NONASSOCIATIVE ALGEBRAS
AND
RELATED TOPICS

International Symposium on

NONASSOCIATIVE ALGEBRAS AND RELATED TOPICS

Hiroshima, Japan 30 August – 1 September 1990

Editors:

KIYOSI YAMAGUTI
Department of Mathematics
Hiroshima University

NAOKI KAWAMOTO
Department of Mathematics
Maritime Safety Academy

World Scientific
Singapore • New Jersey • London • Hong Kong

Published by

World Scientific Publishing Co. Pte. Ltd.
P O Box 128, Farrer Road, Singapore 9128
USA office: Suite 1B, 1060 Main Street, River Edge, NJ 07661
UK office: 73 Lynton Mead, Totteridge, London N20 8DH

NONASSOCIATIVE ALGEBRAS AND RELATED TOPICS

ISBN 981-02-0655-0

Printed in Singapore by Utopia Press.

The Organizing Committee

H. Asano	Yokohama City University
Y. Kakiichi	Toyo University
N. Kawamoto	Maritime Safety Academy
M. Kikkawa	Shimane University
O. Maruo	Hiroshima University
K. Yamaguti	Hiroshima University

Preface

The International Symposium on Nonassociative Algebras and Related Topics was held at ANA Hotel in Hiroshima from August 30 to September 1, 1990 as a satellite symposium of the International Congress of Mathematicians, Kyoto, 1990. This volume contains the papers contributed to the Symposium.

In Hiroshima, Prof. K. Morinaga studied the triple systems and their applications to the geometry of matrix spaces. Prof. S. Tôgô investigated Lie algebras, especially infinite-dimensional Lie algebras, and also infinite groups. Thus, it was a memorial event for us to have the Symposium on nonassociative algebras.

We wish to thank the Mathematical Society of Japan for its support of the Symposium. We greatly appreciate Prof. A.A. Sagle and Prof. I. Yokota for their contributions of papers though they could not come to Hiroshima. We are also grateful to the staff of the World Scientific Publishing Co. for their kind cooperation on the publishing of the Proceedings.

Kiyosi Yamaguti and Naoki Kawamoto July, 1991

Contents

NONASSOCIATIVE ALGEBRAS AND RELATED TOPICS

GENERALIZED JORDAN TRIPLE SYSTEMS

HIROSHI ASANO

Department of Mathematics, Yokohama City University
Kanazawa-ku, Yokohama 236, Japan

ABSTRACT

The generalized Jordan triple systems are some generalization of the Jordan triple systems, hence of Jordan algebras. This is a brief review about the classification of simple generalized Jordan triple systems of the first or the second kind.

§1. Some basic results on generalized Jordan triple systems

1.1. Let U be a finite dimensional vector space over a field K of characteristic zero and let $B : U \times U \times U \to U$ be a trilinear mapping. Then the pair (U, B) is called a *triple system* over K. We shall often write (xyz) instead of $B(x, y, z)$. A triple system (U, B) is called a *generalized Jordan triple system* (or shortly *GJTS*) if the following identity

$$(uv(xyz)) = ((uvx)yz) - (x(vuy)z) + (xy(uvz))$$

is satisfied for all $u, v, x, y, z \in U$. Furthermore, if the additional condition

$$(xyz) = (zyx)$$

is satisfied, then (U, B) is called a *Jordan triple system* (or simply *JTS*). Let (U, B) be a GJTS. For subspaces $V_i (1 \leq i \leq 3)$ of U, we denote by $(V_1V_2V_3)$ the subspace spanned by all elements of the form $(x_1x_2x_3)$ for $x_i \in V_i$. A subspace V of U is called an *ideal* (resp. *K-ideal*) of (U, B) if the relation

$$B(V, U, U) + B(U, V, U) + B(U, U, V) \subset V \text{ (resp. } B(V, U, U) + B(U, U, V) \subset V)$$

is valid. A GJTS (U, B) is called *simple* (resp. *K-simple*) if B is not a zero map and if (U, B) has no non-trivial ideal (resp. K-ideal). Of course, every K-simple GJTS is simple. For an element $a \in U$, let us define a bilinear map B_a on U by putting

$$B_a(x, y) = B(x, a, y) \qquad \text{for } x, y \in U.$$

We say that (U, B) *satisfies the condition* (A) if $B_a = 0$ implies $a = 0$. Every simple GJTS satisfies the condition (A) [2].

1.2. Starting from a given GJTS (U,B), Kantor [4] constructed a certain graded Lie algebra (or in short GLA) $\mathcal{L}(B)=\sum_{i=-\infty}^{\infty}\mathcal{G}_i$ as follows:

Let $\mathcal{U}_-=\sum_{i\leq -1}\mathcal{U}_i$ be the free Lie algebra generated by $\mathcal{U}_{-1}=U$. Let $\mathcal{U}_n$ be the space of all $(n+1)$-linear mappings from $U\times\cdots\times U\to U$. Put

$$\mathcal{U}_+=\sum_{n\geq 0}\mathcal{U}_n \qquad \text{and} \qquad \mathcal{U}=\mathcal{U}_-+\mathcal{U}_+=\sum_{i=-\infty}^{\infty}\mathcal{U}_i.$$

Then $\mathcal{U}$ becomes a graded Lie algebra with respect to a suitably defined product. Let U_1 be the subspace of $\mathcal{U}_1$ consisting of all operators B_a $(a\in U)$ and let $\mathcal{L}_0(B)$ be the graded subalgebra of $\mathcal{U}$ generated by $\mathcal{U}_{-1}$ and U_1. Let $\mathcal{D}$ be a maximal graded ideal of $\mathcal{L}_0(B)$ contained in $\sum_{i\leq -2}\mathcal{U}_i$. Then we can obtain a graded Lie algebra

$$\mathcal{L}(B)=\mathcal{L}_0(B)/\mathcal{D}=\sum_{i=-\infty}^{\infty}\mathcal{G}_i,$$

which is called the *Kantor algebra* for (U,B).

We say that (U,B) is *of the n-th kind* $(n>0)$ if $\mathcal{G}_{\pm m}=\{0\}$ for all $m>n$ and $\mathcal{G}_{\pm n}\neq\{0\}$. If (U,B) satisfies the condition (A), there exists a grade-reversing involutive automorphism τ_B of $\mathcal{L}(B)$ such that $\tau_B(a)=B_a$ for $a\in U$ [3]. This automorphism is called the *grade-reversing canonical involution* of $\mathcal{L}(B)$.

Hereafter we assume that all GJTS's are of the first or second kind and satisfy the condition (A).

1.3. In a GJTS (U,B), we define the linear endomorphisms $L(a,b)$ and $S(a,b)$ on U by

$$L(a,b)(x)=(abx), \qquad S(a,b)(x)=(axb)-(bxa).$$

The following theorem is fundamental.

THEOREM 1 [3]. *Let (U,B) be a GJTS and τ_B be the grade-reversing canonical involution of $\mathcal{L}(B)=\sum_{i=-2}^{2}\mathcal{G}_i$. Then*

(1) *$\mathcal{G}_{-2}$ is isomorphic with the subspace of $End(U)$ spanned by all operators $S(a,b)$,*

$\mathcal{G}_{-1}=U, \qquad \mathcal{G}_1=\tau_B(\mathcal{G}_{-1}), \qquad \mathcal{G}_2=\tau_B(\mathcal{G}_{-2}),$

$\mathcal{G}_0$ is the subspace of $End(U)$ spanned by all operators $L(a,b)$.

(2) *We have the following bracket relations in $\mathcal{L}(B)$:*

$$[a,b]=S(b,a), \qquad [L(a,b),\tau_B(c)]=-\tau_B(B(b,a,c)),$$

$$[L(a,b),c]=B(a,b,c), \qquad [\tau_B(a),b]=L(b,a),$$

$$[\tau_B(S(a,b)),c]=\tau_B(S(a,b)c),$$

$$[L(a,b),S(c,d)]=S(L(a,b)c,d)+S(c,L(a,b)d),$$

$$[S(a,b),\tau_B(S(c,d))]=L(S(a,b)c,d)-L(S(a,b)d,c),$$

$$[L(a,b),L(c,d)] = L(L(a,b)c,d) - L(c,L(b,a)d),$$

where $a,b,c,d \in U$.

Obviously a GJTS (U,B) is the 1st kind if and only if it is a JTS. On the other hand, (U,B) is of the 2nd kind if and only if the following identity is valid:

$$S(S(x,y)u,v) = S(x,y)L(u,v) + L(v,u)S(x,y)$$

for all $u,v,x,y \in U$.

For the Kantor algebra $\mathcal{L}(B)=\sum_{i=-2}^{2}\mathcal{G}_i$, we put

$$W = \mathcal{G}_{-1} + \mathcal{G}_1, \qquad \text{and} \qquad V = \mathcal{G}_{-2} + \mathcal{G}_0 + \mathcal{G}_2.$$

Then the space W becomes a Lie triple system (abbreviated as LTS) together with the triple product $\{XYZ\}=[[X,Y],Z]$. Let $L(W,W)$ be the space spanned by the all operators $L(X,Y)$, and let $\mathcal{L}(W)=L(W,W)+W$ be the standard enveloping Lie algebra of the LTS W. Then $\mathcal{L}(B)$ is isomorphic to $\mathcal{L}(W)$ [2].

§2 Non-degenerate generalized Jordan triple systems

2.1. In a GJTS (U,B), we consider the symmetric bilinear form γ_B on U defined by

$$\gamma_B(x,y) = (1/2)\mathrm{Tr}\{2R(x,y) + 2R(y,x) - L(x,y) - L(y,x)\}$$

is called the *trace form* of (U,B), where $\mathrm{Tr}(f)$ means the trace of a linear endomorphism f [9]. We say that a GJTS (U,B) is *non-degenerate* if the form γ_B is non-degenerate. Any simple GJTS is non-degenerate and any non-degenerate GJTS satisfies the condition (A) [2]. We remark that in a non-degenerate GJTS the trace form is reduced to $\gamma_B(x,y) = \mathrm{Tr}\{2R(x,y) - L(x,y)\}$.

PROPOSITION 2 [2]. *Let (U,B) be a GJTS and $\mathcal{L}(B)$ its Kantor algebra.*
(1) If the trace form γ_B is identically zero, then $\mathcal{L}(B)$ is solvable.
(2) (U,B) is nondegenerate if and only if $\mathcal{L}(B)$ is semisimple.

The following theorem gives us a relation between the trace form of (U,B) and the Killing form of $\mathcal{L}(B)$.

THEOREM 3 [2]. *Let (U,B) be a nondegenerate GJTS. Let β be the Killing form of $\mathcal{L}(B)=\sum_{i=-2}^{2}\mathcal{G}_i$ and β_V be the Killing form of the subalgebra V $(=\mathcal{G}_{-2}+\mathcal{G}_0+\mathcal{G}_2)$ of $\mathcal{L}(B)$. Then we have*

$$\begin{aligned}\beta(X_1,X_2)=&\beta_V(S_1,\tau_B(S_2')) + \beta_V(T_1,T_2) + \beta_V(\tau_B(S_1'),S_2)\\ &+ \mathrm{Tr}_U(S_1S_2' + 2T_1T_2 + S_1'S_2) - 2\{\gamma_B(x_1,y_2) + \gamma_B(y_1,x_2)\}\end{aligned}$$

where $X_i = S_i + x_i + T_i + \tau_B(y_i) + \tau_B(S_i') \in \mathcal{L}(B)$, $S_i, S_i' \in \mathcal{G}_{-2}$, $T_i \in \mathcal{G}_0$, x_i, $y_i \in \mathcal{G}_{-1} = U$.

2.2. Let (U, B) be a non-degenerate GJTS and γ_B its trace form. We denote by $\tilde{\psi}$ the adjoint operator of $\psi \in End(U)$ with respect to γ_B. If φ is an automorphism of (U, B), then we get $\tilde{\varphi}^{-1}=\varphi$. The set

$$\Gamma(U, B) = \{\psi \in GL(U) \mid \psi \circ L(x, y) = L(\psi(x), \tilde{\psi}^{-1}(y)) \circ \psi \text{ for } x, y \in U\}.$$

becomes a group together with the composition of mappings and this group contains the automorphism group $\mathrm{Aut}(U, B)$ of (U, B) as subgroup. The group $\Gamma(U, B)$ is called the *structure group* of (U, B). Any element $\psi \in \Gamma(U, B)$ induces a grade-preserving automorphism, denoted by $\mathcal{L}(\psi)$, of $\mathcal{L}(B)$ [1]. Furthermore it satisfies

$$\mathcal{L}(\psi) \circ \tau_B = \tau_B \circ \mathcal{L}(\tilde{\psi}^{-1}).$$

Now we put

$$\mathrm{Aut}_+\mathcal{L}(B) = \{\sigma \in \mathrm{Aut}(\mathcal{L}(B)) \mid \sigma \text{ is grade-preserving } \},$$

$$\mathrm{Aut}_+(\mathcal{L}(B), \tau_B) = \{\sigma \in \mathrm{Aut}_+\mathcal{L}(B) \mid \sigma \circ \tau_B = \tau_B \circ \sigma\}.$$

It is known [1] that the following relations are valid:

$$\mathcal{L}(\Gamma(U, B)) = \mathrm{Aut}_+\mathcal{L}(B), \qquad \mathcal{L}(\mathrm{Aut}(U, B)) = \mathrm{Aut}_+(\mathcal{L}(B), \tau_B),$$

$$\Gamma(U, B) = \{\sigma |_U \mid \sigma \in \mathrm{Aut}_+\mathcal{L}(B)\},$$

$$\mathrm{Aut}(U, B) = \{\sigma |_U \mid \sigma \in \mathrm{Aut}_+(\mathcal{L}(B), \tau_B)\},$$

where $\sigma |_U$ denotes the restriction of σ on U.

2.3. For any involutive automorphism φ of a GJTS (U, B), we define a new triple product B_φ on U by

$$B_\varphi(x, y, z) = B(x, \varphi(y), z).$$

Then (U, B_φ) also becomes a GJTS of the same kind as (U, B) and it satisfies the condition (A). We call (U, B_φ) the *φ-modification* of (U, B). If (U, B) is non-degenerate, then so is (U, B_φ). It is known [1] that $\mathcal{L}(B)$ is isomorphic to $\mathcal{L}(B_\varphi)$ as graded Lie algebra.

PROPOSITION 4 [1]. *If (U, B) is a simple GJTS, then any modification (U, B_φ) is a simple GJTS or the direct sum of two simple ideals, which are mutually transferred by φ.*

THEOREM 5 [1]. *Let (U, B) be a non-degenerate GJTS, and let φ and ψ be involutive automorphisms of (U, B). Then the following three conditions are equivalent;*
(1) Two modifications (U, B_φ) and (U, B_ψ) are isomorphic with each other.
(2) There exists an element $\omega \in \Gamma(U, B)$ satisfying $\tilde{\omega} \circ \psi \circ \omega = \varphi$.

(3) *There exists a grade-preserving automorphism* σ *of* $\mathcal{L}(B)$ *such that*

$$\sigma^{-1} \circ \tau_B \circ \mathcal{L}(\varphi) \circ \sigma = \tau_B \circ \mathcal{L}(\psi).$$

2.4. For a non-degenerate GJTS (U, B), we put

$\mathrm{Inv}(U, B) = \{\varphi \mid \varphi$ is an involutive automorphisms of $(U, B)\}$,

$\mathrm{Inv}(\mathcal{L}(B)) = \{\sigma \mid \sigma$ is an involutive automorphisms of $\mathcal{L}(B)\}$,

$\mathrm{Inv}_-(\mathcal{L}(B), \tau_B) = \{\sigma \in \mathrm{Inv}(\mathcal{L}(B)) \mid \sigma$ is grade-reversing, $\sigma \circ \tau_B = \tau_B \circ \sigma\}$.

For σ, $\rho \in \mathrm{Inv}_-(\mathcal{L}(B), \tau_B)$, we say that σ is *equivalent* to ρ under $\mathrm{Aut}_+\mathcal{L}(B)$ if $\xi^{-1} \circ \sigma \circ \xi = \rho$ for some $\xi \in \mathrm{Aut}_+\mathcal{L}(B)$. We denote by $\mathcal{R}$ a representative system of equivalence classes of $\mathrm{Inv}_-(\mathcal{L}(B), \tau_B)$ under this equivalence relation. Moreover, we put $\mathcal{S} = \{\tau_B \circ \sigma |_U \mid \sigma \in \mathcal{R}\}$.

For $\varphi, \psi \in \mathrm{Aut}(U, B)$, we say that φ is *equivalent* to ψ under $\Gamma(U, B)$ if there exists an element $\omega \in \Gamma(U, B)$ such that $\tilde{\omega} \circ \varphi \circ \omega = \psi$. Then we see that the set $\mathcal{S}$ is a representative system of equivalence classes under $\Gamma(U, B)$ of $\mathrm{Inv}(U, B)$.

§3 Real simple generalized Jordan triple systems.

In this section, we consider about real GJTS's.

3.1. Let (U, B) be a real GJTS. It is said to be *compact* if its trace form γ_B is positive definite.

THEOREM 6 [2]. *Let* (U, B) *be a real nondegenerate GJTS. Then* (U, B) *is compact if and only if the grade-reversing canonical involution* τ_B *is a Cartan involution of* $\mathcal{L}(B)$.

THEOREM 7 [2]. *If* (U, B) *is a compact real simple GJTS, then* $\mathcal{L}(B)$ *is a simple graded Lie algebra.*

Let $\mathcal{G} = \sum_{i=-2}^{2} \mathcal{G}_i$ be a real simple GLA, and τ be a grade-reversing Cartan involution of $\mathcal{G}$. The pair $(\mathcal{G}, \tau)$ is called an *admissible pair*. We say that two admissible pairs $(\mathcal{G}, \tau)$ and $(\mathcal{G}', \tau')$ are *isomorphic*, if there exists a grade-preserving isomorphism φ of $\mathcal{G}$ onto $\mathcal{G}'$ such that $\varphi \circ \tau = \tau' \circ \varphi$. We have the following

THEOREM 8 [3]. *There exists a bijection of the set of isomorphism classes of compact real simple GJTS's onto the set of isomorphism classes of admissible pairs.*

3.2. For a given GJTS (U, B), let us consider the direct sum $(U + U, B + B)$ as triple systems. This direct sum also becomes a GJTS. A linear map $\varphi: U + U \to$

$U+U$ defined by $\varphi((x,y)) = (y,x)$ is an involutive automorphism of $(U+U, B+B)$. Hereafter we shall denote by $(\widetilde{U}, \widetilde{B})$ the *φ-modification* of $(U+U, B+B)$, that is,

$$\widetilde{B}((x_1,x_2),(y_1,y_2),(z_1,z_2)) = (B(x_1,y_2,z_1), B(x_2,y_1,z_2)).$$

We call $(\widetilde{U}, \widetilde{B})$ the *special direct sum* of (U, B). If (U, B) is a compact simple GJTS, then $(\widetilde{U}, \widetilde{B})$ is a simple GJTS of the 2nd kind [1].

3.3. Let (U, B) be a real GJTS. An involutive automorphism φ of (U, B) is called a *Cartan involution* if the φ-modification (U, B_φ) is compact.

THEOREM 9 [1]. *For every non-degenerate real GJTS (U, B), there exists a Cartan involution on it.*

Using this theorem, we have the following

THEOREM 10 [1]. *Any non-compact real simple GJTS is obtained as*
(1) *a modification of a compact simple GJTS by an involutive automorphism, or*
(2) *the special direct sum of a compact simple GJTS.*

§4 Classification of simple generalized Jordan triple systems

4.1. Loos [5] classified the simple Jordan triple systems over an algebraically closed field. Also he [6] gave the classification of the real simple compact Jordan triple systems. The classification of non-compact simple Jordan triple systems was made by Neher [7], [8].

4.2. As for GJTS's of the 2nd kind, by making use of Theorem 8, we classified all compact real classical simple GJTS's up to isomorphisms [3]. For the classification of non-compact real simple GJTS's, by making use of Theorem 10 and Theorem 5, we gave a representative system S of equivalence classes under $\Gamma(U, B)$ of $\mathrm{Inv}(U, B)$ for every compact real classical simple GJTS (U, B) [1].

4.3. At last, we remark about the classification of complex simple GJTS's of the 2nd kind. Now let (U, B) and (U', B') be two simple GJTS's. A linear map $f : U \to U'$ is called a *weak isomorphism* if

$$f(B(x,y,z)) = B'(f(x), (f^*)^{-1}(y), f(z)),$$

where f^* denotes the dual operator of f, that is,

$$\gamma_{B'}(f(x), y') = \gamma_B(x, f^*(y')).$$

In [4], Kantor classified the complex K-simple GJTS's of the 2nd kind up to weak isomorphisms. The complete classification (up to isomorphisms) of complex simple GJTS's of the 2nd kind has not yet published. But, about the classical ones, we can obtain it from our results for compact real simple GJTS's by means of complexification.

References

[1]. H. Asano, *Classification of non-compact real simple generalized Jordan triple systems of the second kind*, Hiroshima Math. J., (to appear)

[2]. H. Asano and S. Kaneyuki, *On compact generalized Jordan triple systems of the second kind*, Tokyo J. Math., **11** (1988), 105-118.

[3]. S. Kaneyuki and H. Asano, *Graded Lie algebras and generalized Jordan triple systems*, Nagoya Math. J., **112** (1988), 81-115.

[4]. I. L. Kantor, *Some generalization of Jordan algebras*, Trudy Sem. Vect. Tenz. Anal., **16** (1972), 407-499.

[5]. O. Loos, Symmetric spaces I, W.A.Benjamin Inc., New York, 1969.

[6]. O. Loos, Bounded symmetric domains and Jordan pairs, Math. Lect., Univ. Calif., Irvine, 1977.

[7]. E. Neher, *Klassifikation der einfachen reellen speziellen Jordan Tripelsysteme*, manuscripta math.,**31**(1980), 197-215.

[8]. E. Neher, *Klassifikation der einfachen reelen Ausnahme Jordan Tripelsysteme*, J. reine angew. Math., **322**(1981), 145-169.

[9]. K. Yamaguti, *On the metasymplectic geometry and triple systems*, Kokyuroku RIMS, Kyoto Univ. **308** (1977), 55-92 (in Japanese).

SIMPLE LIE ALGEBRAS AND PROJECTIVE PLANES IN A WIDER SENSE

KENJI ATSUYAMA
Kumamoto Institute of Technology
Ikeda, Kumamoto 860, Japan

ABSTRACT

A geometry of exceptional Lie groups is studied by the Chen-Nagano's method of polar sets (M_+, M_-). The generalized projective planes are introduced and classified.

Introduction

This paper is divided into three parts. First we give a unified construction of real simple Lie algebras. Secondly, by making use of these algebras, we obtain some series of symmetric spaces which contain the four projective planes, that is, real, complex, quaternion and Cayley projective planes. The study of the connections between these Lie algebras and projective planes leads us to the notion of projective spaces in a wider sense. For example, in a wider sense the Grassmann manifold $G(4,4n)$ is an n-dimmensional projective space and $E_8/Ss(16)$ is a projective plane. These spaces have "lines" as geometric objects. In the last section we give the classification of these planes and also give the intersection numbers of two lines. These are our main subjects. In general two lines intersect at finite points. However, if they are in a singular position, the intersection becomes a symmetric subspace in the each projective plane as a symmetric space according to their position.

§1. Real simple Lie algebras.

We give here a unified construction of real simple Lie algebras.

Definition of $L(A^{(1)}, M^n, A^{(2)})$ (cf. [1]).

$$L(A^{(1)}, M^n, A^{(2)}) := \mathrm{Der}\, A^{(1)} \oplus \underline{M} \oplus \mathrm{Der}\, A^{(2)} \quad \text{(a direct sum)}$$

(1.1) Notations.

$A^{(i)}$ = one of Hurwitz (or composition) algebras.
This algebra has a usual conjugation $-: a \to \bar{a}$.

R, C, Q = real, complex or quaternion numbers respectively.
$\underline{C}$ = Cayley algebra.

Der $A^{(i)}$ = the Lie algebra of inner derivations of $A^{(i)}$.
This algebra is generated by operators $D_{a,b}$, where

$D_{a,b}(c) = [[a,b],c] - 3(a,b,c), \quad (a,b,c \in A^{(i)}),$
$[a,b] = ab-ba$ and $(a,b,c) = (ab)c - a(bc)$.

M^n = $n \times n$ matrix algebra over R.

$A^{(1)} \otimes M^n \otimes A^{(2)}$ = the tensor product of $A^{(1)}$, M^n and $A^{(2)}$.
(1) The elements have the form $\Sigma a \otimes X \otimes u$. ($\Sigma$ means the sum and $a \in A^{(1)}$, $u \in A^{(2)}$ and $X \in M^n$.)
(2) Moreover this space has three operations:

a product: $(a\otimes X\otimes u)(b\otimes Y\otimes v) = ab\otimes XY\otimes uv$,
an involution $-$: $\Sigma a\otimes X\otimes u \rightarrow \Sigma \bar{a}\otimes X^T\otimes \bar{u}$,
a trace Tr : $\Sigma a\otimes X\otimes u \rightarrow \Sigma a\otimes tr(X) I\otimes u$,

where T is the transposed operator of M^n, $tr(X) = (x_{11}+x_{22}+\cdots+x_{nn})/n$ and I is the $n \times n$ unit matrix.

$\underline{M}$ = the vector space which is generated by all elements in $A^{(1)} \otimes M^n \otimes A^{(2)}$ that have trace Tr 0 and the skew-symmetric form with respect to the involution $-$.

(1.2) A Lie algebra structure in $L(A^{(1)}, M^n, A^{(2)})$.

Let $D^{(1)} + a\otimes X\otimes s + D^{(2)} \in \text{Der } A^{(1)} \oplus \underline{M} \oplus \text{Der } A^{(2)}$.

(1) $[D^{(i)}, D^{(j)}] = \begin{cases} \text{the Lie product of Der } A^{(i)} & (i=j), \\ 0 & (i \neq j), \end{cases}$

(2) $[D^{(1)} + D^{(2)}, a\otimes X\otimes s] = (D^{(1)}a)\otimes X\otimes s + a\otimes X\otimes (D^{(2)}s)$,

(3) For $x = a\otimes X\otimes s$ and $y = b\otimes Y\otimes t$ in $\underline{M}$,

$$[x,y] = (X,Y)(s,t)D_{a,b} + (xy - yx - Tr(xy - yx)) + (X,Y)(a,b)D_{s,t},$$

where $(X,Y) = tr(XY)$ and $(s,t) = (st + \overline{st})/2$.

(1.3) Results.

The Lie algebras $L(A^{(1)}, M^n, A^{(2)})$ give the Freudenthal's magic cube (or the magic square). Hurwitz algebras consist of seven types. They are R, C, Q, $\underline{C}$ of non-split types and are C_s, Q_s, $\underline{C}_s$ of split types. If we use R, C, Q, $\underline{C}$ as $A^{(i)}$, our Lie algebras always become compact types. Especially, in the case of n=3, we can obtain all exceptional Lie algebras. The first aim of our study was to find some geometry for E_8 by making use of $L(\underline{C}, M^3, \underline{C})$.

Examples. $F_4 = L(R, M^3, \underline{C})$, $E_6 = L(C, M^3, \underline{C})$, $E_7 = L(Q, M^3, \underline{C})$, $E_8 = L(\underline{C}, M^3, \underline{C})$.
$G_2 = L(R, M^1, \underline{C})$.

Remark. $L(A^{(1)}, M^n, A^{(2)}) \cong L(A^{(2)}, M^n, A^{(1)})$.

(Fig.1)

($n \geq 2$)

	R	C	Q
R	$B_{(n-1)/2}$ or $D_{n/2}$	A_{n-1}	C_n
C	A_{n-1}	$A_{n-1} \oplus A_{n-1}$	A_{2n-1}
Q	C_n	A_{2n-1}	D_{2n}

Freudenthal's magic square

(n=3)

	R	C	Q	$\underline{C}$
R	B_1	A_2	C_3	F_4
C	A_2	$A_2 \oplus A_2$	A_5	E_6
Q	C_3	A_5	D_6	E_7
$\underline{C}$	F_4	E_6	E_7	E_8

§2. Some orbits.

We construct some series of symmetric spaces. These can be obtained as the orbits of some two automorphisms α and β in $L(A^{(1)}, M^n, A^{(2)})$. The orbits of β's contain the real, complex, quaternion and Cayley projective planes which are denoted by RP_2, CP_2, QP_2 and $\underline{C}P_2$ respectively.

(2.1) Involutive automorphisms α and β.

(1) β is the automorphism in M^n defined by

$$\begin{pmatrix} a_{11} & a_{12} & \cdots & a_{1n} \\ a_{21} & a_{22} & \cdots & a_{2n} \\ \cdot & \cdot & \cdots & \cdot \\ a_{n1} & a_{n2} & \cdots & a_{nn} \end{pmatrix} \longrightarrow \begin{pmatrix} a_{11} & -a_{12} & \cdots & -a_{1n} \\ -a_{21} & a_{22} & \cdots & a_{2n} \\ \cdot & \cdot & & \cdot \\ -a_{n1} & a_{n2} & \cdots & a_{nn} \end{pmatrix}.$$

This map can be extended to $L(A^{(1)}, M^n, A^{(2)})$ easily as an automorphism of the Lie algebra by

$$\beta:\ D^{(1)} + a \otimes X \otimes u + D^{(2)} \longrightarrow D^{(1)} + a \otimes (\beta X) \otimes u + D^{(2)}.$$

(2) α is the canonical automorphism in C, Q, or $\underline{C}$ defined by, respectively,

$$
\begin{array}{llcl}
(C): & e_0, e_1 & \longrightarrow & e_0, -e_1 \\
(Q): & e_0, e_1, e_2, e_3 & \longrightarrow & e_0,\ e_1, -e_2, -e_3 \\
(\underline{C}): & e_0, e_1, e_2, e_3, & \longrightarrow & e_0,\ e_1,\ e_2,\ e_3, \\
 & e_4, e_5, e_6, e_7 & & -e_4, -e_5, -e_6, -e_7,
\end{array}
$$

where $\{e_i\}$ is a basis in the each algebra. This map can be also extended to $L(A^{(1)}, M^n, A^{(2)})$ by

$$\alpha:\ D_{a,b}{}^{(1)} + c \otimes X \otimes u + D^{(2)} \longrightarrow D_{\alpha a, \alpha b}{}^{(1)} + \alpha c \otimes X \otimes u + D^{(2)}.$$

(2.2) The orbits of β under the automorphism groups of compact Lie algebras $L(A^{(1)}, M^n, A^{(2)})$.

Notations. $\underline{C}P_2$ = FⅡ = $F_4/Spin(9)$, EⅢ = $E_6/((Spin(10) \times T)/Z_4)$, EⅧ = $E_8/Ss(16)$,
EⅥ = $E_7/((Spin(12) \times SU(2))/Z_2)$, $G^C(2,4) = SU(6)/S(U_2 \times U_4)$,
$G(4,8)$ = the adjoint space of $SO(12)/SO(4) \times SO(8)$.

Each symmetric space in Fig. 2 corresponds to the each Lie algebra in Fig. 1. For instance, the orbit of β becomes $G(4,4n-4)$ in the case of $D_{2n} = L(Q, M^n, Q)$ and becomes EⅧ in the case of $E_8 = L(\underline{C}, M^3, \underline{C})$. The orbits of β are always compact connected symmetric spaces and they are irreducible except in the case of $L(C, M^n, C)$. They are also the adjoint spaces of each type.

(Fig. 2)

RP_{n-1}	CP_{n-1}	QP_{n-1}
CP_{n-1}	$CP_{n-1} \times CP_{n-1}$	$G^c(2,2n-2)$
QP_{n-1}	$G^c(2,2n-2)$	$G(4,4n-4)$

RP_2	CP_2	QP_2	$\underline{C}P_2$
CP_2	$CP_2 \times CP_2$	$G^c(2,4)$	EⅢ
QP_2	$G^c(2,4)$	$G(4,8)$	EⅥ
$\underline{C}P_2$	EⅢ	EⅥ	EⅧ

In the exceptional Lie algebras the orbits of α and β give the following series of symmetric spaces:

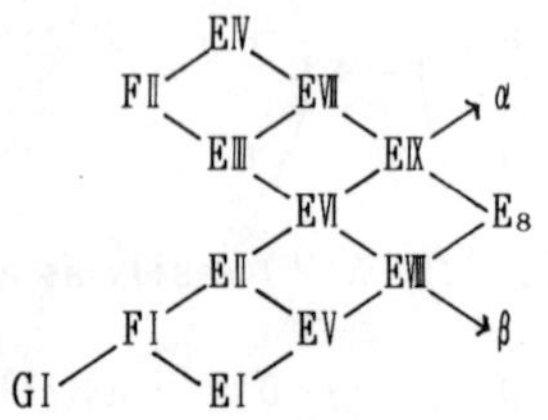

§3. Projective planes in a wider sense.

Since the four projective planes can be obtained as the orbits of β under the automorphism groups of Lie algebras, we can expect that the orbit in each case of E_6, E_7 and E_8 also have the similar structures to the projective planes. In fact, we can find such structures and we describe them in this section. The orbits of β will be called projective planes in a wider sense.

(3.1) Notations.

$M=G/K$: Let M be a compact symmetric space of which the isometry group is G. Then $M=G/K$ holds, if K is an isotropy subgroup of G for some point o in M.

s_p : Let s_p be the (geodesic) symmetry at p in M.

$M_+{}^\circ$: Let p be an antipodal point of o on a closed smooth geodesic. Then the orbit $K(p)$ is called a *polar* of o and we denote it by $M_+{}^\circ(p)$ or by $M_+{}^\circ$.

$M_-{}^\circ$: There exists an orthogonal subspace of M for $M_+{}^\circ(p)$. We denote it by $M_-{}^\circ(p)$ or by $M_-{}^\circ$. They are orthogonal at p. (cf. Lemma 3.5, [4]).

(3.2) Definition.

M is called a <u>projective plane in the wider sense</u> with respect to $M_+{}^\circ(p)$ if

(1) for $q, r \in M$, $s_q=s_r$ is equivalent to $q=r$,

(2) for any $q \in M_+{}^\circ$, there exists $r \in M_+{}^\circ$ such that $s_o s_q = s_r$.

Remark 3.3. The condition (1) means that M is the adjoint space (= the bottom space) by Theorem 4.1, [4]. The condition (2) is equivalent to $M_+{}^\circ(p)\cong M_-{}^\circ(p)$ by Theorem 4.3, [4].

We call $M_+{}^\circ$ a *line* (in the sense of a projective geometry) and denote it by $L(o)$ for simplicity. Let M^L be the manifold consisting of all lines, that is, $M^L = \{ L(o) \mid \text{any } o \in M \}$. The incidence relation is introduced into M by the inclusion of sets.

We define a bijective map L in the plane M. L maps any point p to $L(p)$ there and, conversely, it maps $L(p)$ to p.

Theorem 3.4. The map L gives a polarity, that is, L satisfies

(1) L^2 = identity,

(2) $p \in L(q)$ is equivalent to $q \in L(p)$.

Proof. We know the surjectivity of L from the definition of L. In order to see the injectivity of L, we may show that, for two point $o, r \in M$, $L(o)=L(r)$ implies $o=r$. Since $L(o)=K(p)=M_+{}^o(p)$ holds for some antipodal point p of o, there exists $q \in L(o)$ such that $s_o s_p = s_q$ by the definition of M. Since $M_+{}^o(p) \cong M_-{}^o(p)$ holds in projective planes in the wider sense, L(o) has the same rank as M as a symmetric space. Hence any lines have the same rank as M.

Since $p \in L(o)$ holds (hence $p \in L(r)$ also), the symmetries s_o and s_r leave p fixed. In the tangent space T_pM at p to M, we have a direct sum decomposition $T_pM = T_pM_+{}^o(p) \oplus T_pM_-{}^o(p)$. The subspaces are (±1)-eigenspaces of the differential $(s_o)_*$. We know $T_pM_+{}^o(p)=T_pM_+{}^r(p)$ from $L(o)=L(r)$. Hence $T_pM_-{}^o(p)=T_pM_-{}^r(p)$ holds. These give that the differentials $(s_o)_*$ and $(s_r)_*$ are equal as linear transformations in T_pM. Therefore, by Lemma 11.2, (p.62[5]), we have $s_o=s_r$. Since M is the bottom space, it means o=p. □

Corollary 3.5. The polarity gives a duality in M.

Corollary 3.6. M^L has the same symmetric structure as M.

Corollary 3.7. The intersection $L(p) \cap L(q)$ is diffeomorphic to the manifold which consists of all lines passing through p and q.

Let A be a maximal torus in M passing through o. For $u \in A$, put

$$N(u) = \{ v \in M \mid o \in L(v) \text{ and } u \in L(v) \}.$$

Let U(u) be the identity component of the isotropy group which leaves both u and o fixed. Moreover we define two subsets in M and in L(o) respectively;

$$S := \cap\ L(u) \quad (u \text{ ranges over } A),$$
$$S_o := A \cap L(o).$$

Theorem 3.8.
$$N(u) = \{g \cdot q \mid q \in S \text{ and } g \in U(u)\},$$
$$\cong \{g \cdot q \mid q \in S_o \text{ and } g \in U(u)\}.$$

Proof. See [3] for the proof. By the second isomorphism, we can calculate the number of lines which pass through o and u, that is, we can do it for any two points in M.

Remark 3.9. N(u) is a finite set generally. This means that two lines intersect at finite points in general.

Remark 3.10. There are some singular positions for two points generally. In the case of EIII, there exist many lines passing through two points p and q if and only if they are on a closed geodesic with the minimum length. In EIII, we say that such two points are in the singular position. There exists only one line passing through two points which are in the general (= non-singular) position (cf. [2]).

Problem 3.11. Does hold the harmonicity in EIII generally?

Problem 3.12. Find a good definition for n-dimensional projective spaces in a wider sense. Those examples are $G(4,4n)$ and $G^{C}(2,2n)$.

Problem 3.13. Find finite projective spaces in a wider sense for finite simple (Lie) groups.

(3.14) Classification of projective planes in the wider sense.
(Chen [4], Leung [6], Atsuyama [3])

M_+ := the type of polar sets which play a role of "lines" in the sense of projective plane.

$\#(M_+)$:= the number of points in the intersection set of two lines which are in general position. (This number can be given by Theorem 3.4 and 3.8.)

Let $p, q \in M$. We consider the following two statements (a) and (b);
(a): p and q are in the singular position in the sense of symmetric spaces.
(b): p and q are in the singular position in the sense of projective plane in the wider sense.

Then in general it holds that (b) $\Rightarrow$ (a), but the converse does not hold. So, if (a) $\Rightarrow$ (b) holds, we call M of *type I* and, if not so, we call M of *type II*.

Example 3.15. We consider $GI=G_2/SO(4)$ as M. The line L(o) is a symmetric subspace of M which is diffeomorphic to $S^2 \cdot S^2$ (a semi-direct product of two spheres). Let A be a maximal torus in M which passes through o. Let $\underline{A}$ be the tangent space T_oA at o to A. We can regard it as two-dimensional Euclidean space, that is, $\underline{A} = \{ (a,b) \mid a, b \in R\}$, since the rank of M is 2. Let Δ be the set of roots of M with respect to A. Then, we have the following;

The fundamental roots: λ_1, λ_2,
The highest root: $\lambda_3 = 2\lambda_1 + 3\lambda_2$,

A fundamental domain (cell):

$$D = \{ x \in \underline{A} \mid \lambda_1(xi) \geqq 0,\ \lambda_2(xi) \geqq 0 \text{ and } \lambda_3(xi) \leqq \pi \},$$

The vertexes of D corresponding to λ_1 and λ_2: v_1, v_2,
The extended Dynkin diagram: ⊚—○⇛○, $((-\lambda_3)$—$\lambda_1 \Rrightarrow \lambda_2)$

First we (always) regard the extended Dynkin diagram of the symmetric space M as that of a Lie group G ($G=G_2$ in this case by chance). Next we denote by K_i the maximal subgroups of G which correspond to the roots λ_i (i=1,2). Now $K_1=SO(4)$ and $K_2=SU(3)$. The relations between K_i and the diagram are given by the following:

◎—○⇛○	type G_2	$\#(W(G_2))=12$	$c_0=1$,
◎ ○	type $A_1 \times A_1$	$\#(W(K_1))=4$	$c_1=1$,
◎—○	type A_2	$\#(W(K_2))=6$	$c_2=1$,

where $W(K_i)$ means the Weyl group which corresponds to the group K_i(i=1,2), and $\#(W(K_i))$ is the order of the Weyl group. These give the number of cells around the vertexes v_i in $\underline{A}$. We denote by c_i(resp. n_i) the number of conjugate points of the vertex v_i in $\underline{A}$ (resp. in D) under the affine Weyl group. Then we have

$$\begin{aligned}\text{the number of all cells in A} &= (\#(W(G_2)\times 1)/c_0 = 12\\ &= (\#(W(K_1)\times n_1)/c_1 = 4\times n_1\\ &= (\#(W(K_2)\times n_2)/c_2 = 6\times n_2 .\end{aligned}$$

From these equations we obtain $n_1=3$ and $n_2=2$. This means the cardinal number of $L(o)\cap A$ is 3 ($=n_1$) since the orbit of $\exp v_1 \in A$ becomes $L(o)$. Therefore we can say that there exist three lines which pass through any two points in the general position.

For this model it holds that $n_1 = \#(W(G_2))/\#(W(K_1))$. Hence n_1 = the Euler number of GI.

We define three sets (see p.63, [6]):

$$U(\lambda) = \text{kernel } \chi = \{ u=\exp H \in A \mid \chi(u) = \exp\lambda(H) = 1 \}, \text{ for } \lambda \in \Delta,$$

$$\Delta_o = \{ \lambda \in \Delta \mid S_o \subset U(\lambda) \} \quad \text{and} \quad \Lambda_u = \{ \lambda \in \Delta \mid u \in U(\lambda) \}.$$

Let two points o and u in A. We say that o and u (resp. two lines L(o) and L(u)) are in the general position if $\Lambda_u \subset \Delta_o$. Conversely, if $\Lambda_u \cap (\Delta - \Delta_o) \neq 0$, we say that they are in the singular position. In the case of M=GI there does not exist $\lambda \in \Delta$ such that $S_o \subset U(\lambda)$. Therefore the notion that o and u are in the singular position is equivalent to the notion that u is a singular point for o in the symmetric space.

Classification in the case that M is an irreducible compact symmetric space.

M	M_+	$\#(M_+)$	Type
(Exceptional spaces)			
EII	$S^2\cdot G^C(3,3)$	12	I
EIII	$G^{OR}(2,8)$	1	II
EV*	$AI(8)/Z_4$	36	I
EVI	$G^{OR}(4,8)$	3	II
	$S^2\cdot DIII(6)$	12	I
EVII*	$(T\cdot EIV)/Z_2$	4	I
EVIII	$G(8,8)^{\#}$	135	I
EIX	$S^2\cdot EVII$	12	I
FI	$S^2\cdot CI(3)$	12	I
FII	S^8	1	II
GI	$S^2\cdot S^2$	3	I
(Classical spaces)			
AIII $G(2p,q)$ $(2p\neq q,\ p\leqq q)$	$G(p,p)\times G(p,q-p)$	${}_{2p}C_p$, $(2p<q)$ ${}_{q}C_p$, $(2p>q)$	II
$G(2p,2p)^*$	$G(p,p)\cdot G(p,p)$	${}_{2p}C_p/2$	II
$G(p,p)^*$	$U(p)/Z_2$	2^{p-1}	I $(p\neq 1)$
BI $G(2p,q)$ $(2p\neq q,\ p\leqq q,\ q$:odd$)$	$G(p,p)\times G(p,q-p)$	${}_{2p}C_p$, $(2p<q)$ ${}_{q}C_p$, $(2p>q)$	I
$CI(n)^*$	$UI(n)/Z_2$	2^{n-1}	I $(n\neq 1)$
CII $G(2p,q)$ $(2p\neq q,\ p\leqq q)$	$G(p,p)\times G(p,q-p)$	${}_{2p}C_p$, $(2p<q)$ ${}_{q}C_p$, $(2p>q)$	II
$G(2p,2p)^*$	$G(p,p)\cdot G(p,p)$	${}_{2p}C_p/2$	II
DI $G(2p,q)$ $(2p\neq q,\ p\leqq q,\ q$:even$)$	$G(p,p)\times G(p,q-p)$	${}_{2p}C_p$, $(2p<q)$ ${}_{q}C_p$, $(2p>q)$ (The type of $G(4,2)$ is II)	I
$G^{OR}(2p,2p)^*$	$G(p,p)\cdot G(p,p)$	${}_{2p}C_p/2$	I
$DIII(2n)^*$	$UII(2n)/Z_2$	2^{n-1}	I

References

[1] K. Atsuyama, Another construction of real simple Lie algebra, Kodai Math. J., 6 (1983), 122-133.

[2] K. Atsuyama, The connection between the symmetric space $E_6/SO(10)\cdot SO(2)$ and projective planes, ibid., 8 (1985), 236-248.

[3] K. Atsuyama, The projective spaces in a wider sense. I, (to appear).

[4] B. Y. Chen, A new approach to compact symmetric spaces and applications (a report on joint work with Professor T. Nagano), Katholieke Universiteit Leuven, August 1987.

[5] S. Helgason, Diffential Geometry, Lie groups, and Symmetric Spaces, Academic Press, New York, 1978.

[6] D. S. P. Leung, Reflective submanifolds. II. Congruency of isometric reflective submanifolds and corrigenda to the classification of reflective submanifolds, J. Differential Geometry, 14(1979), p.167-177.

Recent Developments in the Cohomology Theory of Modular Lie Algebras

Rolf FARNSTEINER

Department of Mathematics, University of Wisconsin
Milwaukee, WI 53201, U.S.A.

Abstract

This paper adapts techniques recently introduced into the cohomology theory of modular Lie algebras in order to study two particular cases.

We first investigate the connection between the cohomology theory of a finite dimensional modular Lie algebra L and that of its p-envelope G. By establishing a Morita equivalence between certain enveloping algebras, we obtain a reduction formula that allows to compute extension functors of G in terms of cohomology groups of L.

Motivated by questions concerning the cohomology of graded Cartan type Lie algebras, the second part of the paper studies graded Lie algebras. The general techniques involving Frobenius extensions are refined to yield information concerning the gradation of the cohomology groups. The concluding examples serve as an illustration for the utility of our methods.

§0. Introduction

From the beginning on, the cohomology theory of Lie algebras was devised for the study of Lie algebras over fields of characteristic 0. In fact, the foundational paper by Chevalley and Eilenberg [8] was motivated by the connection between the singular homology theory of a semisimple, connected compact Lie group and the cohomology theory of its Lie algebra. Up to today the most powerful results of the ordinary cohomology theory pertain to Lie algebras related to certain Lie groups or, in the modular case, algebraic groups (cf. [1,23]).

After the discovery of non-classical simple modular Lie algebras by E. Witt in 1937 and Jacobson's subsequent theory of restricted Lie algebras [28], there slowly arose some interest in cohomology groups of modular Lie algebras. The initial results, such as those in [6], already indicated that this part of the theory would differ markedly from its classical precursor. Not only do the results concerning complex Lie algebras usually lose their validity in this setting, but even for simple Lie algebras the classical averaging techniques related to invariant forms are frequently not available (cf. [13]). In view of these conceptual difficulties many of the early results valid for modular Lie algebras employed inductive arguments related to spectral sequences (cf. [3,4,10]) that were independent of the characteristic of the underlying base field. Later, several deep results were established for restricted Lie algebras associated to certain algebraic groups. The cohomology theory of arbitrary modular Lie algebras was first studied by Dzhumadil'daev in [11]. He showed that the existence of a central p-polynomial in the universal enveloping algebra $\mathcal{U}(L)$ of a Lie algebra L forces the vanishing of all those cohomology groups for which it operates invertibly on the coefficient module. The success of his approach, which amounts to an extension of Chevalley's and Eilenberg's methods involving the natural operation of $\mathcal{U}(L)$ on the Koszul complex, rest on the fact that p-polynomials are primitive elements of the Hopf algebra $\mathcal{U}(L)$. This circumstance suggested the existence of more general vanishing theorems valid in arbitrary characteristic and in a more general setting. The utility of ring theoretic methods was already observed by M.P. Malliavin in her paper [33], where she reformulated and extended Dixmier's results [10].

In this article we shall give a survey and new applications of recently developed methods for the cohomology theory of modular Lie algebras. We shall confine our attention to those methods that do not require the machinery of algebraic groups. Since the results discussed below frequently afford a reduction of general problems to those concerning the cohomology theory of classical Lie algebras, they complement those of [1,22], and [23].

The author would like to take this opportunity to thank Professors Kawamoto and Yamaguti for their hospitality.

§1 Review of Recent Results

In this section we shall recall basic results that have proven to be effective in the cohomology theory of modular Lie algebras. The two major tools are vanishing theorems involving operator theoretic techniques, and reduction theorems related to the theory of Frobenius extensions.

Let F be a field of arbitrary characteristic, A an associative F-algebra. In the commutator algebra A^- we consider the left multiplication effected by the element u: $a \mapsto [u,a]$. This mapping is customarily denoted by (adu). For a left A-module M and $u \in A$, we put $u_M: M \to M; m \mapsto u.m$.

THEOREM 1.1 [18]: Let $u \in A$ be an element such that adu: $A \to A$ is locally nilpotent. Given A-modules M and N, the following statements hold:

(1) If u_M is locally nilpotent and u_N is invertible, then $\mathrm{Ext}^n_A(M,N) = (0)\ \forall\ n\geq 0$

(2) If u_M is invertible and u_N is nilpotent, then $\mathrm{Ext}^n_A(M,N) = (0)\ \forall\ n\geq 0$.

(3) If u_M is locally finite and injective and u_N is locally nilpotent, then $\mathrm{Ext}^n_A(M,N) = (0)\ \forall\ n\geq 0$. □

This result, which can be refined considerably by introducing weight space decompositions [17], applies in a variety of contexts ranging from the cohomology theory of rings to the restricted cohomology of restricted Lie algebras of positive characteristic (cf. [17,19]).

Let A be an F-algebra. Given an algebra homomorphism α of A and a left A-module M we shall denote by ${}_\alpha M$ the left A-module with underlying vector space M and operation $r\cdot m := \alpha(r)m$.

Let $B \subset A$ be a subalgebra of A, α an automorphism of B.

DEFINITION: We say that A:B is an *α-Frobenius extension* if

(a) A is a finitely generated projective left B-module

(b) There exists an isomorphism φ: ${}_A A_B \longrightarrow \mathrm{Hom}_B(A, {}_\alpha B)$ of (A,B)-bimodules.

EXAMPLES: (a) Let G be a group, $H \subset G$ a subgroup of finite index. Then $F[G]:F[H]$ is an $id_{F[H]}$-Frobenius extension.

(b) Every finite dimensional Hopf algebra is an id_F-Frobenius extension of its underlying base field F (cf. [30]).

(c) Let L be a Lie superalgebra with enveloping algebra $\mathcal{U}(L)$, $L_0 \subset L$ the subalgebra of even elements. Then $\mathcal{U}(L):\mathcal{U}(L_0)$ is a Frobenius extension. The latter fact seems to explain some of the conceptual similarities between Lie superalgebras and restricted Lie algebras of positive characteristic (cf. [5]).

Let M,N be two left A-modules. The extension A:B defines elements $x_1,\dots,x_n$, $y_1,\dots,y_n \in A$ such that for $f \in \mathrm{Hom}_B(M, {}_\alpha N)$ the mapping $\mathrm{Tr}(f)\colon M \to N$

$$\mathrm{Tr}(f)(m) = \Sigma_i x_i f(y_i m) \quad \forall\ m \in M$$

belongs to $\mathrm{Hom}_A(M,N)$.

Let $\mathrm{Res}_n:\mathrm{Ext}^n_A(M,N) \to \mathrm{Ext}^n_B(M,N)$ denote the canonical change of rings map. In addition to these maps, the Frobenius extension induces corestriction mappings

$$\mathrm{Cor}_n\colon \mathrm{Ext}^n_B(M, {}_\alpha N) \to \mathrm{Ext}^n_A(M,N)$$

for every $n \geq 0$.

THEOREM 1.2 [5]: Let A:B be an α-Frobenius extension and suppose that X,M and N are left A-modules. Then the following statements hold:

(1) $\mathrm{Cor}_0 = \mathrm{Tr}$

(2) If $f \in \mathrm{Hom}_B(M, {}_\alpha N)$, then $\mathrm{Ext}^n_A(id_X, \mathrm{Tr}(f)) = \mathrm{Cor}_n \circ \mathrm{Ext}^n_B(id_X, f) \circ \mathrm{Res}_n\ \forall\ n \geq 0$.

This result, which can be used to retrieve the well-known semisimplicity criterion for finite dimensional Hopf algebras [40], allows the introduction of "Casimir operators" in case α can be extended to a homomorphism of A (cf. [19]).

We now turn to the study of modular Lie algebras. Accordingly, F is assumed to be a field of positive characteristic p. Let L be a finite dimensional Lie algebra over F, $K \subset L$ a subalgebra as well as $\{e_1,\dots,e_k\}$ a cobasis of K in L. The universal enveloping algebras of L and K will be denoted $\mathcal{U}(L)$ and $\mathcal{U}(K)$, respectively. According to Jacobson's refinement of the Poincaré-Birkhoff-Witt Theorem there exist elements $z_1,\dots,z_k \in \mathcal{Z}(\mathcal{U}(L))$, the center of $\mathcal{U}(L)$, such that

(i) $\mathcal{R} := F[z_1,\dots,z_k] \subset \mathcal{U}(L)$ is a polynomial ring in k indeterminates

(ii) $\mathcal{O}(L,K) := \mathrm{alg}_F(K\cup\{z_1,\dots,z_k\})$ is isomorphic to $\mathcal{R}\otimes_F\mathcal{U}(K)$

(iii) $\mathcal{U}(L)$ is a free $\mathcal{O}(L,K)$-module of finite rank.

Owing to (ii) there exists exactly one automorphism α of $\mathcal{O}(L,K)$ such that

$$\alpha(z_i) = z_i \quad 1\le i\le k, \qquad \alpha(x) = x - \mathrm{tr}(\mathrm{ad}L/_K(x))1 \quad \forall\, x \in K,$$

where $\mathrm{ad}L/_K$ denotes the natural representation of K on $L/_K$.

THEOREM 1.3. [21]: The extension $\mathcal{U}(L):\mathcal{O}(L,K)$ is an α-Frobenius extension. □

Property (ii) ensures that every K-module V can be given the structure of an $\mathcal{O}(L,K)$-module by defining $z_i\cdot V = (0)$ $1\le i\le k$. Henceforth we shall regard K-modules $\mathcal{O}(L,K)$-modules in this fashion.

DEFINITION: Let V be a K-module. Then we call $\mathrm{Ind}_K(V) := \mathcal{U}(L)\otimes_{\mathcal{O}(L,K)}V$ the *generalized reduced Verma module* defined by V. The coinduced module $\mathrm{Hom}_{\mathcal{O}(L,K)}(\mathcal{U}(L),V)$ will be denoted $\mathrm{Coind}_K(V)$.

Generalized reduced Verma modules unify various conventional constructions in the representation theory of solvable, classical and Cartan type Lie algebras. If $(L,[p])$ is restricted, K is a p-subalgebra and V a K-module with character $S \in L^*$, then the choice $z_i := e_i^p - e_i^{[p]} - S(e_i)^p 1$ yields natural isomorphisms $\mathrm{Ind}_K(V) \cong u(L,S)\otimes_{u(K,S|_K)}V$ (cf. [21]). Furthermore, if L is one of the graded Lie algebras of Cartan type, then mixed product modules (cf. [37]) can be shown to be isomorphic to generalized reduced Verma modules. Consequently, Shen's representation theory [37,38] can be completely described in terms of these modules. This not only obviates many of the technical considerations pertaining to mixed product, but also affords an extension of Shen's approach to the Lie algebras of Cartan type.

The following reduction theorem is the key to many recent results in the cohomology theory of modular Lie algebras. Its proof combines features from the theory of Frobenius extensions with the product theory of extension functors of (cf. §3).

For convenience we shall view the polynomial ring $\mathcal{R}$ as the universal enveloping algebra of the abelian Lie algebra with underlying F-vector space $L/_K$.

THEOREM 1.4. [18]: Let M be an L-module, V a K-module. Assume there exists a finitely generated $\mathcal{R}$-module P and a $\mathcal{U}(K)$-module Q such that $M \cong P \otimes_F Q$ as an $\mathcal{O}(L,K)$-module. Then there exist isomorphisms

$$\mathrm{Ext}^u_{\mathcal{U}(L)}(M,\mathrm{Ind}_K(V) \cong \bigoplus_{p+q=n} H^p(L/_K,P^*) \otimes_F \mathrm{Ext}^q_{\mathcal{U}(K)}(Q,{}_\alpha V).$$

The seemingly technical condition of this result obtains in many standard situations. If M is, for instance, irreducible, then either each z_i operates trivially on M, or there is a z_i acting invertibly. Since all of the z_i act trivially on $\mathrm{Ind}_K(V)$, Theorem 1.1 implies the triviality of $\mathrm{Ext}^n_{\mathcal{U}(L)}(M,\mathrm{Ind}_K(V))$ in the latter case. Alternatively, we may choose P = F and Q = M and Theorem 1.4 applies.

One consequence of the preceding result is a non-vanishing theorem that illustrates the dichotomy between the modular and non-modular cohomology theory of Lie algebras.

COROLLARY 1.5. [18]: Let $M \neq (0)$ be a finite dimensional $\mathcal{U}(L)$-module. Then there exists a finite dimensional completely reducible L-module N such that $\mathrm{Ext}^n_{\mathcal{U}(L)}(M,N) \neq (0) \neq \mathrm{Ext}^n_{\mathcal{U}(L)}(N,M) \quad 0 \le n \le \dim_F L$.

Restricted Lie algebras possess a restricted cohomology theory that takes into account the existence of a p-power operator. While one cannot expect this theory to be very closely related to ordinary cohomology (because the underlying associative rings are vastly different), Hochschild's early work [25] and more recent results by Friedlander and Parshall [22,23,24] reveal important connections. The following result, which extends [24,(5.3)], exploits certain features of the extension $\mathcal{U}(L):\mathcal{O}(L,K)$ to obtain a description of the ordinary Lie algebra cohomology in terms of the restricted one. Given a restricted Lie algebra (L,[p]) and a restricted L-module M, we define, following Hochschild,

$$H^n_*(L,M) = \mathrm{Ext}^n_{u(L)}(F,M),$$

where u(L) denotes the restricted enveloping algebra of L.

THEOREM 1.6. [20]: Let $K \subset L$ be a p-ideal of a restricted Lie algebra $(L,[p])$ and suppose that M is a restricted L-module. Then there exists a third quadrant spectral sequence

$$\bigoplus_{i+j=q} \Lambda^i(L/K)\otimes_F H_*^p(L/K, H^j(K,M)) \Rightarrow H^n(L,M),$$

where $\Lambda^i(L/K)$ denotes the i-fold wedge product of L/K. If L/K is a torus, then the above sequence collapses to isomorphisms

$$\bigoplus_{i+j=n} \Lambda^i(L/K)\otimes_F H^j(K,M)^L \cong H^n(L,M).$$

Note that this result also generalizes Theorem 13 of [26] in this particular context.

§2 The Cohomology of p-Envelopes

Throughout this section K is assumed to be a finite dimensional Lie algebra over a field F of positive characteristic p. We shall employ the concepts delineated in the previous section to determine the relationship between the cohomology theory of K and that of the restricted Lie algebra $(L,[p])$ generated by it. Suppose that $(L,[p])$ is a restricted Lie algebra and let $f = \Sigma_{i=0}^n \alpha_i X^{p^i} \in F[X]$ be a p-polynomial. For $x \in L$, we put $\lambda_x(f) = \Sigma_{i=0}^n \alpha_i x^{[p]^i}$. Given a subalgebra $H \subset L$, we let H_p denote the p-subalgebra generated by H. If H_p coincides with L, then L is called a *p-envelope of H*. We recall that any finite dimensional Lie algebra K *possesses* a finite dimensonal p-envelope (cf. [39,(5.3)p.93]).

LEMMA 2.1.: Let $(L,[p])$ be a finite dimensional restricted Lie algebra, $K \subset L$ a subalgebra of codimension k such that $K_p = L$. Then there exist elements $x_1,\dots,x_k$ of K and p-polynomials $f_1,\dots,f_k \in F[X]$ such that

(a) $\{\lambda_{x_i}(f_i);\ 1\le i\le k\}$ is a cobasis of K in L.

(b) The central elements $c_i := f_i(x_i)-\lambda_{x_i}(f_i)$ $1\le i\le k$ along with K span a Lie subalgebra G of $\mathcal{U}(L)$ of dimension $\dim_F L$.

Proof. Let $\{y_1,\dots,y_n\}$ be a basis of K. Owing to [39, (1.3) p.66] the space L is linearly generated by the set $\{y_i^{[p]^j};\ 1\le i\le n,\ j\ge 0\}$. By choosing a subset whose cosets form a basis of $L/_K$, we define a subset $J \subset \{1,\ldots,n\}\times\mathbb{N}$ of cardinality k and p-monomials $f_{ij} = X^{p^j}$ for $(i,j)\in J$ such that $\{\lambda_{y_i}(f_{ij});$ $(i,j) \in J\}$ is a cobasis of K in L. For $(i,j) \in J$ we put $x_{(i,j)} := y_i$. To verify property (b) consider the central elements

$$c_{ij} := f_{ij}(y_i) - \lambda_{y_i}(f_{ij}) \quad ; \quad (i,j) \in J.$$

Note that $c_{ij} \equiv f_{ij}(y_i) \bmod(\mathcal{U}(L)_{(1)})$, where $(\mathcal{U}(L)_{(i)})_{i\geq 0}$ denotes the canonical filtration of $\mathcal{U}(L)$. Hence, if $\Sigma_{ij}\alpha_{ij}c_{ij}\in K$, then $\Sigma_{ij}\alpha_{ij}f_{ij}(y_i) \equiv 0 \bmod(\mathcal{U}(L)_{(1)})$, and the Poincaré-Birkhoff-Witt Theorem forces $\alpha_{ij} = 0$ for all $(i,j) \in J$. It now follows that $G = K \oplus \bigoplus_{ij}Fc_{ij}$ has the requisite property. □

For the remainder of this section, we adopt the following conventions. The restricted Lie algebra $(L,[p])$ is assumed to be a finite dimensional p-envelope of K such that the codimension of K in L is k. In accordance with (2.1) we select $x_1,\ldots,x_k \in K$ and p-polynomials $f_1,\ldots,f_k \in F[X]$ such that (a) and (b) hold. Furthermore, we let G denote the Lie algebra generated by K and the c_i.

THEOREM 2.2: The algebras $\mathcal{U}(L)$ and $F[X_1,\ldots,X_k]\otimes_F\mathcal{U}(K)$ are Morita equivalent. In particular, there is an isomorphism $\mathcal{Z}(\mathcal{U}(L)) \cong F[X_1,\ldots,X_k]\otimes_F\mathcal{Z}(\mathcal{U}(K))$.

<u>Proof</u>. The universal enveloping algebra of G isomorphic to $F[X_1,\ldots,X_k]\otimes_F\mathcal{U}(K)$ and it thus suffices to establish a Morita equivalence between $\mathcal{U}(G)$ and $\mathcal{U}(L)$. Let ${}_{U(L)}\mathcal{M}$ and ${}_{U(G)}\mathcal{M}$ denote the categories of $\mathcal{U}(L)$-modules and $\mathcal{U}(G)$-modules, respectively. Given a representation $\sigma: \mathcal{U}(L) \to \mathrm{End}_F(M)$ we define a representation, $\rho: \mathcal{U}(G) \to \mathrm{End}_F(M)$ via

$$\rho(x) = \sigma(x) \quad \forall\ x \in K,\ \rho(c_i) := \sigma(c_i) \quad 1\leq i\leq k.$$

We denote the resulting $\mathcal{U}(G)$-module by $\mathcal{F}(M)$ and thereby obtain a functor $\mathcal{F}: {}_{U(L)}\mathcal{M} \to {}_{U(G)}\mathcal{M}$.

In order to define a functor $\mathcal{G}: {}_{U(G)}\mathcal{M} \to {}_{U(L)}\mathcal{M}$ we consider a representation $\rho: G \to \mathfrak{gl}(N)$. For $x,y \in K$ and p-polynomials $f,g \in F[X]$ one readily verifies the identity

$$\rho([\lambda_x(f),\lambda_y(g)]) = [f(\rho(x)),g(\rho(y))].$$

Using (2.1) we can therefore extend ρ linearly to a representation σ of L by defining

$$\sigma(\lambda_{x_i}(f_i)) := f_i(\rho(x_i)) - \rho(c_i) \quad 1\leq i\leq k.$$

We denote the corresponding $\mathcal{U}(L)$-module by $\mathcal{G}(N)$.

Since $\mathcal{F}\circ\mathcal{G}$ clearly coincides with the identity functor on $_{U(G)}\mathcal{M}$, we consider a representation $\sigma\colon \mathcal{U}(L) \to \mathrm{End}_F(M)$ whose restriction to G defines a representation $\rho\colon \mathcal{U}(G) \to \mathrm{End}_F(\mathcal{F}(M))$. Let $\pi\colon \mathcal{U}(L) \to \mathrm{End}_F(\mathcal{G}(\mathcal{F}(M)))$ denote its extension. For $u \in G$ we obtain, observing the set-theoretic identity $\mathcal{G}(\mathcal{F}(M)) = M$, $\pi(u) = \rho(u) = \sigma(u)$. This in conjunction with the definition of π implies

$$\pi(\lambda_{x_i}(f_i)) = f_i(\rho(x_i)) - \rho(c_i) = f_i(\rho(x_i)) - \sigma(c_i) = f_i(\sigma(x_i)) - \sigma(c_i)$$

$$= \sigma(f_i(x_i)) - \sigma(c_i) = \sigma(\lambda_{x_i}(f_i)).$$

Consequently, $\pi = \sigma$ and $\mathcal{G}\circ\mathcal{F}$ is the identity functor.

Standard Morita theory (cf.[2, (21.10)]) now provides an isomorphism between the centers of $\mathcal{U}(L)$ and $\mathcal{U}(G)$. Since $\mathcal{Z}(\mathcal{U}(G))$ can be identified with $\mathcal{R}\otimes_F\mathcal{Z}(\mathcal{U}(K))$, the last assertion follows. □

Corollary 2.3. Let $\mathcal{O}(L,K)$ be constructed as in §1. Then $\mathcal{U}(L):\mathcal{O}(L,K)$ is a Frobenius extension with winding automorphism $\alpha=\mathrm{id}_{\mathcal{O}(L,K)}$. Moreover, the algebras $\mathcal{U}(L)$ and $\mathcal{O}(L,K)$ are Morita equivalent.

Proof. The first statement follows directly from the fact that K is an ideal in L. Since $\mathcal{O}(L,K)$ is isomorphic to $F[X_1,\dots,X_k]\otimes_F\mathcal{U}(K)$ an application of (2.2) yields the second assertion. □

Corollary 2.4.: Let M and N be L-modules.

(1) $\mathrm{Ext}^n_{U(L)}(M,N) \cong \mathrm{Ext}^n_{U(G)}(M,N) \quad \forall \quad n\geq 0.$

(2) If $c_iM = (0) = c_iN \quad 1\leq i\leq k$. Then there are natural isomorphisms

$$\mathrm{Ext}^n_{U(L)}(M,N) \cong \bigoplus_{p+q=n}\Lambda^p(L/K)\otimes_F\mathrm{Ext}^q_{U(K)}(M,N).$$

Proof. (1) Since the functor $\mathcal{F}$ defines a Morita equivalence, it is a full and faithful functor. There results a natural equivalence

$$\mathrm{Hom}_{U(L)}(M,N) \cong \mathrm{Hom}_{U(G)}(M,N).$$

Let $\mathcal{P}: \dots \to P_n \to P_{n-1} \to \dots \to P_0 \to M \to (0)$ be a projective resolution of the $\mathcal{U}(L)$-module M. Since $\mathcal{F}$ is exact and maps projectives to projectives, $\mathcal{P}$ is also a projective resolution relative to the algebra $\mathcal{U}(G)$. The above equivalence thus gives rise to an isomorphism

$$\mathrm{Hom}_{U(L)}(\mathcal{P},N) \cong \mathrm{Hom}_{U(G)}(\mathcal{P},N)$$

of complexes. The assertion now follows by passing to the cohomology groups.

(2) According to (1) the left-hand term reduces to $\mathrm{Ext}^n_{U(G)}(M,N)$. Since the c_i operate trivially on M and N the product theory of extension functors (cf. [7, p.205f] provides the following isomorphisms

$$\begin{aligned}\mathrm{Ext}^n_{U(G)}(M,N) &\cong \mathrm{Ext}^n_{F[X_1,\dots,X_k]\otimes_F U(K)}(F\otimes_F M, F\otimes_F N)\\ &\cong \bigoplus_{p+q=n}\mathrm{Ext}^p_{F[X_1,\dots,X_k]}(F,F)\otimes_F \mathrm{Ext}^q_{U(K)}(M,N)\\ &\cong \bigoplus_{p+q=n} H^p(L/K,F)\otimes_F \mathrm{Ext}^q_{U(K)}(M,N)\\ &\cong \bigoplus_{p+q=n} \Lambda^p(L/K)\otimes_F \mathrm{Ext}^q_{U(K)}(M,N). \qquad \square\end{aligned}$$

Remark. Since the isomorphism $\mathcal{U}(G) \cong \mathcal{O}(L,K)$ maps c_i onto z_i, the reduction formula above retains its validity if $z_iM = (0) = z_iN$ $1 \le i \le k$.

Example: We shall show that the assumption concerning the operation of the c_i is essential for the validity of the preceding result. Consider the Zassenhaus algebra $W(1;\underline{2}) =: K$. This algebra is not restrictable and has a p-envelope L of the form $L = W(1;\underline{2})\oplus Fx^{[p]}$, where $x \in W(1,\underline{2})$. We let the central element $c: = x^p - x^{[p]}$ operate on the trivial K-module F by postulating $c.1: = 1$. The trivial K-module F obtains the structure of a nontrivial L-module with $x^{[p]}.1: = -1$. Theorem 1.1 ensures the vanishing of $H^n(L,F)$ for all $n \ge 0$. On the other hand, it is well-known (cf. [15, (3.2)]) that $H^2(W(1;\underline{2}),F)$ does not vanish.

Note that, in analogy with earlier observations, every K-module M obtains the structure of a $\mathcal{U}(G)$-module by letting the c_i act trivially. Hence M can be given the structure of an L-module such that the c_i annihilate M.

§3 Cohomology groups of graded Lie algebras

In this section we shall be studying the cohomology of finite dimensional $\mathbb{Z}$-graded Lie algebras. We begin with several basic observations that are valid in a slightly more general setting. Recall that if R is a $\mathbb{Z}$-graded ring, M a finitely generated $\mathbb{Z}$-graded R-module and N an arbitrary $\mathbb{Z}$-graded module, then $\mathrm{Hom}_R(M,N)$ is $\mathbb{Z}$-graded by means of

$$\mathrm{Hom}_R(M,N)_k := \{f \in \mathrm{Hom}_R(M,N);\ f(M_i) \subset N_{i+k}\ \forall\ i \in \mathbb{Z}\}.$$

Suppose $\mathcal{X} = (X,d)$ is a complex. We say that $\mathcal{X}$ is a *graded complex* if each component is a $\mathbb{Z}$-graded R-module and the differentiation $d_n : X_n \to X_{n-1}$ has degree 0 for every $n \in \mathbb{Z}$. In that case the homology groups $H_n(\mathcal{X})$ inherit a grading via $H_n(\mathcal{X})_i := {}^{X_{n,i}\cap \ker d_n}/_{X_{n,i}\cap \operatorname{im} d_{n+1}}$. Consequently, finitely generated $\mathbb{Z}$-graded modules affording a $\mathbb{Z}$-graded projective resolution $\mathcal{P} = (P_n)_{n\geq 0}$ such that P_n is finitely generated for every $n\geq 0$ allow us to introduce a $\mathbb{Z}$-grading on the extension functors. Moreover, if R is noetherian, then the category of all finitely generated $\mathbb{Z}$-graded R-modules is a good category with enough projectives. This is the proper conceptual framework for most of what will follow, yet we shall not dwell on this aspect here. We shall require the following:

Lemma 3.1: Let $R = \bigoplus_{i\in\mathbb{Z}} R_i$ be a $\mathbb{Z}$-graded ring, $f\colon M \to N$ a degree 0 homomorphism between two $\mathbb{Z}$-graded R-modules. Suppose that $\mathcal{P} = (P_n)_{n\geq 0}$ and $\mathcal{Q} = (Q_n)_{n\geq 0}$ are $\mathbb{Z}$-graded projective resolutions of M and N, respectively. If P_n is finitely generated for every $n\geq 0$, then there exists a degree 0 mapping $\varphi\colon \mathcal{P} \to \mathcal{Q}$ over f.

Proof. The Comparison Theorem (cf. [36, p. 179]) provides a mapping $\Gamma\colon \mathcal{P} \to \mathcal{Q}$ over f. For each $n\geq 0$, we consider $\Gamma_n\colon P_n \to Q_n$. Since P_n is finitely generated, Γ_n decomposes into its homogeneous components. One readily verifies that the maps $\varphi_n := (\Gamma_n)_0$ define a degree 0 chain map over f. □

Remark. Suppose $R = \bigoplus_{i\in\mathbb{Z}} R_i$ is a noetherian F-algebra with supplementation $\epsilon\colon R \to F$ such that $\epsilon(R_i) = (0)$ for $i\neq 0$. Then F obtains the structure of a

$\mathbb{Z}$-graded $\mathbf{R}$-module via $r.\alpha := \epsilon(r)\alpha$. If $\mathbf{M}$ is a graded $\mathbf{R}$-module, then (3.1) in conjunction with standard arguments shows that the grading on the cohomology groups

$$H^n(\mathbf{R},\mathbf{M}) := \mathrm{Ext}^n_{\mathbf{R}}(F,\mathbf{M})$$

can be computed from any $\mathbb{Z}$-graded projective resolution $\mathcal{P}: \to P_n \to P_{n-1} \to \dots \to P_0 \to F \to (0)$ of F for which P_n is finitely generated for every $n\geq 0$.

For future reference we collect the following basic facts:

(3.2) Let $\mathbf{R}$ be a $\mathbb{Z}$-graded ring, $S \subset \mathbf{R}$ a graded subring of $\mathbf{R}$. Suppose that (a) $\mathbf{M}$ is a finitely generated graded $\mathbf{R}$-module,
(b) $\mathbf{R}$ is a finitely generated S-module, and
(c) N is a graded S-module,
then the natural isomorphism $\mathrm{Hom}_{\mathbf{R}}(\mathbf{M},\mathrm{Hom}_S(\mathbf{R},N)) \cong \mathrm{Hom}_S(\mathbf{M},N)$ is a map of degree 0.

(3.3) Let $\mathbf{R}$ and S be graded rings, $\mathbf{M},P$ graded $\mathbf{R}$-modules, N,Q graded S-modules. If $\mathbf{M}$ and N are projective and finitely generated, then $\mathbf{M}\otimes_F N$ is finitely generated over $\mathbf{R}\otimes_F S$ and the natural isomorphism

$$\mathrm{Hom}_{\mathbf{R}}(\mathbf{M},P)\otimes_F\mathrm{Hom}_S(N,Q) \cong \mathrm{Hom}_{\mathbf{R}\otimes_F S}(\mathbf{M}\otimes_F N,P\otimes_F Q)$$

is an isomorphism of degree 0.

(3.4) If X and Y are graded complexes, then $X\otimes_F Y$ is a graded complex and the Künneth isomorphism $\bigoplus_{p+q=n} H_p(X)\otimes_F H_q(Y) \cong H_n(X\otimes_F Y)$ has degree 0.

Throughout this section we shall be studying finite dimensional $\mathbb{Z}$-graded Lie algebras $L = \bigoplus_{i=-r}^{s} L_i$, where $1\leq r,s$. The subalgebras $\oplus_{i<0} L_i$ and $\oplus_{i>0} L_i$ will be denoted by L^- and L^+, respectively. The universal enveloping algebra $\mathcal{U}(L)$ inherits a $\mathbb{Z}$-gradation and the $\mathbb{Z}$-graded L-modules are just the $\mathbb{Z}$-graded $\mathcal{U}(L)$-modules.

Given an L-module V, we consider the ordinary cochain complex

$$C(L,V) := \oplus_{n\geq 0} C^n(L,V)$$

along with its coboundary operator $\delta: C(L,V) \to C(L,V)$ of degree 1,

where $\delta(f)(x_0,\dots,x_n) = \Sigma_{0\leq i<j\leq n}(-1)^{i+j} f([x_i,x_j], x_1,\dots,\hat{x}_i,\dots,\hat{x}_j,\dots,x_n) +$
$\Sigma_{i=0}^{n}(-1)^i x_i\cdot f(x_0,\dots,\hat{x}_i,\dots,x_n)$ for $f \in C^n(L,V)$.

As usual, the circumflexes indicate the omission of the corresponding variable. Recall that $C^n(L,V) = \mathrm{Hom}_F(\Lambda^n(L),V)$, and thereby $C(L,V)$ carries the natural L-module structure via

$$(x\cdot f)(x_1,\dots,x_n) := x\cdot f(x_1,\dots,x_n) - \Sigma_{i=1}^{n} f(x_1,\dots,[x,x_i],\dots,x_n)$$

Owing to [8, (23.6)]), δ is a homomorphism of L-modules, and L acts trivially on the cohomology groups

$$H^n(L,V) := H^n(C(L,V)) \quad n\geq 0.$$

One customarily writes $Z^n(L,V)$ and $B^n(L,V)$ for the spaces of n-cocycles and n-coboundaries, respectively. For a subalgebra $K \subset L$, we define the invariant subspace $V^K := \{v \in V;\ x\cdot v = 0\ \forall\ x \in K\}$. Given $f \in C^n(L,V)$, we define its *radical* via $\mathrm{rad}(f) = \{x \in L;\ f(x,y_1,\dots,y_{n-1}) = 0\ \ \forall\ y_1,\dots,y_{n-1} \in L\}$. If f is an n-cocycle, then rad(f) is a subalgebra of L.

In the sequel we shall be studying the cohomology of L with coefficients in $\mathbb{Z}$-graded L-modules. In case V is an irreducible L-module of finite dimension [17, (3.4)] and [9, (1.1)] show that $H^n(L,V)$ is trivial unless V admits a $\mathbb{Z}$-grading. Let $V = \oplus_{j\in\mathbb{Z}} V_j$ a $\mathbb{Z}$-graded L-module. Since L^* is finite dimensional $C^n(L,V)$ inherits a $\mathbb{Z}$-grading by setting $C^n(L,V)_k \cong \mathrm{Hom}_F(\Lambda^n(L),V)_k$. It readily follows that δ is a degree 0 map, so that the cohomology groups are also graded via

$$H^n(L,V)_k \cong {}^{\ker\delta\cap C^n(L,V)_k}/_{\mathrm{im}\delta\cap C^n(L,V)_k}.$$

We note that this gradation is obtained by considering the Koszul complex $\mathcal{X} := (\mathcal{U}(L)\otimes_F\Lambda^n(L))_{n\geq 0}$ over F. This complex inherits a gradation from $\mathcal{U}(L)$ and $\Lambda^n(L)$. The boundary operator ∂ on $\mathcal{X}$ is easily seen to have degree 0. Moreover, $\mathcal{U}(L)\otimes_F\Lambda^n(L)$ is finitely generated. Hence $\mathrm{Hom}_{\mathcal{U}(L)}(\mathcal{X},V)$ is a graded complex which is isomorphic to $C(L,V)$ by a chaim map of degree 0. Owing to (3.1) and its succeeding remark, the above gradation is computable from any finitely generated $\mathbb{Z}$-graded projective resolution of F.

The relative cohomology groups $H^n(L,L^-,V)$ are those of the subcomplex $C(L,L^-,V)$, where $C^n(L,L^-,V) := \{f \in C^n(L,V);\ x.f = 0 \quad \forall\ x \in L^-$ and $L^- \subset \mathrm{rad}(f)\}$. Note that $C^n(L,L^-,V)$ is a homogeneous subspace of $C^n(L,V)$. We shall consider the canonical homomorphisms $\Psi_2: H^2(L,L^-,V) \to H^2(L,V)$, $\Phi_2: H^2(L,V) \to$

$H^2(L^-,V)$ which have degree 0. It is obvious that $\Phi_2 \circ \Psi_2 = 0$.

The ensuing result provides a reduction method which has proven to be useful for the computation of the second cohomology groups.

THEOREM 3.5.: Let $k \in \mathbb{Z}$ be such that $H^1(L^-,V)_{k+j} = (0) \quad 0 \leq j \leq s$.
Then the following statements hold:

(1) Let $f \in Z^2(L,V)_k$ be a homogeneous cocycle of degree k such that the restriction of f to $L^- \times L^-$ cobounds. Then there is $g \in C^1(L,V)_k$ such that $L^- \subset \mathrm{rad}(f-\delta(g))$.

(2) The sequence $H^2(L,L^-,V)_k \to H^2(L,V)_k \to H^2(L^-,V)_k$ is exact and $\dim_F H^2(L,V)_k \leq \dim_F H^2(L^-,V)_k + \dim_F H^2(L,L^-,V)_k$.

<u>Proof</u>: (1) By assumption we can find $\tau \in C^1(L^-,V)_k$ such that $f|_{L^- \times L^-} - \delta(\tau) = 0$. We extend τ to a 1-cochain of L by setting $\tau(L^+ + L_0) = (0)$. Then $\tau \in C^1(L,V)_k$ and the 2-cocycle $\zeta := f - \delta(\tau)$ is homogeneous of degree k, and vanishes on $L^- \times L^-$.

We put $L_{[j]} := \Sigma_{i \leq j} L_i \, (j \geq -1)$ and proceed by defining inductively linear maps $g_j : L_{[j]} \to V$ of degree k such that

(a) $g_{i+1}|_{L_{[j]}} = g_j \quad j \geq -1$

(b) $e \cdot g_j(x) = g_j([e,x]) + \zeta(e,x) \quad \forall\, e \in L^-;\ \forall\, x \in L_{[j]}$.

Put $g_{-1} := 0$ and note that (b) is satisfied by definition of ζ. Now suppose that g_j has already been defined. For $x \in L_{j+1}$ we consider the linear mapping

$$h_x : L^- \to V;\ h_x(e) := g_j([e,x]) + \zeta(e,x) \quad \forall\, e \in L^-.$$

Clearly, h_x is homogeneous of degree k+j+1. Let e_1, e_2 be two elements of L^-. Then we obtain, observing $\delta(\zeta) = 0$,

$$\begin{aligned} h_x([e_1,e_2]) &= g_j([[e_1,e_2],x]) + \zeta([e_1,e_2],x) \\ &= g_j([e_1,[e_2,x]]) + g_j([[e_1,x]e_2]) + \zeta([e_1,x],e_2) - \zeta([e_2,x],e_1) + \\ &\quad + e_1 \cdot \zeta(e_2,x) - e_2 \cdot \zeta(e_1,x) + x \cdot \zeta(e_1,e_2). \end{aligned}$$

The last term vanishes and the inductive hypothesis entails that the first and fourth term combine to $e_1 \cdot g_j([e_2,x])$. By the same token the second and third term sum up to $-e_2 \cdot g_j([e_1,x])$. Consequently,

$$h_x([e_1,e_2]) = e_1\cdot g_j([e_2,x]) - e_2\cdot g_j([e_1,x]) + e_1\cdot\zeta(e_2,x) - e_2\cdot\zeta(e_1,x)$$
$$= e_1\cdot h_x(e_2) - e_2\cdot h_x(e_1),$$

proving that h_x is an element of $Z^1(L^-,V)_{k+j+1}$. The assumption $H^1(L^-,V)_{k+j+1} = (0)$ now implies, for any $x \in L_{j+1}$, the existence of $v(x) \in V_{k+j+1}$ such that $h_x(e) = e\cdot v(x) \quad \forall \ e \in L^-$. Consequently, there exists a linear mapping $\beta\colon L_{j+1} \to V_{k+j+1}$ satisfying $h_x(e) = e\cdot\beta(x)\ \forall\ x\in L_{j+1},\ e \in L^-$. We now define $g_{j+1}\colon L_{[j]}\oplus L_{j+1} \to V$ via $g_{j+1}|_{L_{[j]}} := g_j$ and $g_{j+1}|_{L_{j+1}} := \beta$. This mapping possesses the requisite degree, and obviously satisfies condition (a). Property (b) only needs to be tested for $x \in L_{j+1}$. Given $e \in L^-$ we obtain, observing $[e,x] \in L_{[j]}$,

$$e\cdot g_{j+1}(x) = e\cdot\beta(x) = h_x(e) = g_j([e,x]) + \zeta(e,x) = g_{j+1}([e,x]) + \zeta(e,x).$$

Note that $g_s\colon L \to V$ is a 1-cochain of degree k. Consider the cocycle $\lambda := \zeta - \delta(g_s)$. By construction of g_s, L^- is contained in $\mathrm{rad}(\lambda)$, and $g := \tau + g_s$ satisfies the requirements of (1).

(2): The result readily follows from (1) in conjunction with the observation that every element of $Z^2(L,V)_k$ whose radical comprises L^- is contained in $Z^2(L,L^-,V)_k$, the space of relative 2-cocycles of degree k. □

We shall apply the preceding result to the study of the cohomology of generalized Verma modules. Let M be an L_0-module. As in section 1 we extend the action of L_0 to L_0+L^+ by setting $L^+\cdot M = (0)$. Consider the *generalized Verma module* $V := \mathcal{U}(L)\otimes_{\mathcal{U}(L_0+L^+)}M$. The Poincaré-Birkhoff-Witt Theorem entails that the linear mapping $\xi\colon \mathcal{U}(L^-)\otimes_F M \to V$, which sends $u\otimes m$ to $u\otimes m$, is an isomorphism of $\mathcal{U}(L^-)$-modules, $\mathcal{U}(L^-)$ acting on $\mathcal{U}(L^-)\otimes_F M$ via left multiplication. By setting $V_i := \xi(\mathcal{U}(L^-)_i\otimes_F M)\ \forall\ i\in\mathbb{Z}$ we introduce a negative gradation on V: $V = \bigoplus_{j\le 0} V_j$. Evidently, $L_i\cdot V_j \subset V_{i+j}$ whenever $i>0$. If $i \ge 0$ we obtain for $x\in L_i$, $u\in\mathcal{U}(L^-)_j$, and $m \in M$

$$x\cdot(u\otimes m) = ux\otimes m + [x,u]\otimes m.$$

For $i=0$ we have $[x,u] \in \mathcal{U}(L^-)_j$ while $ux\otimes m = u\otimes x\cdot m \in V_j$. In order to treat the case $i>0$ we recall that the multiplication map of $\mathcal{U}(L)$ induces an isomorphism

$\mathcal{U}(L^-)\otimes_F\mathcal{U}(L_0+L^+) \to \mathcal{U}(L)$ of graded vector spaces. Consequently, the element $[x,u] \in \mathcal{U}(L)_{i+j}$ has a presentation $[x,u] = \Sigma_\ell\Sigma_k a_{\ell k}b_{\ell k}$, where $a_{\ell k} \in \mathcal{U}(L^-)_k$, $b_{\ell k} \in \mathcal{U}(L_0+L^+)_{i+j-k}$. The above identity then yields, observing $ux\otimes m = u\otimes x\cdot m = 0$, $x\cdot(u\otimes m) = \Sigma_\ell\Sigma_k \; a_{\ell k}\otimes b_{\ell k}\cdot m$. For $k \neq i+j$, the element $b_{\ell k}$ lies in $\mathcal{U}(L_0).(\ker\epsilon)$, where $\epsilon: \mathcal{U}(L^+) \to F$ denotes the canonical supplementation. As a result, $b_{\ell k}\cdot m = 0$ whence

$$x\cdot(u\otimes m) = \Sigma_\ell a_{\ell i+j}\otimes b_{\ell i+j}\cdot m \in V_{i+j}.$$

The author is indebted to G. Hochschild for providing him with the following result.

Lemma 3.6.: Let A be an abelian group, written additively. Suppose that $\{a_1,\ldots,a_n\}$ is a set of commuting endomorphisms of A such that for each $i<n$ we have

$$\bigcap_{j>i} \ker(a_i\circ a_j) = \ker(a_i) + \bigcap_{j>i} \ker(a_j).$$

If $f_i \in a_i(A)$ $1\leq i\leq n$ are given such that $a_j(f_i) = a_i(f_j)$ $1\leq i,j\leq n$, then there is an element $f \in A$ such that $a_i(f) = f_i$ $1\leq i\leq n$.

<u>Proof</u>: We proceed by induction on n. Noting the triviality of the case n=1 we suppose that n>1. By inductive hypothesis there is $g \in A$ such that $a_i(g) = f_i$ $2\leq i\leq n$. We write $f_1 = a_1(g_1)$ and consider the element $h: = g - g_1$. For $i > 1$ we have $a_1\circ a_i(h) = a_1(f_i) - a_i(f_1) = 0$. Consequently, there are elements $u,v \in A$ such that $a_1(u) = 0$, $a_j(v) = 0$ for $j>1$ and $h = u + v$. The element $f: = g - v$ is easily seen to possess the requisite property. □

If one defines a_i to be the left multiplication effected by X_i, then the following choices satisfy the requirements of (3.6):

a) the polynomial ring $F[X_1,\ldots,X_n]$ in n variables

b) the divided power algebra $\mathcal{A}(n;\underline{m})$ of dimension $p^{m_1+\ldots+m_n}$

THEOREM 3.7.: Let L be a finite dimensional graded Lie algebra and consider $V: = \mathcal{U}(L)\otimes_{\mathcal{U}(L_0+L^+)}M$, where M is an L_0-module. Suppose that $L^- = L_{-1}$ is n-dimensional. Then the following statements hold:

(1) $H^1(L^-,V)_\ell = (0)$ unless $\ell = n = 1$.

(2) If $n \geq 2$, then the sequence $H^2(L,L^-,V) \to H^2(L,V) \to H^2(L^-,V)$ is exact.

Proof: (1) By assumption L^- is abelian and $\mathcal{U}(L^-)$ is isomorphic to a polynomial ring in n variables. The L^--module V is, as was noted earlier, canonically isomorphic as an L^--module to the $\dim_F M$ - fold direct sum of $\mathcal{U}(L^-)$. Let $\varphi: L^- \to \mathcal{U}(L^-)$ be a homogeneous derivation of degree ℓ. Since $\mathcal{U}(L^-)$ is negatively graded and $L^- = L_{-1}$, we may assume that $\ell \leq 1$. Let $\{X_1, \dots, X_n\}$ be a basis of L^- over F and put $f_i: = \varphi(X_i)$ $1\leq i\leq n$. Then $f_i \in \mathcal{U}(L^-)_{\ell-1} \cong F[X_1,\dots,X_n]_{1-\ell}$. Note that the f_i satisfy the condition $X_if_j = X_jf_i$. We write $f_i = \Sigma_{q\geq 0}X_i^qh_{i,q}$; $h_{i,q} \in F[X_1,\dots,X_{i-1},X_{i+1},\dots,X_n]_{1-\ell-q}$ $0\leq q$, $1\leq i\leq n$. Then $X_jh_{i,o} = 0$ $\forall$ $j \neq i$. Consequently, $h_{i,o} = 0$ unless $n = 1$. In that case we have $f_1 = h_{1,o}X_1^{1-\ell}$. Hence f_1 is a multiple of X_1 unless $\ell = 1$. We now apply (3.6) in order to see that φ is an inner derivation in all other cases.

(2) This is a direct consequence of (1), and (2) of (3.5). □

Turning to the study of modular Lie algebras we shall henceforth assume that F is a field of positive characteristic p. We consider $K: = L_0+L^+$. An L_0-module V is given the structure of a $\mathbb{Z}$-graded K-module, by letting L^+ operate trivially and by setting $V = V_\mu$ for $\mu \in \mathbb{Z}$. Let $\{e_1,\dots,e_k\}$ be a basis of L^- consisting of homogeneous elements. Since the e_i operate nilpotently on L, there exist $m_1,\dots,m_k \in \mathbb{N}$ such that $z_i: = e_i^{p^{m_i}} \in \mathcal{U}(L^-)\cap\mathcal{Z}(\mathcal{U}(L))$. By construction the subalgebra $\mathcal{O}(L,K)$ is a graded subalgebra of $\mathcal{U}(L)$ and the winding automorphism α: $\mathcal{O}(L,K) \to \mathcal{O}(L,K)$ defined in section 1 is homogeneous of degree 0. Given a subalgebra $A \subset \mathcal{U}(L)$, we let A^+ be its intersection with the kernel of the canonical supplementation ϵ: $\mathcal{U}(L) \to F$. The Poincaré-Birkhoff-Witt Theorem ensures the validity of the following inclusions.

(i) $\mathcal{U}(K) = \bigoplus_{i\geq 0}\mathcal{U}(K)_i$; $\mathcal{U}(K)_i \subset \mathcal{U}(L_0)\mathcal{U}(L^+)^+$ $\forall$ $i>0$.

(ii) $\mathcal{U}(L^-) = \bigoplus_{i\leq 0}\mathcal{U}(L^-)_i$; $\mathcal{U}(L^-)_0 = F\cdot 1$

(iii) $\mathcal{O}(L,K)_i \subset \mathcal{Z}\mathcal{U}(L_0)\mathcal{U}(L^+)^+ \quad \forall\ i>0; \quad \mathcal{O}(L,K)_i \subset \mathcal{Z}^+\mathcal{U}(K) \quad \forall\ \ i<0,$

where $\mathcal{Z} = F[z_1,\dots,z_k]$ is a polynomial ring in k indeterminates.

For $a \in \mathbb{N}_0^k$ we write $\|a\|: = \Sigma_{i=1}^k a_i \deg(e_i)$. The k-tuple $(p^{m_1}-1,\dots,p^{m_k}-1)$ will be denoted τ.

PROPOSITION 3.8.: The following statements hold:

(1) Put $\mathrm{Ind}_K(V)_i: = \langle\{u\otimes v;\ u \in \mathcal{U}(L)_{i-\mu},\ v \in V\}\rangle\ \forall\ i \in \mathbb{Z}$. Then $\mathrm{Ind}_K(V) = \bigoplus_{i=\mu+\|\tau\|}^{\mu} \mathrm{Ind}_K(V)_i$ is a $\mathbb{Z}$-graded $\mathcal{U}(L)$-module.

(2) Put $\mathrm{Coind}_K(V)_i: = \{f; f(\mathcal{U}(L)_j) = (0) \quad \forall\ j \neq \mu-i\}$. Then $\mathrm{Coind}_K(V) = \bigoplus_{i=\mu}^{\mu-\|\tau\|} \mathrm{Coind}_K(V)_i$ is a $\mathbb{Z}$-graded $\mathcal{U}(L)$-module.

(3) There exists an isomorphism φ: $\mathrm{Ind}_K(V) \to \mathrm{Coin}_K({}_{\alpha}V)$ of degree $-\|\tau\|$.

<u>Proof</u>. (1) Using the multi-index notation of [39, p. 51], we recall that $\mathcal{U}(L)$ is a free left and right $\mathcal{O}(L,K)$-module on the basis $\mathcal{B}: = \{e^a;\ 0\leq a\leq\tau\}$. Since $\mathcal{O}(L,K)$ is a graded subalgebra, $\mathcal{U}(L)$ is a $\mathbb{Z}$-graded $\mathcal{O}(L,K)$-module. Note that $e^a \in \mathcal{U}(L)_{\|a\|}$ for every $a \in \{0,\dots,\tau\}$. Now let u be an element of $\mathcal{U}(L)_{i-\mu}$ and write $u = \Sigma_{0\leq a\leq\tau} e^a x_a$, where $x_a \in \mathcal{O}(L,K)_{i-\mu-\|a\|}$. For $v \in V$ we obtain, observing (iii),

$$u\otimes v = \Sigma_{0\leq a\leq\tau} e^a\otimes x_a v = \Sigma_{\|a\|=i-\mu} e^a\otimes x_a v.$$

Consequently, $\mathrm{Ind}_K(V)_i = \langle\{e^a\otimes v; \|a\| = i-\mu,\ v \in V\}\rangle$. This shows that $\mathrm{Ind}_K(V)$ has the desired vector space decomposition. By construction we have $\mathcal{U}(L)_i \mathrm{Ind}_K(V)_j \subset \mathrm{Ind}_K(V)_{i+j}$, proving that $\mathrm{Ind}_K(V)$ is a $\mathbb{Z}$-graded $\mathcal{U}(L)$ module.

(2) This is an immediate consequence of [18, (2.4)].

(3) Since $\alpha(x) = x \quad \forall \quad x \in L^+$, the winding automorphism α: $\mathcal{O}(L,K) \to \mathcal{O}(L,K)$ has degree 0. This implies that ${}_{\alpha}V$ is a $\mathbb{Z}$-graded K-module with $L^+\,{}_{\alpha}V = (0)$ and ${}_{\alpha}V = ({}_{\alpha}V)_{\mu}$. Following [21], we define an $\mathcal{O}(L,K)$-linear map π: $\mathcal{U}(L) \to \mathcal{O}(L,K)$ via

$$\pi(\Sigma_{0\leq a\leq\tau} x_a e^a) = x_\tau.$$

If u is an element of $\mathcal{U}(L)_i$, $u = \Sigma_{0\leq a\leq\tau} x_a e^a$ with $x_a \in \mathcal{O}(L,K)_{i-\|a\|}$, then $\pi(u) = x_\tau \in \mathcal{O}(L,K)_{i-\|\tau\|}$. According to [21, (1.4)] the map φ: $\mathrm{Ind}_K(V) \to \mathrm{Coind}_K({}_{\alpha}V)$ which is given by

$$\varphi(u\otimes v)(u') = \pi(u'u)v$$

is an isomorphism of $\mathcal{U}(L)$-modules. To verify the assertion concerning its degree, let u be an element of $\mathcal{U}(L)_{i-\mu}$ so that $u\otimes v \in \mathrm{Ind}_K(V)_i$. Assuming $u' \in \mathcal{U}(L)_j$, we obtain $\pi(u'u) \in \mathcal{O}(L,K)_{i+j-\mu-\|\tau\|}$. Property (iii) now shows that $\varphi(u\otimes v)(u') = 0$ unless $i+j-\mu-\|\tau\| = 0$. Hence $\varphi(u\otimes v) \in \mathrm{Coind}_K({}_{\alpha}V)_{i-\|\tau\|}$, as required. □

Following Shen [38] we call a $\mathbb{Z}$-graded L-module $M = \bigoplus_{j=\ell}^{t} M_j$ *transitive* if $M_\ell = \{m \in M;\ L^- m = (0)\}$. The Lie algebra L is said to have *bounded representation type* if there exists a natural number n such that $\dim_F M \leq n$ for every indecomposable L-module M.

Corollary 3.9.: (1) Let V be an irreducible L_0-module. Then $\mathrm{Ind}_K(V)$ is indecomposable and not projective.
(2) The Lie algebra L is not of bounded representation type.

Proof. (1) Setting $V = V_0$ we obtain from Proposition 3.8 an isomorphism

$$\mathrm{Ind}_K(V) \cong \mathrm{Coind}_K({}_{\alpha}V)$$

of $\mathbb{Z}$-graded L-modules. Owing to [18, (2.4)] the latter module is transitively graded and the L_0-modules ${}_{\alpha}V$ and $\mathrm{Coind}_K({}_{\alpha}V)_0$ are isomorphic. Consequently, $\mathrm{Ind}_K(V)$ has a transitive grading and $\mathrm{Ind}_K(V)_{\|\tau\|}$ is an irreducible L_0-module.

Now let $X \neq (0)$ be an L-submodule of $\mathrm{Ind}_K(V)$. Since L^- operates on X by nilpotent transformations, the Engel-Jacobson Theorem ensures that $X_0 := \{x \in X;\ L^- x = (0)\}$ is a nonzero subspace of $\mathrm{Ind}_K(V)$. Hence X_0 is a non-trivial L_0-submodule of $\mathrm{Ind}_K(V)_{\|\tau\|}$ and X therefore contains $\mathrm{Ind}_K(V)_{\|\tau\|}$. This shows that $\mathrm{Ind}_K(V)$ is indecomposable. As $\mathrm{Ind}_K(V)$ is not torsion-free, it is not projective.

(2) Let n be given and pick $\ell \geq m_1$ such that $n < p^\ell$. Put $z_1' := e_1^{p^\ell}$ and consider the algebra $\mathcal{O}_\ell(L,K) := \mathrm{alg}_F(K \cup \{z_1', z_2, \ldots, z_k\})$. According to (1) the module $\mathrm{Ind}_K(F)$ is indecomposable of dimension $\mathrm{rk}_{\mathcal{O}_\ell(L,K)}\mathcal{U}(L) \geq p^\ell > n$. □

Let $G = \bigoplus_{i=1}^{k} Fz_i$ be the k-dimensional abelian $\mathbb{Z}$-graded Lie algebra with $Fz_i = G_{p^{m_i}\deg(e_i)}$.

THEOREM 3.10.: Let $V = V_\mu$ be an L_0-module. Then there exist isomorphisms

$$H^n(L,\mathrm{Ind}_K(V))_i \cong \bigoplus_{p+q=n} \bigoplus_{r+s=i-\|\tau\|} \Lambda^p(G)_r \otimes_F H^q(K,{}_\alpha V)_s.$$

Proof. Proposition 3.8 readily yields an isomorphism $H^n(L,\mathrm{Ind}_K(V))_i \cong H^n(L,\mathrm{Coind}_K({}_\alpha V))_{i-\|\tau\|}$. We consider the Koszul complex $(\mathcal{X}_q)_{q\geq 0}$ of F over $\mathcal{U}(L)$, where $\mathcal{X}_q = \mathcal{U}(L)\otimes_F \Lambda^q(L)$ and $u\cdot(u'\otimes v) = uu'\otimes v$. As mentioned before, the L-module $\mathcal{X}_q$ inherits a $\mathbb{Z}$-grading from $\mathcal{U}(L)$ and $\Lambda^q(L)$ and thereby obtains the structure of a finitely generated $\mathbb{Z}$-graded $\mathcal{U}(L)$-module. Owing to (3.2) there are natural degree 0 isomorphisms $\mathrm{Hom}_{\mathcal{U}(L)}(\mathcal{X}_q,\mathrm{Coind}_K({}_\alpha V) \cong \mathrm{Hom}_{\mathcal{O}(L,K)}(\mathcal{X}_q,{}_\alpha V)$.
Now let G′ be the abelian $\mathbb{Z}$-graded Lie algebra with basis $\{z_1,\dots,z_n\}$ and $\deg(z_i) := -p^{m_i}\deg(e_i)$. Then $\mathcal{O}(L,K)$ is the universal enveloping algebra of the direct sum $G'\oplus K$. Since X_q is a finitely generated $\mathbb{Z}$-graded projective $\mathcal{O}(L,K)$-module for every q, (3.1) and its succeeding remark ensure the existence of a degree 0 isomorphism in cohomology

$$H^n(L,\mathrm{Coind}_K({}_\alpha V))_{i-\|\tau\|} \cong H^n(G'\oplus K,{}_\alpha V)_{i-\|\tau\|}.$$

To determine the latter module we consider the Koszul complexes $\mathcal{X}$ and $\mathcal{Y}$ over F with respect to G′ and K, respectively. By virtue of [7, (2.7) p. 166] $\mathcal{X}\otimes_F\mathcal{Y}$ is a finitely generated $\mathbb{Z}$-graded projective resolution of F over $\mathcal{O}(L,K) \cong \mathcal{U}(G'\oplus K)$ and (3.3) provides a degree 0 isomorphism of graded complexes

$$\mathrm{Hom}_{\mathcal{O}(L,K)}(\mathcal{X}\otimes_F\mathcal{Y},F\otimes_F({}_\alpha V)) \cong \mathrm{Hom}_{\mathcal{Z}}(\mathcal{X},F)\otimes_F\mathrm{Hom}_{\mathcal{U}(K)}(\mathcal{Y},{}_\alpha V).$$

By passing to cohomology, while observing (3.4), we obtain isomorphisms

$$\begin{aligned} H^n(G'\oplus K,{}_\alpha V)_{i-\|\tau\|} &\cong \Big(\bigoplus_{p+q=n} H^p(G',F)\otimes_F H^q(K,{}_\alpha V)\Big)_{i-\|\tau\|} \\ &\cong \bigoplus_{p+q=n} \bigoplus_{r+s=i-\|\tau\|} H^p(G'F)_r \otimes_F H^q(K,{}_\alpha V)_s. \end{aligned}$$

Since $H^p(G',F)_r \cong \mathrm{Hom}_F(\Lambda^p(G'),F)_r \cong \Lambda^p(G)_r$, the asserted identity follows. □

Suppose now that $K = \oplus_{i\geq 0} K_i$ is graded and generated by $K_0\oplus K_1\oplus \dots \oplus K_t$ $(t>0)$. We let V be a K_0-module and extend the given action to K by setting $K^+V = (0)$.

Note that, for $\mu \in \mathbb{Z}$, this gives V the structure of a graded K-module such that $V = V_\mu$.

Lemma 3.11.: We have $H^1(K,V) = \bigoplus_{i=\mu-t}^{\mu} H^1(K,V)_i$, where $H^1(K,V)_i \cong \mathrm{Hom}_{K_0}(K_{\mu-i}/[K^+,K^+]_{\mu-i}, V)$ for $\mu-t\leq i<\mu$ and $H^n(K,V)_\mu \cong H^n(K_0,V)$ $\forall$ $n\geq 0$.

<u>Proof</u>. Let $f \in Z^1(K,V)$ be a 1-cocycle of degree ℓ. If $\ell<\mu-t$, then f obviously annihilates the generating set of L, proving that f=0. For $\ell>\mu$ we arrive at the same conclusion. This verifies the first assertion. Suppose that $\mu-t\leq\ell<\mu$. If $x \in K_0$ and $y \in K_{\mu-\ell}$, then we obtain

$$f([x,y]) = x.f(y) - y.f(x) = x.f(y),$$

thereby qualifying $f|_{K_{\mu-\ell}}$ as an element of $\mathrm{Hom}_{K_0}(K_{\mu-\ell},V)$. If $x \in K_i$ and $y \in K_j$ with $i,j\geq 1$ and $i+j = \mu-\ell$, then $f([x,y]) = 0$, as K^+ annihilates V. Noting that $B^1(K,V)_\ell = (0)$ we obtain a mapping

$$\Gamma\colon H^1(K,V)_\ell \to \mathrm{Hom}_{K_0}(K_{\mu-\ell}/[K^+,K^+]_{\mu-\ell}, V).$$

Let ψ be an element of $\mathrm{Hom}_{K_0}(K_{\mu-\ell}/[K^+,K^+]_{\mu-\ell}, V)$, φ the corresponding K_0-linear map that vanishes on $[K^+,K^+]_{\mu-\ell}$. We define f via $f|_{K_i} = \delta_{i,\mu-\ell}\varphi$. A case by case analysis shows that f is a 1-cocycle of degree ℓ which is a preimage of ψ under Γ. Since Γ is obviously an injective map, we have established the asserted isomorphism.

In case $\ell=\mu$ we note that the canonical restriction map $H^n(K,V) \to H^n(K_0,V)$ induces a mapping $H^n(K,V)_\mu \to H^n(K_0,V)$. For an element γ of $C^n(K_0,V)$ we define an n-cochain $\Phi_n(\gamma) \in C^n(K,V)_\mu$ by requiring that $\Phi_n(\gamma)(x_1,\ldots,x_n) = (0)$ whenever one of the x_i is not contained in K_0. The Φ_n are readily seen to define a chain map $C(K_0,V) \to C(K,V)$. The resulting homomorphisms $H^n(K_0,V) \to H^n(K,V)_\mu$ invert the above mappings. □

We return to our general situation, where $L = \bigoplus_{i=-r}^{s} L_i$, with $r,s>0$.

Corollary 3.12. Let $V = V_\mu$ be an L_0-module.

(1) There is an embedding $H^n(L_0,{}_aV) \hookrightarrow H^n(L,\mathrm{Ind}_K(V))_{\mu+\|\tau\|}$ $\forall$ $n\geq 0$.

(2) There are isomorphisms $H^1(L,\mathrm{Ind}_K(V))_i \cong G_{i-\mu-\|\tau\|}\otimes_F({}_aV)^{L_0} \oplus H^1(K,{}_aV)_{i-\|\tau\|}$.

(3) There is an embedding $\Lambda^n(G)_{i-\mu-\|\tau\|}\otimes_F({}_aV)^{L_0} \hookrightarrow H^n(L,\mathrm{Ind}_K(V))_i$

Proof. (1): Lemma 3.11 provides an isomorphism $H^n(L_0,{}_aV) \cong H^n(K,{}_aV)_\mu$. The latter space is, owing to (3.10) a direct summand of $H^n(L,\mathrm{Ind}_K(V))_{\mu+\|\tau\|}$.
(2),(3): These are direct consequences of the reduction formula (3.10). □

Corollary 3.13.: Suppose that F is algebraically closed and let $V = V_\mu$ be an irreducible L_0-module of dimension $\dim_F V > 1$.

(1) Suppose that ${}_aV$ is isomorphic to $L_{\mu-i+\|\tau\|}/[L^+,L^+]_{\mu-i+\|\tau\|}$. Then $H^1(L,\mathrm{Ind}_K(V))_i$ is one dimensional if $\mu-s+\|\tau\|\leq i<\mu+\|\tau\|$, isomorphic to $H^1(L_0,{}_aV)$ for $i=\mu+\|\tau\|$, and trivial otherwise

(2) If $\dim_F V \geq \dim_F L_{\mu-i+\|\tau\|}/[L^+,L^+]_{\mu-i+\|\tau\|}$ and ${}_aV$ is not isomorphic to $L_{\mu-i+\|\tau\|}/[L^+,L^+]_{\mu-i+\|\tau\|}$, then $H^1(L,\mathrm{Ind}_K(V))_i = (0)$ for $i \neq \mu+\|\tau\|$ and $H^1(L,\mathrm{Ind}_K(V))_{\mu+\|\tau\|} \cong H^1(L_0,{}_aV)$.

Proof. (1) Since the irreducible L_0-module ${}_aV$ is not one-dimensional, its invariant space is trivial. The isomorphism of (3.12) thus reduces to

$$H^1(L,\mathrm{Ind}_K(V))_i \cong H^1(K,{}_aV)_{i-\|\tau\|}.$$

Setting t=s in (3.11) the latter groups are seen to be isomorphic to $\mathrm{Hom}_{L_0}(L_{\mu-i+\|\tau\|}/[L^+,L^+]_{\mu-i+\|\tau\|},{}_aV)$ if $\mu-s+\|\tau\|\leq i<\mu+\|\tau\|$. Our present assumption in conjunction with Schur's Lemma now yields the desired result. Applying (3.11) again we obtain the assertion for the remaining values of i.

(2) The assumptions ensure that $\mathrm{Hom}_{L_0}(L_{\mu-i+\|\tau\|}/[L^+,L^+]_{\mu-i+\|\tau\|},{}_aV) = (0)$ and the arguments of (1) now yield the desired result. □

§4 APPLICATIONS.

In this section we shall provide some examples that illustrate the utility of our techniques. Most of the results have already been dealt with elsewhere, yet with considerably more computational effort.

Before studying concrete examples, we provide a subsidiary result concerning dimensionally nilpotent Lie algebras. These were studied in a series of papers by J. M. Osborn. In particular, in [34] he gives various examples for these algebras. By using cohomological methods we shall augment his main result [34, (3.1)].

LEMMA 4.1.: Let $\rho\colon L \to gl(V)$ be an ℓ-dimensional representation of a k-dimensional Lie algebra L. Suppose that $x \in L$ is ad-nilpotent. If

(a) $(\text{ad } x)^{k-1} \neq 0$

(b) $\rho(x)$ is nilpotent

(c) there is $\psi \in \text{Der}_F(L,V)$ such that $\rho(x)^{\ell-1}(\psi(x)) \neq 0$,

then $f\colon L\oplus V \to L\oplus V$; $f(y+v) = [x,y]+\psi(y)+x\cdot v$ is a nilpotent derivation of the semidirect product $L\oplus V$ such that $f^{\ell+k-1} \neq 0$.

REMARK. Condition (c) necessitates the non-triviality of $H^1(L,V)$. For any inner derivation $\psi = \text{ad } v$ one obtains $\rho(x)^{\ell-1}(\psi(x)) = \rho(x)^{\ell}(v) = 0$.

<u>Proof</u>. The fact that f is a nilpotent derivation can be verified by direct computation. Owing to condition (a) $Fx = \ker(\text{ad}x) = \text{ad}^{k-1}(L)$ is one dimensional, thereby ensuring that there is $y \in L$ such that $x = (\text{ad}x)^{k-1}(y)$. It follows that $f^{k-1}(y) = x + w$ for some $w \in V$. Hence $f^k(y) = \psi(x) + \rho(x)(w)$ and $f^{k+t}(y) = \rho(x)^t(\psi(x)) + \rho(x)^{t+1}(w)\ \forall\ t\geq 1$. In particular, we have $f^{k+\ell-1}(y) = \rho(x)^{\ell-1}(\psi(x)) \neq 0$, as desired. □

EXAMPLE 4.2. Let $\underline{\underline{r}} < \underline{\underline{s}}$ be two k-tuples with positive integers as entries. We consider the divided power algebra $\mathcal{A}(k;\underline{\underline{s}}) = \langle\{x^{(a)};\ 0\leq a\leq\tau\}\rangle$, where $\tau := (p^{s_1}-1,\dots,p^{s_k}-1)$. The Jacobson-Witt algebra $W(k;\underline{\underline{r}})$ operates naturally on $\mathcal{A}(k;\underline{\underline{s}})$. Moreover, $\mathcal{A}(k;\underline{\underline{s}}) = \bigoplus_{j=0}^{\mu}\mathcal{A}(k;\underline{\underline{s}})_j$ is a transitively graded $W(k;\underline{\underline{r}})$-module with $\mu = |\tau| = -\|\tau\|$. Consider the graded $W(k;\underline{\underline{r}})_0$-module $V = V_\mu = \mathcal{A}(k;\underline{\underline{s}})_\mu = Fx^{(\tau)}$. We put $K := \Sigma_{i\geq 0}\ W(k;\underline{\underline{r}})_i$ and set $z_i := \partial_i^{p^{s_i}}$ $1\leq i\leq k$. By

construction, the z_i are central annihilators of $\mathcal{A}(k;\underline{s})$. Since $W(k;\underline{r})^{+}V = (0)$ there exists a degree 0 homomorphism φ: $\mathrm{Ind}_K(V) \to \mathcal{A}(k;\underline{s})$ such that $\varphi(u\otimes v) = u.v$. As $x^{(\tau)}$ is a generator of $\mathcal{A}(k;\underline{s})$ and both modules have the same dimension φ is an isomorphism.

The winding automorphism α of $\mathcal{O}(W(k;\underline{r}),K)$ is the identity on K^+ and sends the basis element $x^{(\epsilon_i)}\partial_j$ of $W(k;\underline{r})_0$ to $x^{(\epsilon_i)}\partial_j + \delta_{ij}1$. As a result, ${}_{\alpha}V$ is the trivial one-dimensional K-module F. It now follows from (3) of (3.12) that $H^n(W(k;\underline{r}),\mathcal{A}(k;\underline{s}))$ does not vanish for $0\leq n\leq k$. Since $W(k;\underline{r})_0$ is isomorphic to $gl(k)$ part (1) of (3.12) specializes to an embedding $H^n(gl(k),F) \hookrightarrow H^n(W(k;\underline{r}),\mathcal{A}(k;\underline{s}))_0$. The former groups can, in principle, be determined by the methods employed in [23]. Moreover, by applying (3.12(2)) and (3.11) consecutively one can compute the first cohomology groups entirely. For future reference we confine our attention to derivations of degree $\mu+1$. Owing to (3.12 (3)) there is an embedding $G_{1-\|\tau\|} \hookrightarrow H^1(W(k;\underline{r}),\mathcal{A}(k;\underline{s}))_{\mu+1}$. Hence the latter groups will not vanish if k=1.

Example 4.3: An n-dimensional Lie algebra is referred to a *dimensionally nilpotent* if it possesses a nilpotent derivation D such that $D^{n-1} \neq 0$.

We shall show that the semidirect product $L: = W(1;\underline{r})\oplus\mathcal{A}(1;\underline{s})$ is dimensionally nilpotent. According to our previous example, there exists a (necessarily outer) non-zero derivation ψ: $W(1;\underline{r}) \to \mathcal{A}(1;\underline{s})$ of degree $\mu+1$. Since ψ annihilates the subalgebra K, we have $\psi(\partial) \neq 0$. Consequently, $\partial^{\mu}\cdot\psi(\partial) \neq 0$. Noting that ∂ operates nilpotently on $\mathcal{A}(1;\underline{s})$ and $W(1;\underline{r})$, and $(\mathrm{ad}\partial)^{q-1} \neq 0$, where $q: = \dim_F W(1,\underline{r})$, we obtain the assertion directly from (4.1).

Example 4.4: Our final example concerns the second cohomology group of the p-dimensional Witt-algebra $W(1;\underline{1})$ with coefficients in its (p-1)-dimensional irreducible $\mathbb{Z}$-graded module $V: = \mathcal{A}(1;\underline{1})/F\cdot 1$. Our result corrects an error in [12, Thm. 2]. We shall assume that $p\geq 5$.

Let $f \in Z^2(W(1;\underline{1}),V)$ be a homogeneous cocycle of degree k. Since $V = \bigoplus_{i=1}^{p-1} Fv_i$, f evidently vanishes for $k\geq 2p$ and $k\leq -2p$. As $x\partial.f = kf$ is a boundary, we only need to consider the cases $k = -p,0,p$. If $k = -p$, then

$\{\partial, x^{(3)}\partial\}$ is easily seen to be contained in rad(f). Since the former set generates $W(1;\underline{1})$, we obtain $f = 0$.

Now assume that $k=p$. Writing $e_i := x^{(i+1)}\partial$ for $-1 \le i \le p-2$ we define an alternating bilinear mapping $f_p: W(1;\underline{1}) \times W(1;\underline{1}) \to V$ of degree p via $f_p(e_i, e_j) = (i-j)\delta_{-1,i+j}v_{p-1}$. Consequently, $\delta(f_p)$ has degree p and we have $\delta(f_p)(e_i, e_j, e_\ell) = 0$ unless $-1 \le i < j < \ell$, $i+j+\ell \le -1$. Hence $\delta(f_p) = 0$ and f_p is a cocycle with $f_p(e_0, e_{-1}) = v_{p-1}$. Since every coboundary of degree p vanishes on (e_0, e_{-1}), it follows that f_p defines a non-zero element of $H^2(W(1;\underline{1}), V)$ of degree p. Returning to our originally chosen cocycle f, we put $f(e_0, e_{-1}) = \alpha v_{p-1}$ and consider $\psi := f - \alpha f_p$. Then $\psi(e_0, e_{-1}) = 0$ and since ψ has degree p we obtain $\{e_{-1}, e_2\} \subset \mathrm{rad}(\psi)$, whence $\psi = 0$ and $f = \alpha f_p$.

Finally, we consider the case where $k = 0$. By virture of (3.5) there exists a linear mapping $g: W(1;\underline{1}) \to V$ of degree 0 such that e_{-1} is contained in the radical of $\varphi := f - \delta(g)$. Let $f_0: W(1;\underline{1}) \times W(1;\underline{1}) \to V$ denote the alternating map of degree 0 defined via $f_0(e_i, e_j) := (i+j)^{-1}(i+1)(j+1)(j-i)v_{i+j}$ $1 \le i+j \le p-1$. Then f_0 is a cocycle such that $f_0(e_0, e_{p-2}) = -v_{p-2}$. Hence, if $\varphi(e_0, e_{p-2}) = \lambda v_{p-2}$, then $\varphi + \lambda f_0$ annihilates (e_0, e_{p-2}). As $e_{-1} \in \mathrm{rad}(\varphi + \lambda f_0)$ it follows that $Fe_0 \oplus Fe_{-1} \subset \mathrm{rad}(\varphi + \lambda f_0)$. Since the latter space has even codimension in $W(1;\underline{1})$, it also contains e_1. Consequently, $\varphi + \lambda f_0$ vanishes. Since every coboundary of degree 0 annihilates (e_0, e_{p-2}), f_0 gives rise to a non-trivial element of $H^2(W(1;\underline{1}), V)$ of degree 0. This proves that $H^2(W(1;\underline{1}), V)$ has dimension 2.

REFERENCES

[1] Andersen, H.H., Jantzen, J.C.: Cohomology of Induced Representations for Algebraic Groups. Math. Ann. **269** (1984) 487-525

[2] Anderson, F.W.; Fuller, K.: Rings and Categories of Modules. GTM 13 Springer Verlag New York Heidelberg Berlin 1974

[3] Barnes, D.W.: On the Cohomology of Soluble Algebras. Math. Z. **101** (1967) 343-349

[4] Barnes, D.W.: First Cohomology Groups of Soluble Lie Algebras. J. Algebra **46** (1977) 292-297

[5] Bell, A.D.; Farnsteiner, R.: On the Theory of Frobenius Extensions and its Application to Lie superalgebras. (preprint).

[6] Block, R.E.: On the Extensions of Lie Algebras. Canad. J. Math. 20 (1968) 1439-1450

[7] Cartan, H.; Eilenberg, S.: Homological Algebra. Princeton Univ. Press, Princeton, N.J. 1956

[8] Chevalley, C.; Eilenberg, S.: Cohomology Theory of Lie Groups and Lie Algebras. Trans. Amer. Math. Soc. **63** (1984) 85-124.

[9] Chiu, S.; Shen, G.: Cohomology of Graded Lie Algebras of Cartan Type of Characteristic p. Abh. Math. Sem. Univ. Hamburg **57** (1986) 139-156

[10] Dixmier, J.: Cohomologie des Algèbres de Lie Nilpotentes. Acta Math. (Szeged) **16** (1955) 246-250

[11] Dzhumadil′daev, A.S.: On the Cohomology of Modular Lie Algebras. Math. USSR Sbornik **47** (1984) 127-143

[12] Dzhumadil′daev, A.S.: Abelian Extensions of Modular Lie Algebras. Algebra and Logic **24** (1985) 1-6

[13] Farnsteiner, R.: The Associative Forms of the Graded Cartan Type Lie Algebras. Trans. Amer. Math. Soc. **295** (1986) 417-427

[14] Farnsteiner, R.: Central Extensions and Invariant Forms of Graded Lie Algebras. Algebras, Groups, Geom. **3** (1986) 431-455

[15] Farnsteiner, R.: Dual Space Derivations and $H^2(L,F)$ of Modular Lie Algebras. Canad. J. Math. **39** (1987) 1078-1106

[16] Farnsteiner, R.: On the Vanishing of Homology and Cohomology Groups of Associative Algebras. Trans. Amer. Math. Soc. **306** (1988) 651-665

[17] Farnsteiner, R.: Cohomology Groups of Infinite Dimensional Algebras. Math. Z. **199** (1988) 407-423

[18] Farnsteiner, R.: Extension Functors of Modular Lie Algebras. Math. Ann. (to appear)

[19] Farnsteiner, R.: On the Cohomology of Ring Extensions. Adv. in Math. (to appear)

[20] Farnsteiner, R.: Cohomology Groups of Reduced Enveloping Algebras. Math. Z. (to appear)

[21] Farnsteiner, R.; Strade, H.: Shapiro's Lemma and its Consequences in the Cohomology Theory of Modular Lie Algebras. Math. Z. (to appear)

[22] Friedlander, E.M.; Parshall, B.J.: Cohomology of Infinitesimal and Discrete Groups. Math. Ann. **273** (1986) 353-374

[23] Friedlander, E.M.; Parshall, B.J.: Cohomology of Lie Algebras and Algebraic Groups. Amer. J. Math. **108** (1986) 235-253

[24] Friedlander, E.M.; Parshall, B.J.: Modular Representation Theory of Lie Algebras. Amer. J. Math. **110** (1988) 1055-1094

[25] Hochschild, G.P.: Cohomology of Restricted Lie Algebras. Amer. J. Math. **76** (1954) 555-580

[26] Hochschild, G.P.; Serre, J.P.: Cohomology of Lie Algebras. Ann. of Math. **57** (1953) 591-603

[27] Humphreys, J.: Restricted Lie Algebras (and Beyond). Contemp. Math. **13** (1982) 91-98

[28] Jacobson, N.: Abstract Derivation and Lie Algebras. Trans. Amer. Math. Soc. **42** (1937) 206-224

[29] Jantzen, J.C.: Restricted Lie Algebra Cohomology. Algebraic Groups, Utrecht 1986. Springer Lecture Note **1271** (1987) 91-108

[30] Larson, R.G.; Sweedler, M.: An Orthogonal Bilinear Form for Hopf Algebras. Amer. J. Math. **91** (1986) 75-94

[31] Leger, G.; Luks, E.: Cohomology Theorems for Borel-Like Solvable Lie Algebras in Arbitrary Characteristic. Canad. J. Math. **6** (1972) 1019-1026

[32] Loupias, M.: Représentations Indécomposables de Dimension Finie des Algèbres de Lie. Manuscripta Mat. **6** (1972) 365-379

[33] Malliavin, M.P.: Cohomologie d'Algèbres de Lie Nilpotentes, et Caractéristiques d'Euler-Poincaré. Bull. Sc. Math. 100 (1976) 269-287

[34] Osborn, J.M.: Examples of Dimensionally Nilpotent Lie Algebras in Prime Characteristic. Algebras, Groups, Geom. (to appear)

[35] Pareigis, B.: Kohomologie von Lie-p-Algebren. Math. Z. **104** (1968) 281-336

[36] Rotman, J.: An Introduction to Homological Algebra. Academic Press, San Francisco 1979

[37] Shen, G.: Graded Modules of Graded Lie Algebras of Cartan Type I. Scientica Sinica **29** (1986) 570-581

[38] Shen, G.: Graded Modules of Graded Lie Algebras of Cartan Type II. Scientica Sinica **29** (1986) 1009-1019

[39] Strade, H.; Farnsteiner, R.: Modular Lie Algebras and their Representations. Marcel Dekker vol. 116 New York 1988

[40] Sweedler, M.: Integrals for Hopf Algebras. Ann. Math. **89** (1969) 323-335

Derivations and central extensions of a generalized Witt algebra

Toshiharu IKEDA

Department of Mathematics, Hiroshima University, Hiroshima, 730, Japan

Abstract

In this paper we describe the derivation algebra and central extensions of a finitely generated generalized Witt algebra. Especially explicit forms of a derivation of degree zero and a universal covering algebra are given.

1. Introduction

Let k be a field of characteristic zero, and let G be an additive subgroup of $\prod_{i\in I} k_i^+$, where I is a non-empty index set and k_i^+ $(i \in I)$ are copies of the additive group k^+. A generalized Witt algebra $W(G,I)$ is a Lie algebra which have a basis $\{w(g,i) \mid g \in G,\ i \in I\}$ over k in bijective correspondence with $G \times I$, subject to the Lie multiplication

$$[w(g,i), w(h,j)] = g_j w(g+h,i) - h_i w(g+h,j),$$

where $i, j \in I$ and $g = (g_i)_{i\in I}$, $h = (h_i)_{i\in I} \in G$. This algebra is a natural generalization of a Witt algebra, whose several generalizations have been considered by many authors(cf. [1], [8], [10], [11], [12], [14]). If $|I| = 1$ and $G = \mathbb{Z}$, then $W(G,I)$ is the well known centerless Virasoro algebra or the Witt algebra $W_{\mathbb{Z}}$.

In a recent paper [7] the author and Kawamoto have shown that any derivation of $W(G,I)$ is a sum of a locally inner derivation and a derivation of degree zero, in particular if $|I| = n$ and $G = \mathbb{Z}^n$ then the derivations of $W(G,I)$ are inner. In this paper we sharpen these results and make a more detailed study of the derivations and the central extensions of a finitely generated Lie algebra $W(G,I)$. In section 2 we determine the derivations of degree zero for a simple algebra $W(G,I)$. By making use of this result, we investigate the derivation algebra of a general finitely generated Lie algebra $W = W(G,I)$ and show that the dimension of $H^1(W,W)$ is equal to $\text{rank}G - |I| + (|I|+2)\dim\zeta(W)$. In section 3 we study the central extensions of W. We describe an explicit form of a universal covering algebra, and determine the dimension of $H^2(W,k)$.

We simply write W instead of $W(G,I)$ if there would be no confusion. For every $g \in G$ let W_g be the subspace of W spanned by $\{w(g,i) \mid i \in I\}$. A derivation δ of $W(G,I)$ is said to have degree a if $W_g\delta \subset W_{g+a}$ for any $g \in G$, and hence every W_g is invariant under a derivation of degree zero. We call a derivation δ of W locally inner as in [6] if for any finite-dimensional subspace F of W there exists $x \in W$ such that $\delta|_F = \text{ad}x|_F$. We denote by $\text{Der}(W)$, $\text{Inn}(W)$, $\text{Lin}(W)$ and $\text{Der}(W)_0$ respectively the derivations of W, the inner derivations of W, the locally inner derivations of W and the derivations of W of degree zero.

2. Derivations of $W(G,I)$

We begin by stating some fundamental results on the structure of $W(G,I)$, which have been shown essentially by Kawamoto in [11].

LEMMA 1. (1) *$W = W(G,I)$ has the unique maximal abelian ideal R, and W is a split extension of R by a simple subalgebra S of W, where S is isomorphic to $W(H,J)$ for some $J(\subset I)$ and $H(\simeq G)$.*
(2) *If W is finitely generated then there exist $n \in \mathbf{N}$ and a finitely generated subgroup G of $\oplus_{i=1}^n k_i^+$ such that $W \simeq W(G,I_n)$, where $I_n = \{1,2,\cdots,n\}$.*
(3) *$W(G,I)$ is simple if and only if G is total, that is, $\oplus_{j\in J}k_j$ is spanned by $p_J(G)$ for any finite-subset J of I, where $p_J : \oplus_{i\in I}k_i \to \oplus_{j\in J}k_j$ is the canonical projection.*

The following lemma makes it easy to calculate the derivations and central extensions of W.

LEMMA 2. *Let G, H be additive subgroups of k^n and suppose that $G = HX$ for some $X = (x_{ij}) \in GL(n,k)$. Then $W(G,I_n) \simeq W(H,I_n)$.*

PROOF. Let $\{w(g,i) \mid g \in G,\ i \in I_n\}$ and $\{w'(h,i) \mid h \in H,\ i \in I_n\}$ be canonical bases of $W(G,I_n)$ and $W(H,I_n)$ respectively. Since X is nonsingular and $G = HX$, we can define a linear isomorphism $\varphi : W(G,I_n) \to W(H,I_n)$ such that $w(hX,i)\varphi = \sum_{j=1}^{n} x_{ji} w'(h,j)$ for $h \in H$, $i \in I_n$. It is straightforward to see that φ preserves the Lie brackets. Thus φ is a Lie isomorphism.

REMARK. It is known that $W(G,I_1) \simeq W(H,I_1)$ if and only if $G = xH$ for some $x \in k \setminus \{0\}$(see [1, Proposition 10.3.8]). However, for $n \geq 2$ we do not know whether the condition of Lemma 2 is satisfied or not even if $W(G,I_n) \simeq W(H,I_n)$.

We first study the derivations of a finitely generated and simple algebra $W(G,I)$. By Lemma 1(2) G is a finitely generated free abelian group. The free rank of G is denoted by $\mathrm{rank}(G)$. For any additive mapping $\alpha : G \to k$, we define a linear endomorphism φ_α of W by $x_g \varphi_\alpha = \alpha(g) x_g$ for every $g \in G$ and $x_g \in W_g$. It is easy to see that every φ_α is a derivation of W of degree zero.

LEMMA 3. *Let* $W = W(G,I_n)$ *be finitely generated and simple. Then* $\mathrm{Der}(W)_0 = \{\varphi_\alpha |\ \alpha$ *is additive on* $G\}$ *and* $\mathrm{Der}(W)/\mathrm{Inn}(W)$ *is an abelian Lie algebra of dimension* $\mathrm{rank} G - n$.

PROOF. Using Lemma 1(3) and Lemma 2, if necessary, we may assume that G contains $\sum_{i=1}^{n} \mathbf{Z}\varepsilon_i$, where ε_i is the element of $\oplus_{i=1}^{n} k_i$ whose i-th component is 1 and others are 0. Let δ be a derivation of W of degree zero and suppose that

$$w(g,i)\delta = \sum_{j=1}^{n} c(g,i,j) w(g,j) \quad (g \in G, i \in I_n),$$

where $c(g,i,j) \in k$. As in the same manner described in [7, (3.1)-(3.18)] we can show that $c(g,i,j) = 0$ for $j \neq i$ and each $c(g,i) = c(g,i,i)$ is additive with respect to g. Furthermore, applying δ to the identity $[w(g,i), w(g,j)] = g_j w(2g,i) - g_i w(2g,j)$, we have

$$(c(g,i) + c(g,j))(g_j w(2g,i) - g_i w(2g,j)) = 2g_j c(g,i) w(2g,i) - 2g_i c(g,j) w(2g,j),$$

whence $c(g,i) = c(g,j)$ for any i, $j \in I$. Thus δ is of the form φ_c for some additive mapping $c : G \to k$, and so the space spanned by all φ_α coincides with $\mathrm{Der}(W)_0$. Hence $\mathrm{Der}(W)_0$ is abelian and $\dim \mathrm{Der}(W)_0$ is equal to $\mathrm{rank} G$. In [7] it has been shown that $\mathrm{Der}(W) = \mathrm{Lin}(W) + \mathrm{Der}(W)_0$ and $\mathrm{Lin}(W) = \mathrm{Inn}(W)$ if W is finitely generated. Since W is simple, we see that $\dim(\mathrm{Inn}(W) \cap \mathrm{Der}(W)_0) = n$ and $\dim(\mathrm{Der}(W)/\mathrm{Inn}(W)) = \mathrm{rank}(G) - n$.

Now for a general $W(G,I)$ we define several typical derivations of degree zero. For $f \in \mathrm{Hom}_k(W_0, W_0)$ such that $W_0 f \subset \zeta(W)$ we put $\psi_f = \sum_{g \in G} p_g \tau_{-g} f \tau_g$, where for each $g \in G$, $p_g : W \to W_g$ is the projection of W onto W_g, and τ_g is the linear automorphism of W such that $w(h,i)\tau_g = w(g+h,i)$ $(h \in G, i \in I)$. For each $z \in \zeta(W)$ let π_z be the linear endomorphism of W such that $w(g,i)\pi_z = g_i z \tau_g$ $(g \in G, i \in I)$. φ_α is the derivation of W we defined above for an additive mapping $\alpha : G \to k$. In virtue of these derivations, we can state one of our main theorems.

THEOREM 1. *Let $W = W(G, I_n)$ be a finitely generated generalized Witt algebra. Then any derivation δ of W of degree zero can be written uniquely of the form*

$$\delta = \varphi_\alpha + \psi_f + \pi_z,$$

for some additive mapping $\alpha : G \longrightarrow k$, $f \in \mathrm{Hom}_k(W_0, \zeta(W))$ and $z \in \zeta(W)$. In particular

$$\dim H^1(W,W) = \mathrm{rank}G - n + (n+2)\dim\zeta(W).$$

PROOF. Let l be the dimension of the subspace of $\oplus_{i=1}^n k_i$ spanned by G. Using Lemma 2, we may assume that G is total in $\oplus_{i=1}^l k_i$ and $G \supset \oplus_{i=1}^l \mathbf{Z}\varepsilon_i$. Let δ be a derivation of W of degree zero and suppose that

$$w(g,i)\delta = \sum_{j=1}^n c(g,i,j)w(g,j) \quad (g \in G, i \in I_n).$$

Then we have $W_0\delta \subset \zeta(W)$. Indeed, for any $i,j \in I_n, g \in G$ we have

$$\begin{aligned}[][w(0,i)\delta, w(g,j)] &= [w(0,i), w(g,j)]\delta - [w(0,i), w(g,j)\delta] \\ &= -g_i w(g,j)\delta + g_i w(g,j)\delta = 0.\end{aligned}$$

Hence we can define a derivation ψ_f for $f = \delta|_{W_0}$. Replacing δ by $\delta - \psi_f$ we may suppose that $W_0\delta = 0$. Applying δ to the identity $[w(-g,i), w(g,j)] = -g_i w(0,j)$ $(1 \leq i \leq l < j \leq n)$ we have

$$\sum_{t=1}^n c(-g,i,t)[w(-g,t), w(g,j)] + \sum_{t=1}^n c(g,j,t)[w(-g,i), w(g,t)] = 0.$$

Hence

$$\sum_{t=1}^n c(g,j,t)g_i w(0,t) = (\sum_{s=1}^l c(g,j,s)(-g_s))w(0,i) + (\sum_{s=1}^l c(-g,i,s)g_s)w(0,j), \quad (2.1)$$

and it follows that $c(g,j,t)=0$ for $t\neq j$. Note that from [11, Lemma 1.2, Theorem 2.3] $\zeta(W)=\sum_{j=l+1}^{n}\langle w(0,j)\rangle$ and $\langle w(g,j)|\ g\in G,\ l+1\leq j\leq n\ \rangle$ is the unique maximal abelian ideal R of W. Therefore R is invariant under δ.

Now δ induces a derivation of W/R of degree zero. From Lemma 3, there exists an additive mapping $\alpha: G\to k$ such that $w(g,i)\delta-\alpha(g)w(g,i)\in R\ \ (g\in G)$. Replacing δ by $\delta-\varphi_\alpha$ again, we may suppose that $W_0\delta=0$ and $W\delta\subset R$. Then $c(g,i,s)=0\ \ (g\in G, i\in I_n, 1\leq s\leq l)$. Hence $R\delta=0$ from (2.1).

We claim that there exist $c_j\in k\ (l+1\leq n)$ such that $c(g,i,j)=g_ic_j$ $(g\in G,\ 1\leq i\leq l)$. Applying δ to the identities $[w(a,i),w(b,i)]=(a_i-b_i)w(a+b,i)$ and $[w(g,i),w(-g,s)]=g_sw(0,i)+g_iw(0,s)$, we have

$$(a_i-b_i)c(a+b,i,j)=a_ic(a,i,j)-b_ic(b,i,j)\quad (a,b\in G,\ 1\leq i\leq l<j\leq n),\tag{2.2}$$

$$g_sc(g,i,j)+g_ic(-g,s,j)=0\quad (g\in G,\ 1\leq i,s\leq l<j\leq n).\tag{2.3}$$

If $g_i\neq 0$ then we have $c(-g,i,j)=-c(g,i,j)$ from (2.3). If $g\neq 0$ and $g_i=0$ then $g_s\neq 0$ for some s, and so $c(g,i,j)=0=c(-g,i,j)$. If $g=0$ then it is clear that $c(g,i,j)=0$. In any case we have $c(-g,i,j)=-c(g,i,j)$. Take $a=2g,b=-g$ in (2.2). Then we obtain that $c(2g,i,j)=2c(g,i,j)$. Furthermore, using (2.2) for $(a,b)=(\varepsilon_i,g)$, (ε_i+g,g) and $(\varepsilon_i,2g)$ we have

$$\begin{aligned}(1-g_i)c(\varepsilon_i+g,i,j)&=c(\varepsilon_i,i,j)-g_ic(g,i,j),\\ c(\varepsilon_i+2g,i,j)&=(1+g_i)c(\varepsilon_i+g,i,j)-g_ic(g,i,j),\\ (1-2g_i)c(\varepsilon_i+2g,i,j)&=c(\varepsilon_i,i,j)-4g_ic(g,i,j).\end{aligned}$$

From these we can deduce that $c(g,i,j)=g_ic(\varepsilon_i,i,j)$. Next, let $l\geq 2$ and $1\leq i\neq s\leq l$. Using (2.3) for $g=\varepsilon_i+\varepsilon_s$, we have $c(\varepsilon_i+\varepsilon_s,i,j)=c(\varepsilon_i+\varepsilon_s,s,j)$. Since $c(\varepsilon_i+\varepsilon_s,i,j)=c(\varepsilon_i,i,j)$ by (2.2), we get $c(\varepsilon_i,i,j)=c(\varepsilon_s,s,j)$. Putting $c_j=c(\varepsilon_1,1,j)$ we obtain that $c(g,i,j)=g_ic_j$ as claimed.

Let $z=\sum_{j=l+1}^{n}c_jw(0,j)$. Then $z\in\zeta(W)$ and it is easy to see that $\delta=\pi_z$. Therefore the initial δ is of the form $\varphi_\alpha+\psi_f+\pi_z$. To show the uniqueness, let $\varphi_\alpha+\psi_f+\pi_z=0$. Then $W_0(\varphi_\alpha+\psi_f+\pi_z)=W_0f=0$. Therefore $f=0$ and so $\psi_f=0$. Since $R\pi_z=0$, we have $w(g,j)(\varphi_\alpha+\pi_z)=\alpha(g)w(g,j)=0$ for any $g\in G,\ j>l$. Hence $\alpha=0$ and $\varphi_\alpha=\pi_z=0$. Finally noting that $\dim(\mathrm{Der}(W)_0\cap\mathrm{Inn}(W))=l$, we have $\dim H^1(W,W)=\dim(\mathrm{Der}(W)_0/(\mathrm{Der}(W)_0\cap\mathrm{Inn}(W)))=\mathrm{rank}G+n(n-l)+(n-l)-l$.

COROLLARY 1. *If $W(G,I)$ and $W(H,I)$ are finitely generated and isomorphic, then* $\mathrm{rank}G=\mathrm{rank}H$.

3. Central extensions of $W(G, I)$

For a central extension $0 \longrightarrow A \longrightarrow \widetilde{L} \longrightarrow L \longrightarrow 0$ of a Lie algebra L, $\widetilde{L}$ is called a covering algebra of L if $\widetilde{L}$ is perfect, that is, $[\widetilde{L}, \widetilde{L}] = \widetilde{L}$. A covering algebra $\widetilde{L}$ is a universal covering algebra if for any other covering $0 \longrightarrow A' \longrightarrow \widetilde{L}' \longrightarrow L \longrightarrow 0$ of L, there are homomorphisms $i : A \to A'$, $j : \widetilde{L} \to \widetilde{L}'$ such that the diagram

$$\begin{array}{ccccccccc} 0 & \longrightarrow & A & \longrightarrow & \widetilde{L} & \longrightarrow & L & \longrightarrow & 0 \\ & & {\scriptstyle i}\big\downarrow & & {\scriptstyle j}\big\downarrow & & {\scriptstyle id}\big\downarrow & & \\ 0 & \longrightarrow & A' & \longrightarrow & \widetilde{L}' & \longrightarrow & L & \longrightarrow & 0 \end{array}$$

is commutative. Every perfect algebra has a universal covering algebra which is unique up to isomorphism. A detailed exposition of this theory is given in [4, §1]. A generalized Witt algebra is perfect and so it has a universal covering algebra.

A Lie algebra is called centrally closed if its every central extension splits. A universal covering algebra of some algebra is centrally closed(see [4, Theorem 1.16]). For example perfect affine Lie algebras are universal covering algebras of twisted or untwisted loop algebras(see [15]), and so they are centrally closed. Furthermore GIM Lie algebras, that generalize Kac-Moody or GCM Lie algebras(see [13]), are centrally closed(see [2, Theorem 2.6]). It is worth while noting, however, that the Kac-Moody algebras in the sense of [9] are not always centrally closed(see [3, Corollary 3.4]).

Let V be the Virasoro algebra with basis $\{L_i | \ i \in \mathbb{Z}\} \cup \{c\}$ and multiplication $[L_i, L_j] = (i-j)L_{i+j} + \frac{1}{12}\delta_{i,-j}(i^3 - i)c$, $[V, c] = 0$. It is well known that V is a universal covering algebra of the Witt algebra $W_{\mathbb{Z}}$ and so V is centrally closed(see [5]). We extend this result to a generalized Witt algebra $W(G, I_n)$. By Lemma 2 we may assume that $\sum_{i=1}^{l} \mathbb{Z}\varepsilon_i \subset G \subset \oplus_{i=1}^{l} k_i$, where l is the dimension of the subspace of $\oplus_{i=1}^{n} k_i$ spanned by G.

THEOREM 2. (1) *If* $l = 1$ *then* $W(G, I_n)$ *has a universal covering algebra* $V(G, I_n)$ *with a basis* $\{L(g, i) | \ g \in G, i \in I_n\} \cup \{\omega_{ij} | \ 2 \le i \le j \le n\} \cup \{\zeta_j | \ 2 \le j \le n\} \cup \{c\}$ *and the Lie multiplication*

$$[L(g, i), L(h, j)] = g_j L(g + h, i) - h_i L(g + h, j) + \delta_{g,-h}\beta_{ij}(g),$$

where

$$\begin{aligned}
&\omega_{ij}(2 \le i \le j \le n),\ \zeta_j(2 \le j \le n),\ c \in \zeta(W),\\
&\beta_{11}(g) = \frac{g^3 - g}{12}c,\ \beta_{1j}(g) = \frac{g^2 - g}{2}\zeta_j,\ \beta_{j1}(g) = -\frac{g^2 + g}{2}\zeta_j\ (2 \le j \le n),\\
&\beta_{ij}(g) = \beta_{ji}(g) = g\omega_{ij}\ (2 \le i \le j \le n).
\end{aligned}$$

In particular

$$\dim H^2(W, k) = \frac{n(n+1)}{2}.$$

(2) *If* $l \ge 2$ *then* $W(G, I_n)$ *is centrally closed.*

PROOF. By Lemma 1(1), we may suppose that $W = W(G, I_l)\dot{+}R$, where $R = \langle w(g,j) |\ g \in G, l+1 \le j \le n \rangle$. Let $\widetilde{W} = W\dot{+}C$ (vector space sum) be a central extension of W by C. The multiplication is given by

$$[x+u, y+v] = [x,y] + \wedge(x,y),$$

where $x, y \in W$, $u, v \in C$ and $\wedge : W \times W \to C$ is a 2-cocycle. By the cocycle property, we have

$$\begin{aligned}
0 = \quad & b_k \wedge (w(a,i), w(b+c,j)) - c_j \wedge (w(a,i), w(b+c,k))\\
&+ c_i \wedge (w(b,j), w(c+a,k)) - a_k \wedge (w(b,j), w(c+a,i))\\
&+ a_j \wedge (w(c,k), w(a+b,i)) - b_i \wedge (w(c,k), w(a+b,j))
\end{aligned} \tag{3.1}$$

for $a, b, c \in G$, $i, j, k \in I_n$.

If $b = c = 0$ in (3.1), then $a_j \wedge (w(0,k), w(a,i)) = a_k \wedge (w(0,j), w(a,i))$. If $a \ne 0$ then $a_k \ne 0$ for some k, and so we can define a central element $z(a,i)$ as follows.

$$z(a,i) = \frac{\wedge(w(0,k), w(a,i))}{a_k} \quad (0 \ne a \in G,\ i \in I_n).$$

Then $\wedge(w(0,j), w(a,i)) = a_j z(a,i)$ for $1 \le i, j \le n$. Setting $a + b + c = 0$, we eliminate c in (3.1) and we obtain

$$\begin{aligned}
0 = \quad & b_k \wedge_{ij}(a) + (a_j + b_j) \wedge_{ik}(a) - (a_i + b_i) \wedge_{jk}(b)\\
&- a_k \wedge_{ji}(b) - a_j \wedge_{ik}(a+b) + b_i \wedge_{jk}(a+b),
\end{aligned} \tag{3.2}$$

where $\wedge_{ij}(a) = \wedge(w(a,i), w(-a,j))$. Since $\wedge$ is anticommutative, $\wedge_{ji}(-a) = -\wedge_{ij}(a)$.

Now we make a change of basis in $\widetilde{W}$ as follows.

$$\begin{aligned} L(a,i) &= w(a,i) - z(a,i) \quad (0 \neq a \in G,\ 1 \leq i \leq n), \\ L(0,i) &= w(0,i) + \frac{1}{2} \wedge_{ii} (\varepsilon_i) \quad (1 \leq i \leq l), \\ L(0,j) &= w(0,j) + \wedge_{1j}(\varepsilon_1) \quad (l+1 \leq j \leq n). \end{aligned}$$

We note here that if $l \geq 2$ then $\wedge_{ij}(a) = a_i \wedge_{1j} (\varepsilon_1)$ for $1 \leq i \leq l < j \leq n$, in particular $\wedge_{ij}(\varepsilon_i) = \wedge_{1j}(\varepsilon_1)$. Indeed, let $j \geq l+1$, $i = k$, $g = -b$ and $g_i = 0$ in (3.2). Then we have $2a_i \wedge_{ij} (g) = 0$ for any $a \in G$. Therefore if $1 \leq i \leq l < j \leq n$ and $g_i = 0$ then $\wedge_{ij}(g) = 0$. Let $1 = k < i \leq l < j \leq n$ and $b = -\varepsilon_1$ in (3.2). Then since $\wedge_{ij}(\varepsilon_1) = 0$, we have $\wedge_{ij}(a) = a_i \wedge_{1j} (\varepsilon_1)$.

Let $\beta(L(a,i), L(b,j)) = [L(a,i), L(b,j)] - a_j L(a+b,i) + b_i L(a+b,j) \in C$ for all $a, b \in G$, $i, j \in I_n$. Extending β bilinearly to the entire span of the $L(a,i)$'s, β is also a 2-cocycle. It is clear that $[L(0,i), L(b,j)] = -b_i w(b,j) + \wedge(w(0,i), w(b,j)) = -b_i L(b,j)$, whence $\beta(L(0,i), L(b,j)) = 0$ for all $b \neq 0$. This holds even for $b = 0$. In fact, putting $a = 0$ in (3.2) we have $a_k \wedge_{ij} (0) = a_i \wedge_{jk} (0)$. If $j \leq l$ then $\wedge_{ij}(0) = (\varepsilon_j)_i \wedge_{jj} (0) = 0$. If $j > l$ then $\wedge_{ij}(0) = (\varepsilon_1)_i \wedge_{j1} (0) = 0$. Thus $\wedge_{ij}(0) = 0$ for any $i, j \in I_n$, whence $\beta(L(0,i), L(0,j)) = [L(0,i), L(0,j)] = 0$. Using the Jacobi identity for $[L(0,k), [L(a,i), L(b,j)]]$ we get

$$(a_k + b_k)\beta(L(a,i), L(b,j)) = 0$$

for all $k, i, j \in I_n$. Hence if $a + b \neq 0$ then $\beta(L(a,i), L(b,j)) = 0$. Therefore we can write

$$\beta(L(a,i), L(b,j)) = \delta_{a,-b}\beta_{ij}(a),$$

where $\beta_{ij}(a) = \beta(L(a,i), L(-a,j))$.

(1) We first treat the case of $l = 1$. Since the identity (3.2) is valid for any 2-cocycle, putting $i = j = k = 1$ we get

$$(a-b)\beta_{11}(a+b) = (a+2b)\beta_{11}(a) - (2a+b)\beta_{11}(b) \tag{3.3}$$

for all $a, b \in G \subset k$. Noting $\beta_{11}(0) = \beta_{11}(1) = 0$, we use (3.3) for $(a,b) = (1,g), (2,g)$ and $(1+g, g)$ to obtain

$$\begin{gathered} (1-g)\beta_{11}(1+g) = -(2+g)\beta_{11}(g), \\ (2-g)\beta_{11}(2+g) = (2+2g)\beta_{11}(2) - (4+g)\beta_{11}(g), \\ g\beta_{11}(2+g) = (3+g)\beta_{11}(1+g). \end{gathered}$$

From these identities we can easily deduce that $6\beta_{11}(g) = (g^3 - g)\beta_{11}(2)$. Hence if we put $c = 2\beta_{11}(2)$ then $\beta_{11}(g) = (g^3 - g)c/12$.

In the same way as above, we produce two more kinds of central elements ζ_k and ω_{ij}. If we put $i = j = 1$ and $k \geq 2$ then we have

$$(a - b)\beta_{1k}(a + b) = (a + b)(\beta_{1k}(a) - \beta_{1k}(b)). \tag{3.4}$$

Substituting $(a, b) = (1, g), (1 + g, -1)$ in (3.4) and noting that $\beta_{1k}(1) = 0$, we get

$$\begin{aligned}(1 - g)\beta_{1k}(1 + g) &= -(1 + g)\beta_{1k}(g),\\ (2 + g)\beta_{1k}(g) &= g(\beta_{1k}(1 + g) - \beta_{1k}(-1)).\end{aligned}$$

From these it follows that $\beta_{1k}(g) = g(g-1)\zeta_k/2$ and $\beta_{k1}(g) = -g(g+1)\zeta_k/2$, where $\zeta_k = \beta_{1k}(-1)$. Next, let $b = 1 \in G$, $k = 1$ and $i, j \geq 2$ in the β-version of (3.2). Then we have $\beta_{ij}(a) = a\beta_{ji}(1)$ and $\beta_{ij}(1) = \beta_{ji}(1)$. Putting $\omega_{ij} = \beta_{ij}(1)$, we obtain $\beta_{ij}(g) = \beta_{ji}(g) = g\omega_{ij}$ for $g \in G$ and $i, j \geq 2$.

Let $V(G, I_n)$ be the Lie algebra given in the statement of the theorem. Then from above it is easy to see that $V(G, I_n)$ is a universal covering algebra of W.

Now let $c_1, z_j, w_{ij} : W \times W \to k$ be 2-cocycles such that $c_1(w(g, 1), w(-g, 1)) = (g^3 - g)/12$, $z_j(w(g, 1), w(-g, j)) = (g^2 - g)/2$, $w_{ij}(w(g, i), w(-g, j)) = g$ and others are zero $(i, j \geq 2)$. From the above argument for a 1-dimensional central extension, we can see that the space $Z^2(W, k)$ are spanned by c_1, z_j's, w_{ij}'s and the coboundaries. Besides, it is straightforward to see that c_1, z_j's, w_{ij}'s are linearly independent modulo $B^2(W, k)$. Thus $H^2(W, k) = n(n + 1)/2$.

(2) Let $l \geq 2$. We shall show $\beta_{ij} = 0$ for all i, j. If $i, j \geq l + 1$, then from (3.2) we have $b_k\beta_{ij}(a) = a_k\beta_{ij}(b)$ for any $a, b \in G$ and $k \in I_n$. Hence $\beta_{ij}(a) = a_1\beta_{ij}(\varepsilon_1) = a_1(\varepsilon_1)_2\beta_{ij}(\varepsilon_2) = 0$. We have already shown that $\wedge_{ij}(a) = a_i \wedge_{1j}(\varepsilon_1)$ for $1 \leq i \leq l < j \leq n$. Therefore $[L(a, i), L(-a, j)] = a_iL(0, j)$, whence $\beta_{ij} = 0$ if $i > l$ or $j > l$.

Finally, we suppose $1 \leq i \neq j \leq l$. Since $[L(\varepsilon_i, i), L(-\varepsilon_i, i)] = 2L(0, i)$, we have $\beta_{ii}(\varepsilon_i) = 0$. In the same way as we computed $\beta_{11}(a)$, we can deduce that $\beta_{ii}(a) = (a_i^3 - a_i)\beta_{ii}(2\varepsilon_i)/6$. Putting $k = j$ in the β-version of (3.2), we have

$$\begin{aligned}&(a_j + 2b_j)\beta_{ij}(a) - a_j\beta_{ji}(b) - a_j\beta_{ij}(a + b)\\ &= (a_i + b_i)\beta_{jj}(b) - b_i\beta_{jj}(a + b)\end{aligned} \tag{3.5}$$

for all $a, b \in G$. Let $i \neq j$, $a_j = 0$ and $b = \varepsilon_j$ in (3.5). Then we obtain $\beta_{ij}(a) = 0$. Therefore $\beta_{ij}(a) = 0$ if $a_i = 0$ or $a_j = 0$. We next set $a = \varepsilon_i, b = 2\varepsilon_j$ $(i \neq j)$. Then we have $\beta_{jj}(2\varepsilon_j) = 0$, whence $\beta_{jj}(a) = 0$ for all $a \in G$ and $1 \le j \le l$. Thus the identity (3.5) becomes

$$(a_j + 2b_j)\beta_{ij}(a) + a_j\beta_{ij}(-b) = a_j\beta_{ij}(a+b). \tag{3.6}$$

Substitute $(a, \varepsilon_j), (a + \varepsilon_j, \varepsilon_j)$ and $(a, 2\varepsilon_j)$ for (a, b) in (3.6). Then from the gotten equations, it is easy to see that $\beta_{ij}(a) = 0$. Thus $\beta_{ij} = 0$ for all $i, j \in I_n$, and therefore $\widetilde{W}$ splits.

References

[1] R.K. Amayo and I. Stewart, *Infinite-dimensional Lie Algebras*, Noordhoff, Leyden, 1974.

[2] G. M. Benkart and R. V. Moody, *Derivations, central extensions and affine Lie algebras*, Algebras Groups Geom. **3**(1986), 456-492

[3] R. Farnsteiner, *Derivations and central extensions of finitely generated graded Lie algebras*, J. Algebra **118**(1988), 33-45.

[4] H. Garland, *The arithmetic theory of loop groups*, Inst. Hautes Études Sci. Publ. Math. **52**(1980), 5-136.

[5] I. M. Gel'fand and D. B. Fuks, *Cohomologies of the Lie algebra of vector fields on the circle*, Funktsional. Anal. i Prilozhen **2**(1968), 92-93 (Russian), translated in Functional Anal. Appl. **2**(1968), 342-343.

[6] T. Ikeda, *Locally inner derivations of ideally finite Lie algebras*, Hiroshima Math. J. **17**(1987), 495-503.

[7] T. Ikeda and N. Kawamoto, *On the derivations of generalized Witt algebras over a field of characteristic zero*, Hiroshima Math. J. **20**(1990), 47-55.

[8] N. Jacobson, *Lie Algebras*, Interscience, New York, 1962.

[9] V. G. Kac, *Infinite-dimensional Lie Algebras - An Introduction*, Progress in Mathematics, Vol. **44**, Birkhäuser, Boston (1983).

[10] I. Kaplansky, *Seminar on simple Lie algebras*, Bull. Amer. Math. Soc. **60**(1954), 470-471.

[11] N. Kawamoto, *Generalizations of Witt algebras over a field of characteristic zero*, Hiroshima Math. J. **16**(1986), 417-426.

[12] R. Ree, *On generalized Witt algebras*, Trans. Amer. Math. Soc. **83**(1956), 510-546.

[13] P. Slodowy, *Beyond Kac-Moody algebras, and inside*, Canad. Math. Soc. Conf. Proc. **5**, Britten, Lemire, Moody, eds., (1986), 361-371.

[14] R. L. Wilson, *Classification of generalized Witt algebras over algebraically closed fields*, Trans. Amer. Math. Soc. **153**(1971), 191-210.

[15] R. L. Wilson, *Euclidean Lie algebras are universal central extensions, Lie Algebras and Related Topics*, D. J. Winter, Ed., Springer Verlag Lecture Notes in Math. **933**(1982), 210-213.

Another construction of Lie (super) algebras by associative triple systems and Freudenthal-Kantor (super) triple systems

Yoshiaki KAKIICHI

Department of Mathematics, Faculty of Engineering, Toyo University
Kawagoe City, Saitama 350, Japan

Abstract

Let A_1, A_2 be the commutative associative triple systems and $U(\varepsilon,\delta)$, $\varepsilon,\delta = \pm 1$, be a Freudenthal-Kantor (super) triple system. It is shown that the tensor product $A_1 \otimes U(\varepsilon,\delta) \otimes A_2$ becomes a Freudenthal-Kantor (super) triple system which induces a Lie (super) triple system $A_1 \otimes U(\varepsilon,\delta) \otimes A_2 \oplus \overline{A_1 \otimes U(\varepsilon,\delta) \otimes A_2}$. Then we obtain a Lie algebra or a Lie superalgebra as a standard embedding of it according to $\delta = 1$ or $\delta = -1$ (cf. [2]).

1. Introduction

Let A_1, A_2 be the commutative associative triple systems and $U(\varepsilon,\delta)$, $\varepsilon,\delta = \pm 1$, be a Freudenthal-Kantor (super) triple system. Then the tensor product $A_1 \otimes U(\varepsilon,\delta) \otimes A_2$ becomes also a Freudenthal-Kantor (super) triple system [9]. On the other hand, it has been shown in [8] thet the vector space direct sum $U(\varepsilon,\delta) \oplus \overline{U(\varepsilon,\delta)}$ becomes the (anti-) Lie triple system by a certain triple product (cf. [10]). It follows from above results that the vector space direct sum $A_1 \otimes U(\varepsilon,\delta) \otimes A_2 \oplus \overline{A_1 \otimes U(\varepsilon,\delta) \otimes A_2}$ becomes the Lie triple system ($\delta = 1$) or the anti-Lie triple system ($\delta = -1$), to which we can associate a Lie algebra or a Lie superalgebra as a standard embedding of it.

We assume that any vector space considered in this paper is finite dimensional and the characteristic of the base field is different from two. The author wishes to express his hearty thanks to Prof. K. Yamaguti for his kind advice and encouragement.

2. Associative triple systems and Freudenthal-Kantor (super) triple systems

A triple system A with a trilinear product $\langle abc\rangle := \ell(a,b)c := m(a,c)b$ is called an associative triple system (ATS) if

(2.1) $\quad \langle ab\langle cde\rangle\rangle = \langle a\langle dcb\rangle e\rangle = \langle\langle abc\rangle de\rangle,$

and is said to be commutative if

(2.2) $\quad \langle abc\rangle = \langle cba\rangle$ for $a,b,c,d,e \in A$ [4, 5, 6].

A Freudenthal-Kantor (super) triple system (FKS) $U(\varepsilon,\delta)$ is a vector space with a triple product $\{xyz\}$ satisfying

(2.3) $\quad [L(x,y), L(u,v)] = L(L(x,y)u,v) + \varepsilon L(u, L(y,x)v),$

(2.4) $\quad K(K(x,y)u,v) = L(v,u)K(x,y) - \varepsilon K(x,y)L(u,v),$

where $L(x,y)u = \{xyu\}$ and $K(x,y)u = \{xuy\} - \delta\{yux\}$ for $x,y,u,v \in U(\varepsilon,\delta)$, $\varepsilon = \pm1$, $\delta = \pm1$ [3,7].

Using (1),(2),(3),(4), we have

PROPOSITION ([9]). *For the pair of commutative associative triple systems A_1 and A_2, and any Freudenthal-Kantor (super) triple system $U(\varepsilon,\delta)$, define a trilinear product in $A_1 \otimes U(\varepsilon,\delta) \otimes A_2$ by*

$$\{a\otimes x\otimes p \quad b\otimes y\otimes q \quad c\otimes z\otimes r\} = \langle abc\rangle \otimes \{xyz\} \otimes \langle pqr\rangle$$

for $a,b,c \in A_1$, $x,y,z \in U(\varepsilon,\delta)$, $p,q,r \in A_2$. Then $A_1 \otimes U(\varepsilon,\delta) \otimes A_2$ becomes an FKS, where $L(a\otimes x\otimes p, b\otimes y\otimes q) = \ell(a,b)\otimes L(x,y)\otimes \ell(p,q)$ and $K(a\otimes x\otimes p, b\otimes y\otimes q) = m(a,b)\otimes K(x,y)\otimes m(p,q)$.

COROLLARY. *If $U(\varepsilon,\delta)$ is a Jordan triple system (or an anti-Jordan triple system), then so is $A_1 \otimes U(\varepsilon,\delta) \otimes A_2$.*

3. Induced (anti-) Lie triple systems and standard embedding Lie (super) algebras

A triple system $T(\delta)$ with a trilinear product $[xyz]$, $\delta = \pm1$, is called a Lie triple system (LTS) if it satisfies the following identities for any elements $x,y,z,u,v \in T(\delta)$

(i) $[xyz] = -\delta[yxz]$,

(ii) $[xyz] + [yzx] + [zxy] = 0$,

(iii) $[xy[uvz]] = [[xyu]vz] + [u[xyv]z] + [uv[xyz]]$.

When $\delta = 1$, $T(1)$ is an ordinary Lie triple system and when $\delta = -1$, $T(-1)$ is an anti-Lie triple system [1].

For the FKS $A_1 \otimes U(\varepsilon, \delta) \otimes A_2$, we consider a vector space direct sum $A_1 \otimes U(\varepsilon, \delta) \otimes A_2 \oplus \overline{A_1 \otimes U(\varepsilon, \delta) \otimes A_2}$, of which element is denoted by a finite sum of the matrix form $\begin{pmatrix} a \otimes x \otimes p \\ b \otimes y \otimes q \end{pmatrix}$ and define a triple product on it by

$$(3.1) \quad \left[\begin{pmatrix} a_1 \otimes x_1 \otimes p_1 \\ a_2 \otimes x_2 \otimes p_2 \end{pmatrix}\begin{pmatrix} b_1 \otimes y_1 \otimes q_1 \\ b_2 \otimes y_2 \otimes q_2 \end{pmatrix}\begin{pmatrix} c_1 \otimes z_1 \otimes r_1 \\ c_2 \otimes z_2 \otimes r_2 \end{pmatrix}\right]$$

$$:= L\left(\begin{pmatrix} a_1 \otimes x_1 \otimes p_1 \\ a_2 \otimes x_2 \otimes p_2 \end{pmatrix}, \begin{pmatrix} b_1 \otimes y_1 \otimes q_1 \\ b_2 \otimes y_2 \otimes q_2 \end{pmatrix}\right)\begin{pmatrix} c_1 \otimes z_1 \otimes r_1 \\ c_2 \otimes z_2 \otimes r_2 \end{pmatrix}$$

$$:= \begin{pmatrix} M_{11} & M_{12} \\ M_{21} & M_{22} \end{pmatrix}\begin{pmatrix} c_1 \otimes z_1 \otimes r_1 \\ c_2 \otimes z_2 \otimes r_2 \end{pmatrix},$$

where

$$M_{11} = \ell(a_1, b_2) \otimes L(x_1, y_2) \otimes \ell(p_1, q_2) - \delta\ell(b_1, a_2) \otimes L(y_1, x_2) \otimes \ell(q_1, p_2),$$
$$M_{12} = \delta m(a_1, b_1) \otimes K(x_1, y_1) \otimes m(p_1, q_1),$$
$$M_{21} = -\varepsilon m(a_2, b_2) \otimes K(x_2, y_2) \otimes m(p_2, q_2),$$
$$M_{22} = \varepsilon\ell(b_2, a_1) \otimes L(y_2, x_1) \otimes \ell(q_2, p_1) - \varepsilon\delta\ell(a_2, b_1) \otimes L(x_2, y_1) \otimes \ell(p_2, q_1),$$

then $A_1 \otimes U(\varepsilon, \delta) \otimes A_2 \oplus \overline{A_1 \otimes U(\varepsilon, \delta) \otimes A_2}$ becomes an LTS with respect to this product (cf. [8]).

Hence we have the following

THEOREM 1. *Let A_1, A_2 be the commutative triple systems as above and $U(\varepsilon, \delta)$ be an* FKS. *Then, from an* FKS $A_1 \otimes U(\varepsilon, \delta) \otimes A_2$ *we obtain an* LTS $A_1 \otimes U(\varepsilon, \delta) \otimes A_2 \oplus \overline{A_1 \otimes U(\varepsilon, \delta) \otimes A_2}$ *which induces the standard embedding Lie algebra ($\delta = 1$) and Lie superalgebra ($\delta = -1$)*

$$\mathfrak{G}(\varepsilon, \delta) := \mathcal{D} \oplus A_1 \otimes U(\varepsilon, \delta) \otimes A_2 \oplus \overline{A_1 \otimes U(\varepsilon, \delta) \otimes A_2},$$

where $\mathcal{D}$ is the Lie algebra of inner derivation in the LTS $A_1 \otimes U(\varepsilon, \delta) \otimes A_2 \oplus \overline{A_1 \otimes U(\varepsilon, \delta) \otimes A_2}$.

Put $\mathfrak{G} = V_0 \oplus V_1$, where $V_0 = \mathcal{D}$, $V_1 = A_1 \otimes U(\varepsilon,\delta) \otimes A_2 \oplus \overline{A_1 \otimes U(\varepsilon,\delta) \otimes A_2}$, then we have the relations $[V_0, V_0] \subset V_0$, $[V_0, V_1] \subset V_1$, $[V_1, V_1] \subset V_0$.

More precisely, let

$\mathfrak{G}_{-2}$ be the vector space spanned by derivations $L\left(\begin{pmatrix} a \otimes x \otimes p \\ 0 \end{pmatrix}, \begin{pmatrix} b \otimes y \otimes q \\ 0 \end{pmatrix}\right)$,

$\mathfrak{G}_0$ be the vector space spanned by derivations $L\left(\begin{pmatrix} a \otimes x \otimes p \\ 0 \end{pmatrix}, \begin{pmatrix} 0 \\ b \otimes y \otimes q \end{pmatrix}\right)$,

$\mathfrak{G}_2$ be the vector space spanned by derivations $L\left(\begin{pmatrix} 0 \\ a \otimes x \otimes p \end{pmatrix}, \begin{pmatrix} 0 \\ b \otimes y \otimes q \end{pmatrix}\right)$,

$\mathfrak{G}_{-1} = A_1 \otimes U(\varepsilon,\delta) \otimes A_2$ and $\mathfrak{G}_1 = \overline{A_1 \otimes U(\varepsilon,\delta) \otimes A_2}$,

where $a, b \in A_1$, $x, y \in U(\varepsilon,\delta)$ and $p, q \in A_2$. Then $V_0 = \mathfrak{G}_{-2} \oplus \mathfrak{G}_0 \oplus \mathfrak{G}_2$, $V_1 = \mathfrak{G}_{-1} \oplus \mathfrak{G}_1$.

By straightforward calculations we have

THEOREM 2. *The Lie algebra or Lie superalgebra obtained in Theorem 1 is the graded Lie algebra or Lie superalgebra of second order such that*

$$\mathfrak{G}(\varepsilon,\delta) = \mathfrak{G}_{-2} \oplus \mathfrak{G}_{-1} \oplus \mathfrak{G}_0 \oplus \mathfrak{G}_1 \oplus \mathfrak{G}_2,$$

$[\mathfrak{G}_i, \mathfrak{G}_j] \subset \mathfrak{G}_{i+j}$ *for* $i, j = 0, \pm 1$ *and* ± 2 *(it being understood that* $\mathfrak{G}_{i+j} = 0$ *if* $i+j$ *is different from* $0, \pm 1$ *and* ± 2 *)* (cf. [7]).

References

[1] J. R. Faulkner and J. C. Ferrar: Simple anti-Jordan pairs. Comm. Algebra, **8**, 993-1013 (1980).

[2] Y. Kakiichi: A construction of Lie algebras and Lie superalgebras by Freudenthal-Kantor triple systems I. Proc. Japan Acad., **61** A, 232-234 (1985).

[3] I. L. Kantor: Models of exceptional Lie algebras. Soviet Math. Dokl., **14**, 254-258 (1973).

[4] O. Loos: Assoziative Tripelsysteme. Manuscripta Math., **7**, 103-112 (1972).

[5] O. Loos: Jordan Pairs. Lecture Notes in Math. **460**, Springer-Verlag, (1975).

[6] K. Meyberg: Lectures on algebras and triple systems. The University of Virginia, Charlottesville (1972).

[7] K. Yamaguti: On the metasymplectic geometry and triple systems. Kôkyûroku RIMS, Kyoto Univ., no.308, 55-92 (1977) (in Japanese).

[8] K. Yamaguti: Constructions of Lie (super) algebras from triple systems. Lecture Notes in Physics, 313, Group theoretical methods in Physics, Proceedings of the XVI International Colloquium, Varna, Bulgaria, June 15-20, 190-197 (1987).

[9] K. Yamaguti: A construction of Lie (super) algebras from associative triple systems and Freudenthal-Kantor (super) triple systems. Proceedings of Gauss Symposium held in Garuja, SP, Brazil, July 24-27 (1989) (in press).

[10] K. Yamaguti and A. Ono: On representations of Freudenthal-Kantor triple systems $U(\varepsilon, \delta)$. Bull. Fac. Sch. Educ. Hiroshima Univ., Part II **7**, 43-51 (1984).

On (ε, δ)– Freudenthal–Kantor triple systems

Noriaki Kamiya
Department of Mathematics, Faculty of Science
Shimane University, 690 Matsue JAPAN

Abstract

In this paper, we investigate a (ε, δ) – Freudenthal – Kantor triple system. And we give a construction of Lie triple systems or anti–Lie triple systems from (ε, δ) –Freudenthal–Kantor triple systems.

Introduction

K.Yamaguti defined identities of algebraic system which is a generalization of a Freudenthal–Kantor triple system.We call the algebraic system a (ε, δ)–Freudenthal–Kantor triple system.

In the case of $\delta = 1$, we constructed all simple Lie algebras over C from (ε, δ)–Freudenthal–Kantor triple systems. In the case of $\delta = -1$, we can construct Lie superalgebras from (ε, δ)–Freudenthal–Kantor triple systems. This is a reason for studying the (ε, δ)–Freudenthal–Kantor triple system.

In this paper,we shall investigate the simplicity and the trace form of a (ε, δ)–Freudenthal–Kantor triple system.

We shall be concerned with algebras and triple systems which are finite dimensional over a field of characteristic different from 2 or 3.

§ 1

For $\varepsilon = \pm 1$ and $\delta = \pm 1$, triple system $U(\varepsilon, \delta)$ with the triple product $< -, -, - >$ is called a (ε, δ)–Freudenthal–Kantor triple system if

$$[L(a, b), L(c, d)] = L(< abc >, d) + \varepsilon L(c, < bad >) \tag{U1}$$

$$K(K(a, b)c, d) - L(d, c)K(a, b) + \varepsilon K(a, b)L(c, d) = 0 \tag{U2}$$

where $L(a, b)c =< abc >$ and $K(a, b)c =< acb > -\delta < bca >$. This algebraic system is defined by K.Yamaguti. [20]

In particular, we shall consider the case of $K(a,b) = 0$ for all $a, b \in U(\varepsilon,\delta)$. We call to be a (ε,δ)–Jordan triple system. That is, the definition of (ε,δ)–Jordan triple system is as follows:

$$<abc>= \delta <cba> \tag{J1}$$

$$<ab<cde>>=<<abc>de>+\varepsilon<c<bad>e>+<cd<abe>> . \tag{U1}$$

If $(\varepsilon,\delta) = (-1,1)$, then this notion reduces a Jordan triple system [7,16] and if $(\varepsilon,\delta) = (1,-1)$, then this notion reduces an anti–Jordan triple system [13].

Next we give some examples of (ε,δ)–Jordan triple system.

Example1.1. *Let W be a vector space equipped with a symmetric bilinear form $B(x,y)$ for $x, y \in W$. Then*

$$<xyz>= B(x,y)z + B(y,z)x - B(z,x)y$$

defines on W a $(-1,1)$–Jordan triple system.

Example 1.2. *Let W be a vector space equipped with an anti–symmetric bilinear form $B(x,y)$ for $x, y \in W$. Then*

$$<xyz>= B(x,y)z + B(y,z)x - B(z,x)y$$

defines on W a $(1,-1)$– Jordan triple system.

Example 1.3. *Let W be a vector space equipped with a symmetric bilinear form $B(x,y)$ for $x, y \in W$. Then*

$$<xyz>= B(x,y)z - B(y,z)x$$

defines on W a $(-1,-1)$–Jordan triple system.

Example 1.4. *Let W be a vector space equipped with an anti-symmetric bilinear form $B(x,y)$ for $x, y \in W$. Then*

$$<xyz>= B(x,y)z - B(y,z)x$$

defines on W a $(1,1)$–Jordan triple system.

If the dimension of vector space in Example 1 and 2 are two, then by using it we discussed a construction of Lie triple system or anti–Lie triple system[13]. We consider $(-\kappa,\kappa)$ Jordan triple system of a two dimensional vector space W defined by

$$<abc>=<a,b>c+<b,c>a-<c,a>b \tag{$\star$}$$

for $a, b, c \in W$, where $< , >$ is a bilinear form satisfying $< a, b >= \kappa < b, a >$, and we denote it by $W(\kappa)$.

Lemma 1.1. *Let $W(\kappa)$ be as above. Then we have the following*
(i) $< ab < cde >>=<< abc > de >$
(ii) $< ab < cde >>= \kappa < c < bad > e >$
(iii) $< ab < cde >>=< cd < abe >>$.

Proof. The case of $\kappa = 1$ reduces to a Jordan triple system. Hence from results in [6], it holds. The case of $\kappa = -1$ reduces to an anti–Jordan triple system. Hence from Section 3 in [13], it holds.

Theorem 1.2. *Let $W(\kappa)$ be as above and let $U(\varepsilon, \delta)$ be any (ε, δ) –Freudenthal–Kantor triple system. Then $W(\kappa) \otimes U(\varepsilon, \delta)$ becomes a $(\kappa\varepsilon, \kappa\delta)$ –Freudenthal–Kantor triple system with respect to the triple product*

$$< a \otimes x, b \otimes y, c \otimes z >=< abc > \otimes < xyz >$$

for $a, b, c \in W(\kappa)$ and $x, y, z \in U(\varepsilon, \delta)$.

Proof. (Proof of U1): By the definition, we have

$$< a \otimes x, b \otimes y, < c \otimes z, d \otimes u, e \otimes v >>$$

$$=< ab < cde >> \otimes < xy < zuv >>$$

$$=< ab < cde >> \otimes (< xyz > uv > + \varepsilon < z < yxu > v > + < zu < xyv >>)$$

$$=< ab < cde >> \otimes << xyz > uv > + \varepsilon < ab < cde >> \otimes < z < yxu > v >$$

$$+ < ab < cde >> \otimes < zu < xyv >> .$$

Hence, by using Lemma 1, we have

$$< a \otimes x, b \otimes y, < c \otimes z, d \otimes u, e \otimes v >>$$

$$=<< abc > de > \otimes << xyz > uv > + \varepsilon\kappa < c < bad > e > \otimes < z < yxu > v >$$

$$+ < cd < abe >> \otimes < zu < xyv >>$$

$$=<< a \otimes x, b \otimes y, c \otimes z >, d \otimes y, e \otimes v > + \varepsilon\kappa < c \otimes z, < b \otimes y, a \otimes x, d \otimes u >, e \otimes v >$$

$$+ < c \otimes z, d \otimes u < a \otimes x, b \otimes y, e \otimes v >> .$$

(Proof of U2): It suffices to show

$$K(K(a \otimes x, b \otimes y)c \otimes z, d \otimes u) - L(d \otimes u, c \otimes z)K(a \otimes x, b \otimes y)$$

$$+ \varepsilon\kappa K(a \otimes x, b \otimes y)L(c \otimes z, d \otimes u) = 0.$$

This is verified using the following relation;

$$<< acb > ed > \otimes (K(K(x,y)z,u) - L(u,z)K(x,y) + \varepsilon K(x,y)L(z,u))v = 0$$

$$for\ a,b,c,d,e \in W(\kappa), and\ x,y,z,u,v \in U(\varepsilon,\delta),$$

which follows by means of Lemma 1.1. This completes the proof.

Corollary. [13] *Let W be a two dimensional anti–Jordan triple system defined by $(\star)$ and J be any Jordan triple system. Then the triple product on $W \otimes J$ defined by*

$$< a \otimes x, b \otimes y, c \otimes z > = < abc > \otimes < xyz >$$

for $a,b,x \in W$ and $z,y,z \in J$ becomes an anti–Jordan triple product on $W \otimes J$.

Corollary. [3] *Let W be a two dimensional Jordan triple system defined by $(\star)$ and U be any generalized Jordan triple system of second order. Then the triple product on $W \otimes U$ defined by*

$$< a \otimes x, b \otimes y, c \otimes z > = < abc > \otimes < xyz >$$

becomes a generalized Jordan triple system of second order.

§ 2

In this section, we shall study a construction of Lie triple systems or anti–Lie triple systems from (ε,δ)–Freudenthal–Kantor triple systems $U(\varepsilon,\delta)$.

First,we recall a Lie triple system (the case of $\delta = 1$) or an anti–Lie triple system (the case of $\delta = -1$) that satisfy the following identities:

(L1) $[xyz] = -\delta[yxz]$

(L2) $[xyz] + [yzx] + [zxy] = 0$

(L3) $[L(x,y), L(z,w)] = L(([xyz],w) + L(z,[xyw])$.

Theorem 2.1. *Let $U(\varepsilon,\delta)$ be a (ε,δ)–Freudenthal–Kan- tor triple system. If P is a linear transformation of $U(\varepsilon,\delta)$ such that $P < xyz\cdot > = < Px, Py, Pz >$ and $P^2 = -\varepsilon\delta Id$, then $(U(\varepsilon,\delta),[-,-,-])$ is a Lie triple system (the case of $\delta = 1$) or an anti–Lie triple system (the case of $\delta = -1$) with respect to the triple product*

$$[xyz] := < xPyz > -\delta < yPxz > +\delta < xPzy > - < yPzx > . \qquad (2-1)$$

Proof. From the definition of $U(\varepsilon,\delta)$, we can obtain this theorem by straightforward but very long calculations and we omit it.

In particular,if $\varepsilon = -1, \delta = 1, K(x,y) = 0$ and $P = Id$ (that is, $U(\varepsilon,\delta)$ is a Jordan triple system), then the triple product becomes

$$[xyz] = <xyz> - <yxz> .$$

Corollary. *Let $V(\varepsilon,\delta)$ be a (ε,δ)–Freudenthal–Kantor triple system. Then the vector space $V(\varepsilon,\delta) \oplus V(\varepsilon,\delta)$ becomes a Lie triple system (the case of $\delta = 1$) or an anti–Lie triple system (the case of $\delta = -1$) with respect to the triple product defined by*

$$[\begin{pmatrix} a \\ b \end{pmatrix} \begin{pmatrix} c \\ d \end{pmatrix} \begin{pmatrix} e \\ f \end{pmatrix}] =$$

$$\begin{pmatrix} L(a,d) - \delta L(c,b) & \delta K(a,c) \\ -\varepsilon K(b,d) & \varepsilon(L(d,a) - \delta L(b,c)) \end{pmatrix} \begin{pmatrix} e \\ f \end{pmatrix} . \qquad (2-2)$$

Proof. Let $U(\varepsilon,\delta) := V(\varepsilon,\delta) \oplus V(\varepsilon,\delta)$.

We put $P = \begin{pmatrix} 0 & \delta Id \\ -\varepsilon Id & 0 \end{pmatrix}$. Then from the above theorem, we obtain the following triple product:

$$[\begin{pmatrix} a \\ b \end{pmatrix} \begin{pmatrix} c \\ d \end{pmatrix} \begin{pmatrix} e \\ f \end{pmatrix}] =$$

$$< \begin{pmatrix} a \\ b \end{pmatrix}, \begin{pmatrix} 0 & \delta Id \\ -\varepsilon Id & 0 \end{pmatrix} \begin{pmatrix} c \\ d \end{pmatrix}, \begin{pmatrix} e \\ f \end{pmatrix} >$$

$$-\delta < \begin{pmatrix} c \\ d \end{pmatrix}, \begin{pmatrix} 0 & \delta Id \\ -\varepsilon Id & 0 \end{pmatrix} \begin{pmatrix} a \\ b \end{pmatrix}, \begin{pmatrix} e \\ f \end{pmatrix} >$$

$$+\delta < \begin{pmatrix} a \\ b \end{pmatrix}, \begin{pmatrix} 0 & \delta Id \\ -\varepsilon Id & 0 \end{pmatrix} \begin{pmatrix} e \\ f \end{pmatrix}, \begin{pmatrix} c \\ d \end{pmatrix} >$$

$$- < \begin{pmatrix} c \\ d \end{pmatrix}, \begin{pmatrix} 0 & \delta Id \\ -\varepsilon Id & 0 \end{pmatrix} \begin{pmatrix} e \\ f \end{pmatrix}, \begin{pmatrix} a \\ b \end{pmatrix} >$$

which implies the relation(2–2). This completes the proof.

The Lie triple system or anti–Lie triple system obtained from this relation (2–2) is called to the Lie triple system or anti–Lie triple system associated with a (ε,δ) –Freudenthal–Kantor triple system.

§ 3

In this section, we shall consider the simplicity of the standard imbedding Lie algebra or Lie superalgebra associated with a (ε,δ)–Freudenthal–Kantor triple system $U(\varepsilon,\delta)$.

From § 2, we obtain the Lie or anti–Lie triple system $U(\varepsilon,\delta)\oplus U(\varepsilon,\delta)$ associated with $U(\varepsilon,\delta)$ and denote it by $T(\varepsilon,\delta)$. According to $\delta=1$ or $\delta=-1$, we have the standard imbedding Lie algebra ($\delta=1$) or Lie superalgebra ($\delta=-1$) as follows:

Lie algebra (the case of $\delta=1$)

$$L(\varepsilon,\delta):=D(T(\varepsilon,\delta),T(\varepsilon,\delta))\oplus T(\varepsilon,\delta),$$

where $D(T(\varepsilon,\delta),T(\varepsilon,\delta))$ is the Lie algebra of inner derivations of Lie triple system $T(\varepsilon,\delta)$.

Lie superalgebra(the case of $\delta=-1$)

$$L(\varepsilon,\delta)=D(T(\varepsilon,\delta),T(\varepsilon,\delta))\oplus T(\varepsilon,\delta),$$

where $D(T(\varepsilon,\delta),T(\varepsilon,\delta))$ is the Lie algebra of inner derivations of anti–Lie triple system $T(\varepsilon,\delta)$.

An element of the Lie algebra or Lie superalgebra $L(\varepsilon,\delta)$ is

$$\begin{pmatrix} L(a,b) & \delta K(c,d) \\ -\varepsilon K(e,f) & \varepsilon L(b,a) \end{pmatrix}\oplus\begin{pmatrix} x \\ y \end{pmatrix}. \qquad (3-1)$$

In the case of the standard imbedding Lie algebra or Lie superalgebra $L(\varepsilon,\delta)$, we have the decomposition

$$L(\varepsilon,\delta)=L_{-2}(\varepsilon,\delta)\oplus L_{-1}(\varepsilon,\delta)\oplus L_0(\varepsilon,\delta)\oplus L_1(\varepsilon,\delta)\oplus L_2(\varepsilon,\delta)$$

where

$$L_{-2}(\varepsilon,\delta)=\textit{the span of all}\begin{pmatrix} 0 & \delta K(c,d) \\ 0 & 0 \end{pmatrix}$$

$$L_{-1}(\varepsilon,\delta)=U(\varepsilon,\delta)\oplus(0)$$

$$L_0(\varepsilon,\delta)=\textit{the linear span of all}\begin{pmatrix} L(a,b) & \\ 0 & \varepsilon L(b,a) \end{pmatrix}$$

$$L_1(\varepsilon,\delta)=(0)\oplus U(\varepsilon,\delta)$$

$$L_2(\varepsilon,\delta)=\textit{the linear span of all}\begin{pmatrix} 0 & 0 \\ -\varepsilon K(e,f) & 0 \end{pmatrix}.$$

The results of above is due to K.Yamaguti.

A (ε,δ)–Freudentahl–Kantor triple system $U(\varepsilon,\delta)$ is said to be unitary if the linear span $\tilde{k}$ of the set $\{K(a,b)|a,b\in U(\varepsilon,\delta)\}$ contains the identity endomorphism Id.

Remark. The case of $\varepsilon\delta = -1$ does not occur in unitary (ε, δ) –Freudenthal –Kantor triple systems $U(\varepsilon, \delta)$.

Let $U(\varepsilon, \delta)$ be a unitary Freudenthal–Kantor triple system. $T(\varepsilon, \delta)$ be the Lie triple system or anti-Lie triple system associated with $U(\varepsilon, \delta)$ and $L(\varepsilon, \delta)$ be the standard imbedding Lie algebra or Lie superalgebra. Since $Id \in \tilde{k}$, the Lie algebra or Lie superalgebra $L(\varepsilon, \delta)$ contains elements

$$E = \begin{pmatrix} 0 & \delta Id \\ 0 & 0 \end{pmatrix} \textit{ and } F = \begin{pmatrix} 0 & 0 \\ -\varepsilon Id & 0 \end{pmatrix}.$$

It also contains the element

$$H = \varepsilon\delta[E, F] = \begin{pmatrix} -Id & 0 \\ 0 & Id \end{pmatrix}.$$

As $[H, E] = -2E$ and $[H, F] = 2F$, we get a three dimensional simple Lie subalgebra of the type $A_1 =< H, E, F >$ and

$$A_1 \leq D(T(\varepsilon, \delta), T(\varepsilon, \delta)) \leq L(\varepsilon, \delta).$$

Since there is an automorphism $\sigma_0 = \begin{pmatrix} 0 & \delta Id \\ -\varepsilon Id & 0 \end{pmatrix}$ of $T(\varepsilon, \delta)$ such that ${\sigma_0}^2 = -\varepsilon\delta$ Id, we extend this automorphism to the Lie algebra or Lie superalgebra $L(\varepsilon, \delta)$ and denote it by $\tilde{\sigma}_0$. That is,

$$\tilde{\sigma}_0 : D(T(\varepsilon), T(\varepsilon)) \oplus T(\varepsilon) \longrightarrow \sigma_0 D(T(\varepsilon), T(\varepsilon))\sigma_0^{-1} \oplus \sigma_0 T(\varepsilon).$$

Then we get

$$\tilde{\sigma}_0(H) = \sigma_0 \circ H \circ {\sigma_0}^{-1} = -H, \tilde{\sigma}_0(E) = F \textit{ and } \tilde{\sigma}_0(F) = E,$$

and we see that $A_1 =< H, E, F >$ is a $\tilde{\sigma}_0$–invariant split three dimensional simple Lie algebra. Thus the standard imbedding Lie algebra or Lie superalgebra $L(\varepsilon, \delta) = D(T(\varepsilon, \delta), T(\varepsilon, \delta)) \oplus T(\varepsilon, \delta)$ contains the simple $\tilde{\sigma}_0$–invariant subalgebra $A_1 =< H, E, F >$. Therefore we obtain a decomposition of $L(\varepsilon, \delta)$ as follows:

$$L(\varepsilon, \delta) = L_{-2}(\varepsilon, \delta) \oplus L_{-1}(\varepsilon, \delta) \oplus L_0(\varepsilon, \delta) \oplus L_1(\varepsilon, \delta) \oplus L_2(\varepsilon, \delta),$$

where $L_i(\varepsilon, \delta)$ is the eigen space for ad H corresponding to the eigenvalue i in $L(\varepsilon, \delta)(-2 \leq i \leq 2)$.

Theorem 3.1. *For a unitary Freudenthal–Kantor triple system $U(\varepsilon, \delta)$, let $T(\varepsilon, \delta)$ be the Lie triple systems or anti–Lie triple system and $L(\varepsilon, \delta)$ be*

the standard imbedding Lie algebra or Lie superalgebra associated with $U(\varepsilon,\delta)$. *The following are equivalent;*
(a) $U(\varepsilon,\delta)$ *is simple.*
(b) $T(\varepsilon,\delta)$ *is simple.*
(c) $L(\varepsilon,\delta)$ *is simple.*

Proof. By the same argument as Theorem 2.1 in [11], we can obtain the proof. Thus we omit it.

§ 4

In this section, we shall define a trace form for a (ε,δ)–Freudenthal–Kantor triple system $U(\varepsilon,\delta)$ and characterize it.

We may define a bilinear form (=trace form) $\gamma(x,y)$ of $U(\varepsilon,\delta)$ by

$$\gamma(x,y) := \frac{1}{2}Tr[2(R(x,y)-\varepsilon R(y,x))+\delta(\varepsilon L(x,y)-L(y,x))] \qquad (4-1)$$

where $L(x,y)z = <xyz>$ and $R(x,y)z = <zxy>$.
Also,from [16] or [18] (resp.[13]), the Killing form $\alpha(t,s)$ of a Lie triple system (resp. anti–Lie triple system) is given as

$$\alpha(t,s) = \frac{1}{2}Tr[R(t,s)+R(s,t)] \qquad (4-2)$$

$$(resp.\ \alpha(t,s) = \frac{1}{2}Tr[R(t,s)-R(s,t)]\ ). \qquad (4-3)$$

Thus we may define as follows:

$$\alpha(t,s) = \frac{1}{2}Tr[R(t,s)+\delta R(s,t)]. \qquad (4-4)$$

By means of relation (2-2), the corespondence between the trace form γ of $U(\varepsilon,\delta)$ and the Killing form α of the Lie triple system or anti–Lie triple system $T(\varepsilon,\delta)$ associated with it is given by the following.

Proposition 4.1.

$$\alpha(\begin{pmatrix} a_1 \\ b_1 \end{pmatrix}, \begin{pmatrix} a_2 \\ b_2 \end{pmatrix}) = \gamma(b_1,a_2)+\delta\gamma(b_2,a_1) \qquad (4-5)$$

On the other hand, we have the following

Proposition 4.2. [13] *Let* α *be the Killing form of an anti–Lie triple system* T *and let* β *be the Killing form* ($\beta(x,y)$ = *strace ad* x *ad* y) *of the standard imbedding Lie superalgebra* $L = D(T,T)\oplus T$.

Then we have

$$\alpha(x,y) = \beta(y,x) \tag{4-6}$$

for $x, y \in T$.

From the correspondence between the Killing form α of a Lie (or anti-) triple system and the Killing form β of the stanadard imbedding Lie algebra (or Lie superalgebra), we have the following.

Proposition 4.3. *For the Killing form* α *of a Lie triple system or an anti–Lie triple system defined by (4–4), we have*
(i) $\alpha([xyz], w) = \delta\alpha(z, [yxw])$
(ii) $\alpha(x, [yzw]) = \delta\alpha([xwz], y)$.

Proposition 4.4. *For the trace form* γ *of a* (ε, δ) *–Freudenthal –Kantor triple system* $U(\varepsilon, \delta)$, *we have*
(i) $\gamma(< xyz >, w) + \varepsilon\gamma(z, < yxw >) = 0$
(ii) $\gamma(< xyz >, w) + \varepsilon\gamma(x, < wzy >) = 0$.
(iii) $\gamma(Dx, y) + \gamma(x, Dy) = 0$
(iv) $\gamma(Bx, y) - \gamma(x, By) = 0$
where $D \in DerU(\varepsilon, \delta)$, $B \in Anti - Der\ U(\varepsilon, \delta)$.

Proof From Proposition 4.1 and Proposition 4.3, we obtain the identities (i) and (ii). We have the following identities in a (ε, δ) –Freudenthal–Kantor triple system:

$$[D, L(x,y)] = L(Dx, y) + L(x, Dy)$$

$$[D, R(x,y)] = R(Dx, y) + R(x, Dy)$$

$$[B, L(x,y)] = L(Bx, y) - L(x, By)$$

$$[B, R(x,y)] = -R(Bx, y) + R(x, By)$$

for $D \in Der\ U(\varepsilon, \delta), B \in Anti - DerU(\varepsilon, \delta)$. Hence we have

$$TrL(Dx,y) = -TrL(x,Dy), \quad TrR(Dx,y) = -TrR(x,Dy),$$

$$TrL(Bx,y) = TrL(x,By), \quad TrR(Bx,y) = TrR(x,By).$$

Thus these imply the identities (iii) and (iv). This completes the proof.

From Proposition 4.1 and Proposition 4.2, we have the final theorem in this article as follows.

Theorem 4.5. *Let* $U(\varepsilon, \delta)$ *be a* (ε, δ)*–Freudenthal–Kantor triple system,* $T(\varepsilon, \delta)$ *be the Lie triple system or anti–Lie triple system and* $L(\varepsilon, \delta)$ *be the Lie algebra or Lie superalgebra associated with* $U(\varepsilon, \delta)$. *Let* γ, α, β *be the respective forms. Then the following statements are equivalent:*
(i) *The trace form* γ *is nondegenerate*

(ii) *The Killing form* α *is nondegenerate*
(iii) *The Killing form* β *is nondegenerate.*

Corollary. *Any derivation of a* (ε, δ) *–Freudenthal–Kantor triple system* $U(\varepsilon, \delta)$ *with nondegenerate trace form* γ *is a finite sum of inner derivations of* $S(a, b) = L(a, b) + \varepsilon L(b, a)$.

References

1. B.N.Allison,: *A construction of Lie algebras from J– ternary algebras*, Amer. J. Math., 98 (1976), 285– 294.
2. B.N.Allison,: *A class of nonassociative algebras with involution containing the class of Jordan algebras*, Math. Ann., 237(1978), 133–156.
3. H.Asano and K.Yamaguti,: *A construction of Lie algebra by generalized Jordan triple systems of second order*, Indag.Math.,42(1980), 249–253.
4. J.R.Faulkner and J.C.Ferrar,: *Simple anti–Jordan pairs*, Comm. Alg., 8 (1980), 993–1013.
5. W.Hein,: *A construction of Lie algebras by triple systems*, Trans. Amer. Math. Soc., 205 (1975), 79– 95.
6. U.Hirzebruch,: *A generalization of Tits' construction of Lie algebras by Jordan algebras to Jordan triple systems*, Indag.Math., 40 (1978), 456–459.
7. N.Jacobson,: *Lie and Jordan triple systems*, Amer. J. Math., 71 (1949), 148– 170.
8. N.Jacobson,: *Structure and representations of Jodan algebras*, Amer. Math. Soc. Colloq. Publ., Vol.39, Providence, R.I., 1968.
9. V.G.Kac,: *Lie superalgebras*, Advances in Math. 26 (1977), 8– 96.
10. N.Kamiya,: *A structure theory of Freudenthal –Kantor triple systems*, J. Algebra, 110 (1987), 108–123,
11. N.Kamiya,: *A structure theory of Freudenthal–Kantor triple systems II*, Comm. Math. Univ. Sancti Pauli, 38 (1989), 41-60.
12. N.Kamiya,: *A structure theory of Freudenthal–Kantor triple systems III*, Mem. Fac. Sci. Shimane Univ., 23 (1989), 33– 51.
13. N.Kamiya,: *A construction of Anti–Lie triple systems from a class of triple systems*, Mem.Fac. Sci. Shimane Univ., 22(1988), 51–62.
14. I.L.Kantor,: *Models of exceptional Lie algebras*, Soviet Math. Dokl., 14 (1973), 254–258.
15. O.Loos,:*Jordan pairs*, Springer Lecture Notes 460,1975.

16. K.Meyberg,:*Lectures on algebras and triple systems*, Lecture notes. The Univ. of Virginia, Charlottesville, 1972.
17. E.Neher,:*On the classification of Lie and Jordan triple systems*, Habilitationsschrift, Wilhelms–Universität zu Münster, 1983.
18. J.S.Ravisankar,:*Some remarks on Lie triple systems*, Kumamoto J. Sci. (Math), 11(1974), 1–8.
19. K.Yamaguti,: *On the metasymplectic geometry and triple systems*, Surikaisekikenkyusho Kokyuroku,306 (1977), 55 –92. Research Institute for Math. Sci., Kyoto Univ., (in Japanese).
20. K.Yamaguti, and A.Ono,: *On representations of Freudenthal–Kantor triple systems* $U(\varepsilon,\delta)$, Bull. Fac. School Ed., Hiroshima Univ., Part II, 7 (1984), 43– 51.

Projectivity of Homogeneous Left Loops

Michihiko KIKKAWA

Department of Mathematics
Shimane University, Matsue, 690 JAPAN

ABSTRACT

After surveying the theory of homogeneous Lie loops, the problem of finding geodesic homogeneous local left Lie loops on any Lie group which are in projective relation with it is treated. Moreover, some special and intrinsic examples of such local left loops are given.

§1. Preliminaries

In this section, some definitions of algebraic concepts concerning loops are given, which will be treated in the other sections as analytic and local concepts on analytic manifolds. Cf. [6], [10], [14] and [22].

1.1. DEFINITION. A <u>loop</u> (G,μ) is a set G equipped with a binary operation $\mu: G\times G \longrightarrow G$ satisfying the followings:

(i) There exists a (two-sided) identity element e;

$$\mu(x,e) = \mu(e,x) = x \quad \text{for any } x \text{ in } G.$$

$(ii)_L$ Any left translation $L_x : G \longrightarrow G$; $L_x y = \mu(x,y)$, is a bijection.

$(ii)_R$ Any right translation R_x is a bijection.

If (G,μ) satisfies (i) and $(ii)_L$, it will be called a <u>left loop</u>. A (left) loop (G,μ) is said to have the <u>left inverse property</u> if $L_x^{-1} = L_{x^{-1}}$ holds for any x in G, where $x^{-1} = L_x^{-1} e$; that is,

$$\mu(x,\mu(x^{-1},y)) = \mu(x^{-1},\mu(x,y)) = y.$$

1.2. PROPOSITION. <u>If a (left) loop</u> (G,μ) <u>has the left inverse property</u>, <u>then the element</u> $x^{-1} = L_x^{-1} e$ <u>is a unique two-sided inverse of</u> x, <u>i.e.</u>,

$$\mu(x^{-1},x) = \mu(x,x^{-1}) = e.$$

1.3. DEFINITION. A (left) loop with the left inverse property is said to be homogeneous if every left inner map

(1.3) $$L_{x,y} := L^{-1}_{\mu(x,y)} \cdot L_x \cdot L_y$$

is an automorphism of the (left) loop (G,μ).

1.4. PROPOSITION. Let (G,μ) be a left loop with the left inverse property and $\eta: G\times G\times G \longrightarrow G$ be the ternary system on G given by

(1.4.1) $$\eta(x,y,z) := L_x\, \mu(L_x^{-1}y, L_x^{-1}z).$$

Then, (G,μ) is homogeneous if and only if η satisfies the following condition:

(1.4.2) $$\eta(x,y,\eta(u,v,w)) = \eta(\eta(x,y,u),\eta(x,y,v),\eta(x,y,w))$$

for any x, y, u, v, w in G ([25]).

1.5. DEFINITION. The ternary system η above is called the homogeneous system of the homogeneous (left) loop (G,μ) (cf. [25], [26], [27], [29], [30], [35]).

1.6. REMARK. Any group (G,μ_0) is a homogeneous loop whose left inner maps are reduced to the identity map of G. In fact, a loop with the left inverse property is reduced to a group if and only if every left inner map is reduced to the identity map. The homogeneous system η_0 of a group (G,μ_0) is given by

(1.6) $$\eta_0(x,y,z) = y\,x^{-1}z.$$

1.7. THEOREM ([22]). Let (G,μ) be a homogeneous left loop and $L(G)$ the left inner mapping group of (G,μ). If a subgroup K of $AUT(G,\mu)$ contains $L(G)$, then the Cartesian product $A = G\times K$ forms a group under the multiplication;

(1.7) $$(x,\alpha)\cdot(y,\beta) = (\mu(x,\alpha(y)),\ L_{x,\alpha(y)}\cdot\alpha\cdot\beta).$$

§2. Loops and 3-Webs

The notion of loops appeared in 1930's under the name of "Normbereich" in the theory of "Gewebe" (cf., e.g., G. Boll [8], [9] or W. Blaschke – G. Bol [7]). The name "loop" was given by A.A. Albert [6].

2.1. Let (G,μ) be a loop with the identity element e. With each pair (x_0,y_0) in the set $W = G\times G$, the following three classes of subsets of W be associated;

$$F_{(1)}(x_0,y_0) = \{x_0\} \times G,$$
$$F_{(2)}(x_0,y_0) = G \times \{y_0\},$$
$$F_{(3)}(x_0,y_0) = \{(x,y) \in W \mid \mu(x,y)=\mu(x_0,y_0)\}.$$

The subsets of W belonging to one of these classes for some "point" (x_0,y_0) in W will be called the "lines" of W. It is easy to see that they satisfy the followings:

(W_1) Every point of W lies in one and only one line of each class.

(W_2) Any two lines of the same class does not meet at any point in W.

(W_3) Any two lines of different classes meet at one and only one point of W.

In general, we can introduce the concept of abstract 3-webs by these conditions; that is,

2.2. DEFINITION. A set of "Points" with the three classes $F_{(1)}$, $F_{(2)}$, $F_{(3)}$ of "lines" is called a 3-web if they satisfy the axioms (W_1), (W_2) and (W_3) above (cf., e.g., G. Bol [9]).

2.3. The 3-web associated with a loop (G,μ) presents a figuring of the loop multiplication μ on the line passing through the origin $E = (e,e)$, for instance $F_{(1)}(E)$, as follows (Fig.1): In the 3-web $W = G\times G$ associated with (G,μ), identify G with the line $F_{(1)}(E)$ by representing each $x \in G$ by the point $X = (e,x) \in F_{(1)}(E)$. Then, for any two points $X=(e,x)$ and $Y=(e,y)$ on $F_{(1)}(E)$, we can find $X' = F_{(3)}(X) \cap F_{(2)}(E) = (x,e)$ on $F_{(2)}(E)$ and $P = F_{(1)}(X') \cap F_{(2)}(Y) = (x,y)$. Since any point (u,v) on $F_{(3)}(P)$ satisfies $\mu(u,v) = \mu(x,y)$ we obtain the point $(e,\mu(x,y)) = \mu(X,Y)$ on the line $F_{(1)}(E)$, which presents the multiplication of X and Y on the line $F_{(1)}(E) = G$.

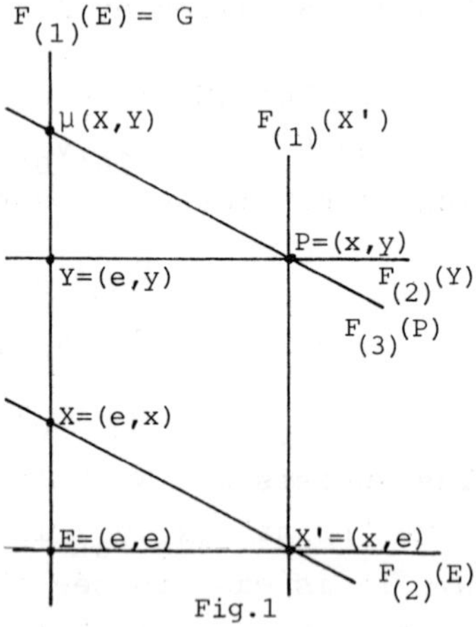

Fig.1

Conversely, any 3-web $\{W; F_{(1)}, F_{(2)}, F_{(3)}\}$ presents on any line, say $F_{(1)}(E)$ through an arbitrarily fixed origin E, a loop multiplication μ with the identity element E by means of the construction in Fig.1.

2.4. CLOSURE CONDITIONS. Various algebraic conditions for loops are figured in the associated 3-web by the corresponding closure conditions of some figures (cf. [9], [53]).

(1) Existence of two-sided inverse; $\mu(x,x^{-1}) = \mu(x^{-1},x) = e$.

$\Longleftrightarrow$

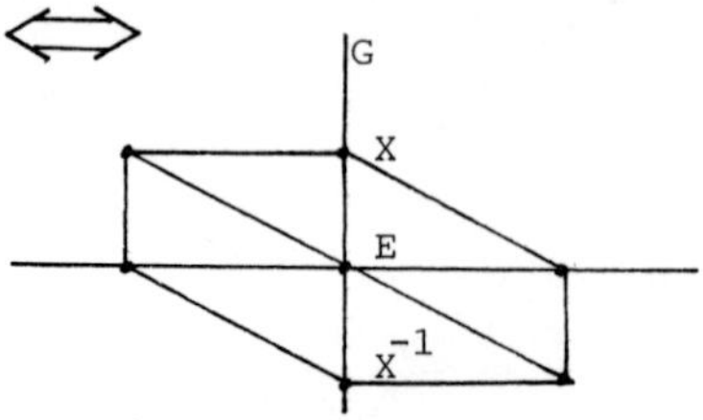

(Hexagonal at E)
Fig.2

(2) The left inverse property; $\mu(x^{-1},\mu(x,y)) = y$.

$\Longleftrightarrow$

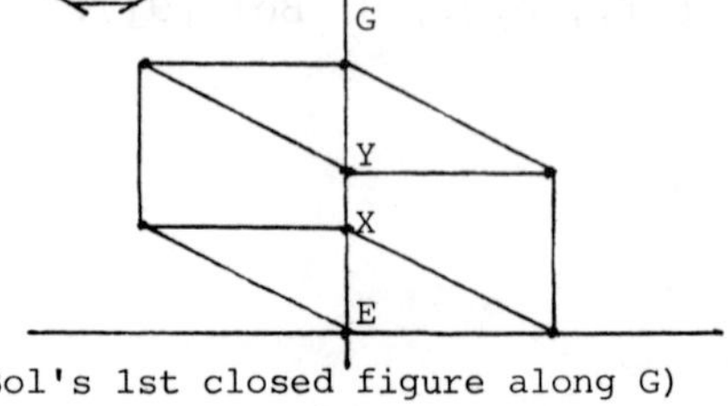

(Bol's 1st closed figure along G)
Fig.3

(3) (G,μ) is associative.

⟺

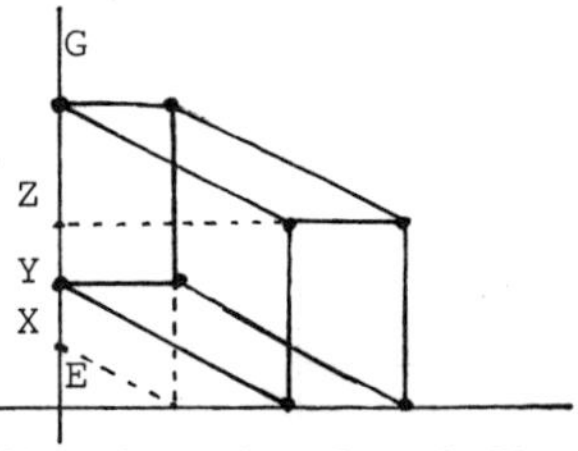

(Reidemeister's closed figure along G)

Fig.4

(4) (G,μ) is commutative.

⟺

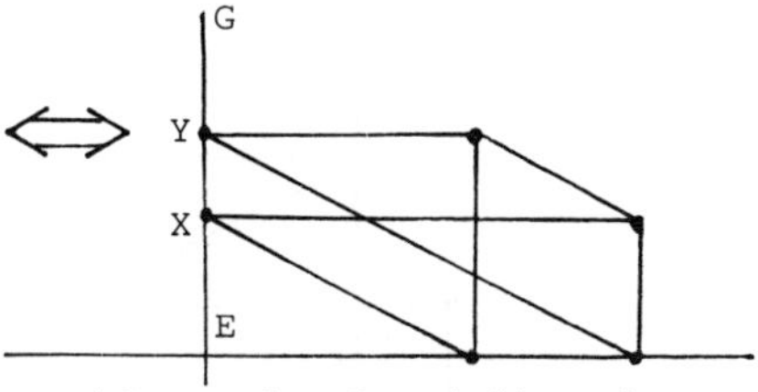

(Thomsen's closed figure)

Fig.5

(5) (G,μ) is a Moufang loop, i.e., $\mu(x,\mu(y,\mu(x,z)))=\mu(\mu(\mu(x,y),x),z)$.

⟺

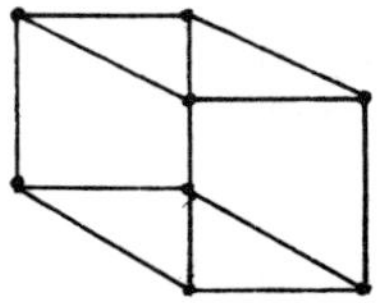

and

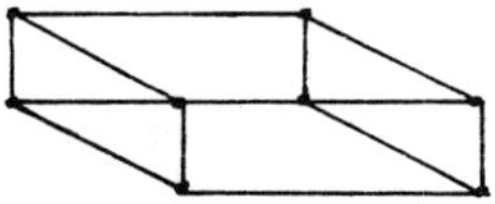

(Bol's 1st and 2nd closed figures holding at everywhere)

Fig.6

(6) (G,μ) is anti-symmetric, i.e., $\mu(x,y)^{-1}=\mu(y^{-1},x^{-1})$.

⟺

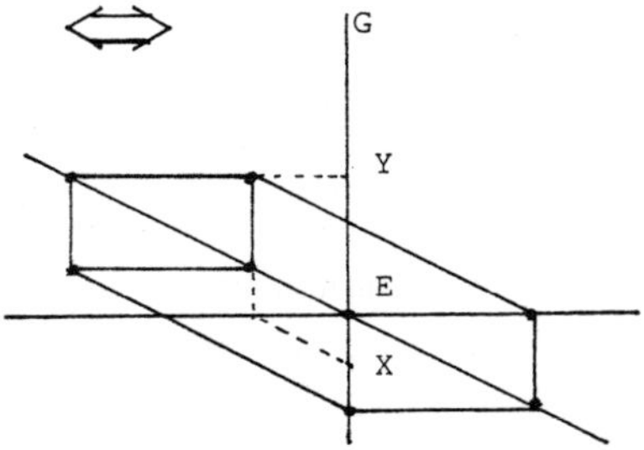

(Bol's 3rd closed figure)

Fig.7

(7) (G,μ) is symmetric ([14],[20]), i.e., $\mu(x,y)^{-1}=\mu(x^{-1},y^{-1})$.

⟺

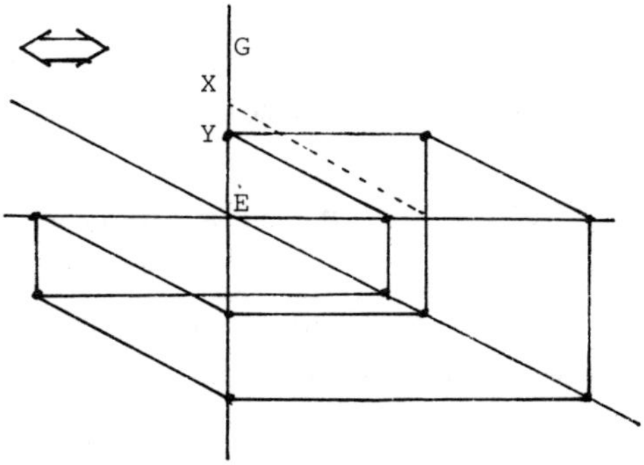

Fig.8

(8) (G,μ) is homogeneous, i.e., $L_{x,y}\mu(z,w) = \mu(L_{x,y}z, L_{x,y}w)$.

$\Longleftrightarrow$

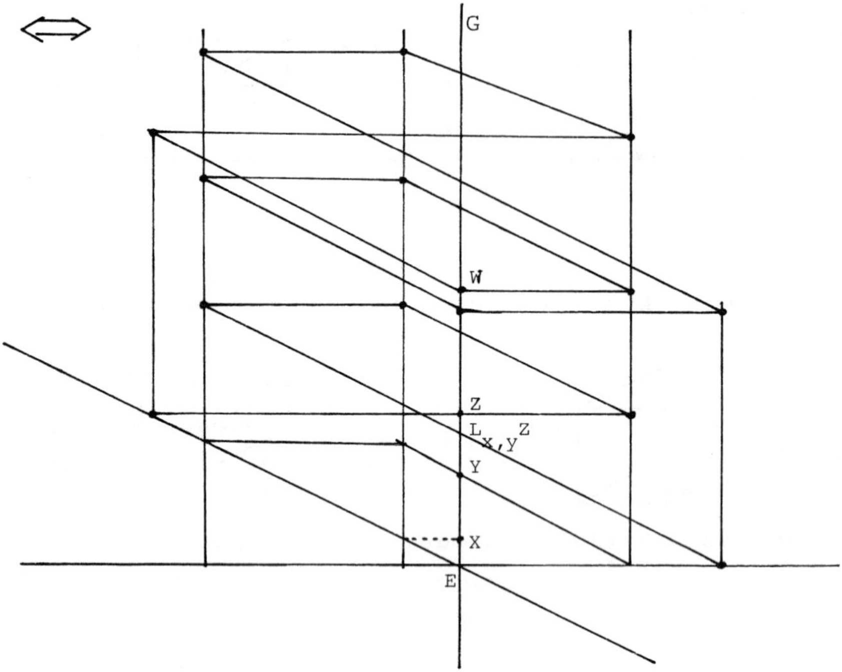

Fig 9

§3. Differentiable Local Loops in 3-Webs

In 1936, S.S. Chern introduced a method of modern differential geometry into web geometry ([12]).

3.1. Let M be a 2n-dimensional differentiable manifold and $\Sigma_{(1)}, \Sigma_{(2)}, \Sigma_{(3)}$ three foliations of codimension n in M, any two of which are independent at each point of M. Then, the conditions (W_1), (W_2), (W_3) for 3-webs are satisfied, in local, at any point E of M, so that an n-dimensional differentiable local loop with the identity element E is induced on the leaf, e.g., $F_{(1)}(E)$ of $\Sigma_{(1)}$, through the point E. Associated with the local differentiable

3-web $\{\Sigma_{(1)}, \Sigma_{(2)}, \Sigma_{(3)}\}$ above, a linear connection on M has been introduced by S.S. Chern [12], which we call the Chern connection of the 3-web.

Let $\omega^1_{(\rho)}, \omega^2_{(\rho)}, \ldots, \omega^n_{(\rho)}$ ($\rho = 1,2,3$) be three involutive families of differential systems in M with which the foliations $\Sigma_{(\rho)}$ is defined by $\omega^k_{(\rho)} = 0$, $k = 1,2,\ldots,n$. Without loss of generality, we can choose the forms $\omega^k_{(3)}$ such that

$$\omega^k_{(3)} = \omega^k_{(1)} + \omega^k_{(2)}, \quad k = 1,2, \ldots, n.$$

Then, the following formulas are obtained:

$$(3.1.1) \qquad d\omega^k_{(1)} = \omega^i_{(1)} \wedge \omega^k_i + a^k_{ij}\omega^i_{(1)} \wedge \omega^j_{(1)}$$

$$(3.1.2) \qquad d\omega^k_{(2)} = \omega^i_{(2)} \wedge \omega^k_i - a^k_{ij}\omega^i_{(2)} \wedge \omega^j_{(2)}$$

$$(3.1.3) \qquad d\omega^k_j = \omega^i_j \wedge \omega^k_i + b^k_{jhm}\omega^h_{(1)} \wedge \omega^m_{(2)},$$

where ω^k_i's are 1-forms and a^k_{ij}, b^k_{jhm} are differentiable functions satisfying $a^k_{ij} + a^k_{ji} = 0$.

3.2. DEFINITION. The Chern connection is the linear connection on the 2n-dimensional manifold M whose connection forms $\tilde{\omega}^\alpha_\beta$, $1 \leqq \alpha,\beta \leqq 2n$, are given by

$$(3.2.1) \qquad \tilde{\omega}^k_i = \tilde{\omega}^{n+k}_{n+i} = \omega^k_i \quad \text{and} \quad \tilde{\omega}^k_{n+i} = \tilde{\omega}^{n+k}_i = 0, \; 1 \leqq i,k \leqq n.$$

The torsion tensor A and the curvature tensor B of the Chern connection are expressed with respect to the base forms $\{\omega^1_{(1)}, \ldots, \omega^n_{(1)}; \omega^1_{(2)}, \ldots, \omega^n_{(2)}\}$ by

$$(3.2.2) \qquad A^k_{ij} = -A^{n+k}_{n+i\, n+j} = -2a^k_{ij} \quad \text{and} \quad A^\alpha_{\beta\gamma} = 0 \text{ otherwise},$$

$$(3.2.3) \qquad B^k_{i\,h\,n+m} = -B^k_{j\,n+m\,h} = B^{n+k}_{n+j\,h\,n+m} = -B^{n+k}_{n+j\,n+h\,m} = -b^k_{jhm}$$

$$\text{and} \quad B^\alpha_{\beta\gamma\delta} = 0 \quad \text{otherwise (cf. [38])}.$$

S.S. Chern described closure conditions of some figures of the differentiable 3-web $\{\Sigma_{(1)}, \Sigma_{(2)}, \Sigma_{(3)}\}$ in terms of the torsion tensor A and the curvature tensor B (cf. [12]).

§4. Tangent Algebras of Differentiable Loops

4.1. MAL'CEV ALGEBRAS. In 1954, A.I.Mal'cev has shown that any analytic Moufang loop (G,μ) is locally characterized by an algebra induced on the tangent space $T_e(G)$ at the identity element e of the loop. This algebra is called Mal'cev algebra which is an anti-symmetric algebra satisfying the relation;

$$(XY)(ZX) + ((XY)Z)X + ((YZ)X)X + ((ZX)X)Y = 0. \tag{4.1}$$

This works seems to be a first access to generalization of the theory of Lie groups and their Lie algebras to the theory of analytic loops. Cf. [46] and [49].

4.2. AKIVIS ALGEBRAS. In 1978, M.A.Akivis considered an algebraic system on the tangent space of differentiable local loops at their identity elements, which is now called an Akivis algebra. An Akivis algebra is a vector space, say $\mathfrak{G}$, equipped with an anti-symmetric bilinear product $<XY>$ and a trilinear product $<X,Y,Z>$ satisfying the following relations:

$$\mathfrak{S}(<X,Y,Z> - <Y,X,Z> - <<XY>Z>) = 0, \tag{4.2}$$

where $\mathfrak{S}$ denotes the cyclic sum with respect to the three vectors X, Y and Z in $\mathfrak{G}$.

4.3. THEOREM(Akivis[2]). For any differentiable loop (G,μ) on a differentiable manifold G, consider the associated 3-web W and express it by three foliations $\Sigma_{(1)}$, $\Sigma_{(2)}$, $\Sigma_{(3)}$ in a neighborhood of the identity element e. Then, the tangent space $T_e(G)$ at e forms an Akivis algebra under the operations

$$<XY> = -2X^i Y^j a^k_{ij} E_k, \tag{4.3.1}$$

$$<X,Y,Z> = X^i Y^j Z^m b^k_{mij} E_k, \tag{4.3.2}$$

where $E_k = \partial_{n+k}$ are the natural basis at e corresponding to the base forms $\omega^1_{(2)},\ldots,\omega^n_{(2)}$, if the loop G is represented by the leaf of $\Sigma_{(1)}$ through the origin, and a^k_{ij}'s b^k_{ijm}'s are constants given by the value of the torsion tensor A and the curvature tensor B of the Chern connection on G×G evaluated at the origin (e,e) (cf. (3.2.2) and (3.2.3)).

§5. Homogeneous Left Lie Loops

In this section, various concepts and results for homogeneous Lie loops referred to [22], [25], [39] and [40] are modified for homogeneous <u>left</u> loops.

5.1. DEFINITION. An <u>analytic left loop</u> or <u>left Lie loop</u> (G,μ) is a left loop μ on an analytic manifold G such that the following maps are analytic;

$$\mu\colon G\times G \longrightarrow G \quad \text{(the multiplication)},$$

$$\lambda\colon G\times G \longrightarrow G,\ \lambda(x,y) = L_x^{-1}y,$$

where L_x denotes the left translation by x.

5.2. CANONICAL CONNECTION. Let (G,μ) be a homogeneous left Lie loop with the identity element e. The <u>canonical connection</u> of (G,μ) is a linear connection ∇ on G given by

(5.2) $$(\nabla_X Y)_x = X_x Y - \eta(x,X_x,Y_x),\quad x\in G$$

for any vector fields X, Y on G. Cf.[39], [40].

5.3. PROPOSITION ([22]). <u>The torsion tensor</u> S <u>and the curvature tensor</u> R <u>of the canonical connection</u> ∇ <u>of a homogeneous left Lie loop</u> (G,μ) <u>are parallel tensor fields on</u> G, <u>i.e.</u>,

(5.3) $$\nabla S = 0 \text{ and } \nabla R = 0 \text{ on } G.$$

5.4. TANGENT LIE TRIPLE ALGEBRAS. Let (G,μ) be a homogeneous left Lie loop and $T_e(G)$ the tangent space to G at the identity element e. The <u>tangent Lie triple algebra</u>

of (G,μ) is the algebraic system on $T_e(G)$ with the following bilinear operation XY and the trilinear operation $[X,Y,Z]$;

(5.4.1) $$XY = S_e(X,Y),$$

(5.4.2) $$[X,Y,Z] = R_e(X,Y)Z$$

for $X,Y,Z \in T_e(G)$, where S_e and R_e denote the value of the torsion tensor S and the curvature tensor R of the canonical connection ∇, respectively, evaluated at e.

5.5. THEOREM ([22]). The tangent Lie triple algebra of a homogeneous left Lie loop (G,μ) forms a general Lie triple system of Yamaguti [66].

Proof. Let ∇ be the canonical connection of (G,μ). The torsion S and the curvature R are defined, respectively, by

(5.5.1) $$S(X,Y) = \nabla_Y X - \nabla_X Y + [X,Y],$$

(5.5.2) $$R(X,Y)Z = \nabla_Y\nabla_X Z - \nabla_X\nabla_Y Z + \nabla_{[X,Y]}Z.$$

Hence the followings are evident in the tangent space $T_e(G)$;

(5.5.3) $$XY = -YX,$$

(5.5.4) $$[X,Y,Z] = -[Y,X,Z].$$

Since $\nabla S = 0$ and $\nabla R = 0$, Bianchi's 1st and 2nd identities (cf. Kobayashi-Nomizu [44]) for vector fields on G;

$$\mathfrak{S}\{R(X,Y)Z + S(S(X,Y),Z) - (\nabla_X S)(Y,Z)\} = 0,$$

$$\mathfrak{S}_{X,Y,Z}\{R(S(X,Y),Z)W - ((\nabla_X R)(Y,Z)W)\} = 0,$$

imply the following relations for tangent vectors at e:

(5.5.5) $$\mathfrak{S}\{[X,Y,Z] + (XY)Z\} = 0,$$

(5.5.6) $$\mathfrak{S}\{[XY,Z,W]\} = 0.$$

Finally, Ricci's identities for S and R;

$$R(X,Y,S(U,V)) = S(R(X,Y)U,V) + S(U,R(X,Y)V) + (R(X,Y)S)(U,V) + (\nabla_{S(U,V)}S)(X,Y)$$

and

$$R(X,Y)R(U,V)W = R(R(X,Y)U,V)W + R(U,R(X,Y)V)W + R(U,V)R(X,Y)W + (\nabla_{R(U,V)W}S)(X,Y),$$

imply the following relations in $T_e(G)$:

(5.5.7) $[X,Y,UV] = [X,Y,U]V + U[X,Y,V]$,

(5.5.8) $[X,Y,[U,V,W]] = [[X,Y,U],V,W] + [U,[X,Y,V],W] + [U,V,[X,Y,W]]$.

Thus the axioms (5.5.3) – (5.5.8) for general Lie triple system are assured for the tangent Lie triple algebra $T_e(G)$ of the homogeneous left Lie loop (G,μ).

q.e.d.

5.6. GEODESIC LOCAL LOOPS. Not depending on 3-webs, has there been considered differentiable local loops on manifolds with linear connections (cf. [17],[22], [23], [54], [55], [56]).

Let M be a differentiable manifold with a linear connection ∇, and e an arbitrarily fixed point of M. Then we can associate with e a differentiable local loop μ_e with the identity element e in the following way (cf. Fig. 5.1): For any two points x and y in a normal neighborhood N of e, let $\alpha(t)$ and $\beta(s)$ be geodesic curves in N joining the origin $e = \alpha(0)=\beta(0)$ to $x = \alpha(t_0)$ and $y = \beta(s_0)$, respectively. By the parallel displacement of the tangent vector $X_0 = \dot{\beta}(0)$ along the geodesic curve α we can obtain a vector X_{t_0} tangent to M at the point $x = \alpha(t_0)$. Then, we can find a geodesic curve γ passing through the point $x = \gamma(0)$ and tangent to the vector X_{t_0}. If the parameter of the geodesic curve γ can be extended in N to the value s_0, we set $\mu_e(x,y) = \gamma(s_0)$. So, we can define a differentiable local loop μ_e in M which will be called a <u>geodesic local loop</u> at e.

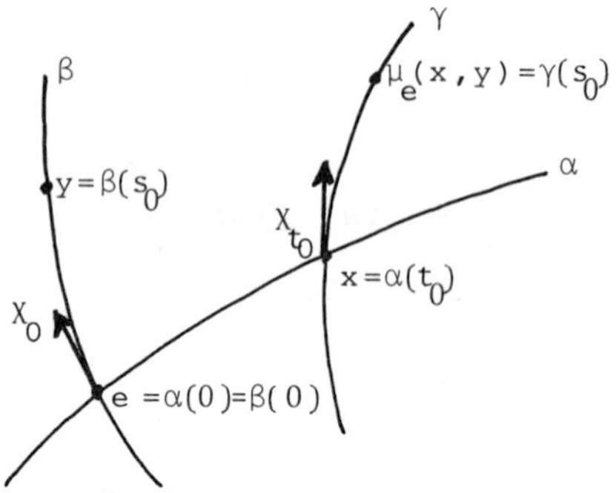

Fig. 5.1

5.7. DEFINITION. Let (G,μ) be a homogeneous left Lie loop with the canonical connection ∇. If any geodesic local loop of ∇ at the identity element e is coinsident with the multiplication of the loop, i.e., $\mu_e = \mu$, (G,μ) is said to be geodesic.

5.8. THEOREM(cf. [22]). Two geodesic homogeneous left Lie loops are locally isomorphic if and only if their tangent Lie triple algebras are isomorphic.

5.9. PROPOSITION. Let (G,μ) be a geodesic homogeneous left Lie loop. Then, any geodesic curve $x(t)$ passing through the identity element $e = x(0)$ is a 1-parameter subgroup (i.e., associative subloop) of (G,μ), and vice versa. Moreover, the geodesic curve $x(t)$ satisfies the following relation; (cf. [22])

$$L_{x(t),x(s)} = \mathrm{Id}. \tag{5.9}$$

5.10. REMARK. For the class of homogeneous Lie loops, K.H. Hofmann and K. Strambach [15] clarified the interrelation between the Akivis algebra and the tangent Lie triple algebra: Let $\{T_e(G); \langle X,Y\rangle, \langle X,Y,Z\rangle\}$ and $\{T_e(G); XY, [X,Y,Z]\}$ be the Akivis algebra and the tangent Lie triple algebra of a homogeneous Lie loop (G,μ). Then, they are related as follows:

$$XY = \langle X,Y\rangle \quad \text{and} \quad [X,Y,Z] = -2\langle X,Y,Z\rangle .$$

This is valid for homogeneous left Lie loops. Hence, the following is obtained from Theorem 5.8.

5.11. COROLLARY. Two geodesic homogeneous left Lie loops are locally isomorphic if and only if their Akivis algebras are isomorphic.

§6. Projectivity of Homogeneous Left Lie Loops

6.1. DEFINITION([41], [43]). On an anlytic manifold G, two geodesic homogeneous left Lie loops μ and $\tilde{\mu}$ with the same identity element e will be said to be in projective relation if the followings are satisfied:

(i) Any 1-parameter subgroup of μ is a 1-parameter subgroup of $\tilde{\mu}$, and vice versa.

(ii) Let η and $\tilde{\eta}$ denote the homogeneous systems of μ and $\tilde{\mu}$, respectively (cf. Definition 1.5). Then, they satisfy the relations;

$$\eta(x,y,\tilde{\eta}(u,v,w)) = \tilde{\eta}(\eta(x,y,u),\eta(x,y,v),\eta(x,y,w)), \tag{6.1.1}$$

$$\tilde{\eta}(x,y,\eta(u,v,w)) = \eta(\tilde{\eta}(x,y,u),\tilde{\eta}(x,y,v),\tilde{\eta}(x,y,w)) \tag{6.1.2}$$

for any x,y,u,v,w in G.

6.2. THEOREM([40], [41]). Let (G,μ) and $(G,\tilde{\mu})$ be geodesic homogeneous left Lie loops which are in projective relation. Then, the following formulas (6.2.1) – (6.2.7) are valid for any vector fields X, Y, Z, W on G, where S and R (resp. $\tilde{S}$ and $\tilde{R}$) are the torsion tensor and the curvature tensor of the canonical connection ∇ (resp. $\tilde{\nabla}$) of μ (resp. $\tilde{\mu}$), respectively, and T denotes the (1,2)-tensor field on G given by

$$T(X,Y) = \nabla_X Y - \tilde{\nabla}_X Y .$$

(6.2.1) $T(X,X) = 0,$

(6.2.2) $T(X,S(Y,Z)) = S(T(X,Y),Z) + S(Y,T(X,Z)),$

(6.2.3) $T(X,R(Y,Z)W) = R(T(X,Y),Z)W + R(Y,T(X,Z))W + R(Y,Z)T(X,W),$

(6.2.4) $T(X,T(Y,Z)) = T(T(X,Y),Z) + T(Y,T(X,Z)),$

(6.2.5) $R(X,Y)T(Z,W) = T(R(X,Y)Z,W) + T(Z,R(X,Y)W),$

(6.2.6) $\tilde{S}(X,Y) = S(X,Y) + 2T(X,Y),$

(6.2.7) $\tilde{R}(X,Y)Z = R(X,Y)Z - T(S(X,Y),Z) - T(T(X,Y),Z).$

Proof. The formulas (6.2.1) – (6.2.5) have shown in Corollary to Proposition 1 in [41]. In this proposition, it has been shown that the tensor fields T and $-T$ are respectively the affine homogeneous structures of the canonical connections ∇ and $\tilde{\nabla}$. Hence, by Proposition 1.1 of [40], we get the formulas (6.2.6) and (6.2.7).

q.e.d.

6.3. LOCAL VERSION ([43]). Let G be an analytic manifold with some base point, say e. The concept of geodesic homogeneous left loops on G with the identity element e can be localized to neighborhoods of e in a natural way, which will be called geodesic homogeneous <u>local</u> left Lie loops. Then, the results mentioned above are valid in a neighborhood of e. It should be noted that any local <u>left</u> loop is reduced to a local (left and right) loop in a neighborhood of the identity.

In particular, the formulas in Theorem 6.2 are valid on the tangent space $T_e(G)$ at the identity, that is, they give some relations in the tangent Lie triple algebras of geodesic homogeneous local left Lie loops.

§7. Projectivity of Lie Groups

In this section, we mention the main theorem of this report, which has been shown in [43].

7.1. LIE GROUPS AS HOMOGENEOUS LIE LOOPS

Let (G,μ_0) be a Lie group. Evidently, μ_0 can be regarded as a geodesic homogeneous left Lie loop. In fact, the canonical connection ∇^0 of μ_0 is coincident with the (−)-connection of E. Cartan whose curvature R^0 vanishes identically and torsion S^0 is given by

$$(7.1)\qquad S^0(X,Y) = [X,Y]_0$$

for any left invariant vector fields X and Y on G, where $[\ ,\ ]_0$ denotes the Lie bracket of the Lie algebra $\mathfrak{G}_0$ of μ_0. Thus we have

7.2. PROPOSITION. <u>The tangent Lie triple algebra of any Lie group</u> (G,μ_0) <u>is reduced to the Lie algebra</u> $\mathfrak{G}_0$ <u>of</u> μ_0.

In the following, we treat the problem of how to find all geodesic homogeneous local left Lie loops on a Lie group (G,μ_0) which are in projective relation with μ_0.

7.3. MAIN THEOREM ([43]). <u>Let</u> (G,μ_0) <u>be a Lie group and</u> $\mathfrak{G}_0 = \{T_e(G);\ [\ ,\ ]_0\}$ <u>its Lie algebra on the tangent space</u> $T_e(G)$ <u>of</u> G <u>at the identity element</u> e. <u>Assume that there is given another Lie algebra structure</u> $\mathfrak{L} = \{T_e(G);\ [\ ,\ ]\}$ <u>on</u> $T_e(G)$ <u>satisfying the condition</u>

$$(7.3.1)\qquad [X,[Y,Z]_0] = [[X,Y],Z]_0 + [Y,[X,Z]]_0,$$

<u>that is,</u> $ad\mathfrak{L} \subset Der\mathfrak{G}_0$. <u>Then, the local multiplication</u> μ <u>given by the following</u> (7.3.2) <u>is a geodesic homogeneous local left Lie loop which is in projective relation with</u> μ_0:

$$(7.3.2)\qquad \mu(\exp X, \exp Y) = \mu_0(\exp X, \exp A(X)Y),$$

<u>where</u> $\exp$ <u>denotes the exponential map of the Lie group and</u> $A(X)$ <u>denotes the endomorphism given by</u>

(7.3.3) $$A(X) = e^{\mathrm{ad}_{\mathbb{L}} X}, \quad X \in T_e(G).$$

Conversely, any geodesic homogeneous local left Lie loop on G which is in projective relation with μ_0 is given by the multiplication μ in (7.3.2) for some Lie algebra $\mathbb{L}$ on $T_e(G)$ satisfying (7.3.1).

Proof. For the detailed proof, we refer to [43]. Here we give a fundamental ideas of the proof of the latter half of the theorem. Assume that μ is a geodesic homogeneous local left Lie loop which is in projective relation with the given Lie group μ_0. Let ∇ (resp. ∇^0) be the canonical connection of μ (resp. μ_0) and denote the torsion tensor and the curvature tensor by S and R (resp. S^0 and R^0), respectively. Then, since $R^0 = 0$, the formulas (6.2.1) – (6.2.5) for the affine homogeneous structure $T = \nabla^0 - \nabla$ of the connection ∇^0 are reduced to the following formulas:

(7.3.4) $$T(X,X) = 0$$

(7.3.5) $$T(X,T(Y,Z)) = T(T(X,Y),Z) + T(Y,T(X,Z))$$

(7.3.6) $$T(X,S^0(X,Y)) = S^0(T(X,Y),Z) + S^0(Y,T(X,Z)).$$

Hence, the value of T at the identity e induces on the tangent space $T_e(G)$ a Lie algebra structure $\mathbb{L} = \{T_e(G); [X,Y] = T_e(X,Y)\}$, which satisfies (7.3.1) by (7.3.6). In this case, the values of the torsion S and the curvature R evaluated at e are given respectively by (6.2.6) and (6.2.7) as follows;

(7.3.7) $$S_e(X,Y) = [X,Y]_0 + 2[X,Y],$$

(7.3.8) $$R_e(X,Y)Z = -[[X,Y]_0,Z] - [[X,Y],Z].$$

If we introduce a local multiplication $\tilde{\mu}$ by (7.3.2) for this Lie algebra $\mathbb{L}$, by using Theorem 5.8, we can show that two local multiplications μ and $\tilde{\mu}$ are coincident in their common domain.

q.e.d.

§8. Examples

In this section, we give some examples of geodesic homogeneous local left Lie loops on Lie groups by means of Theorem 7.3. Above all, Example 8.2 seems to present some essential examples of such local loops that are in projective relation with the Lie groups. (Cf. Added in Proof.)

8.1. PROJECTIVITY ON R^n. If we regard the abelian Lie group $\{R^n, +\}$ as a geodesic homogeneous left Lie loop, then, its Lie algebra is abelian and the condition (7.3.1) is satisfied by every real n-dimensional Lie algebra L set on R^n. The geodesic homogeneous local left Lie loop μ associated with L is given by ([41])

$$(8.1) \qquad \mu(X,Y) = X + A(X)Y, \quad A(X) = e^{\operatorname{ad} X},$$

where $\operatorname{ad} X$ denotes the adjoint representation of L.

8.2. SCALAR MULTIPLE OF LIE BRACKETS. Let $\mathsf{G} = \{V; [\ ,\ ]\}$ be a Lie algebra on a vector space V over a field F. For any fixed element p in F, set a new bracket

$$(8.2.1) \qquad [X,Y]_p = p[X,Y].$$

Then, we get a new Lie algebra $\mathsf{G}_p = \{V; [\ ,\ ]_p\}$ which satisfies the condition (7.3.1) for G, that is,

$$(8.2.2) \qquad [X,[Y,Z]]_p = [[X,Y]_p,Z] + [Y,[X,Z]_p].$$

If G is the Lie algebra of a real Lie group (G,μ_0), then, by Theorem 7.3, a geodesic homogeneous local left Lie loop μ_p is associated with the Lie algebra G_p for any fixed real number p as follows;

$$(8.2.3) \qquad \mu_p(\exp X, \exp Y) = \mu_0(\exp X, \exp A_p(X)Y),$$

where $A_p(X)$ is the endomorphism of the tangent space $T_e(G)$ given by

$$(8.2.4) \qquad A_p(X) = e^{\operatorname{ad}_{\mathsf{G}} pX}.$$

If we denote the group multiplication μ_0 by juxtaposition, i.e., $\mu_0(x,y) = xy$ as usual, then the local multiplication μ_p is expressed by

(8.2.5) $\mu_p(x,y) = x^{p+1} y\, x^{-p}$ for $x = \exp X$, $y = \exp Y$,

where $x^r = \exp rX$ for any real number r.

Now, let ∇ be the canonical connection of the geodesic homogeneous local left Lie loop μ_p and S, R be respectively the torsion tensor and the curvature tensor of ∇. Then, the formulas (7.3.7) and (7.3.8) imply the followings

(8.2.6) $S(X,Y) = (1+2p)[X,Y]$,

(8.2.7) $R(X,Y)Z = -p(1+p)[[X,Y],Z]$

for any left invariant vector fields X, Y and Z on the Lie group (G,μ_0).

8.3. SOME SPECIAL CASES. In the preceding example, if we put for p some special values of real numbers, then we get the following well-known cases.

(1) The case $p = 0$: In this case, the Lie lagebra $\mathfrak{G}_p$ is reduced to an abelian Lie algebra so that $A_p(X) = \mathrm{Id}$, and the multiplication μ_p is reduced to the original multiplication μ_0 of the Lie group. It is evident that the canonical connection ∇ is reduced to the (−)-connection of Cartan whose torsion and curvature are given by;

(8.3.1) $S(X,Y) = [X,Y]_0$ (cf. (7.1)),

(8.3.2) $R(X,Y) = 0$,

for any left invariant vector fields X and Y, which are obtained also by putting $p = 0$ in (8.2.6) and (8.2.7).

(2) The case $p = -1$: In this case the bracket of the Lie algebra $\mathfrak{G}_p$ is reduced to $[X,Y]_p = -[X,Y]_0$ and the associated geodesic homogeneous left Lie loop μ_p is reduced to the Lie group anti-isomorphic to the given Lie group (G,μ_0) because we get from (8.2.5)

(8.3.3) $\mu_{-1}(x,y) = yx.$

By (8.2.6) and (8.2.7) the torsion and the curvature of the canonical connection are expressed as follows:

(8.3.4) $S(X,Y) = -[X,Y]_0,$

(8.3.5) $R(X,Y) = 0,$

which show that the canonical connection of μ_{-1} is reduced to the (+)-connection of Cartan of the Lie group (G,μ_0).

(3) The case $p = -\frac{1}{2}$: In this case, the Lie algebra $\mathfrak{G}_{-\frac{1}{2}}$ has the bracket

(8.3.6) $[X,Y]_{-\frac{1}{2}} = -\frac{1}{2}[X,Y]_0.$

The associated geodesic homogeneous local left Lie loop $\mu_{-\frac{1}{2}}$ is reduced to

(8.3.7) $\mu_{-\frac{1}{2}}(x,y) = x^{\frac{1}{2}}y\, x^{\frac{1}{2}}$ for $x = \exp X$, $y = \exp Y$,

where $x^{\frac{1}{2}} = \exp \frac{1}{2}X$. This local left loop is symmetric (cf. 2.3 (7)), i.e., the local diffeomorphism $j(x) = x^{-1}$ around the identity element e is a local automorphism of $\mu_{-\frac{1}{2}}$. The canonical connection ∇ is a left invariant connection on the Lie group (G,μ_0) whose torsion and curvature are given respectively by

(8.3.7) $S(X,Y) = 0,$

(8.3.8) $R(X,Y)Z = \frac{1}{4}[[X,Y]_0,Z],$

that is, the connection ∇ is the (0)-connection of Cartan on the Lie group.

8.4. ADDED IN PROOF. After the symposium at Hiroshima, Aug. 29 – Sept. 1, Mr. Manabu Sanami, one of the participants of the symposium, found the following fact:

Let $\mathfrak{G}$ be a finite dimensional Lie algebra over the field F = R (real number field) or C (complex number field). If $\mathfrak{G}$ is simple, then the Lie algebra $\mathfrak{L}$ on the underlying vector space of $\mathfrak{G}$ satisfying the condition

(7.3.1); ad $\mathsf{L} \subset Der\ \mathsf{G}$, is just only a Lie algebra $\mathsf{L} = \mathsf{G}_p$ given by (8.2.1) for some $p \in F$. By this result combined with Theorem 7.3, we have;

8.5. THEOREM. Any geodesic homogeneous local left Lie loop on a simple Lie group (G,μ_0) wich is in projective relation with μ_0 must be one of the local Lie loops given by Example 8.2 for some real number p.

BIBLIOGRAPHY

[1] AKIVIS, M.A. Local algebras of a multidimensional web (Russian), Sib. Mat. Zh. 17(1976), 5 – 11.

[2] ________ Geodesic loops and local triple systems in a space with an affine connection (Russian), Sib. Mat. Zh. 19(1987) 243 – 245.

[3] ________ - A.M. SHELEKHOF The computation of the curvateru and torsion tensors of multidimensional three-web and of the associator of the local quasigroup connected with it (Russian), Sib. Mat. Zh. 12(1971), 953 – 960.

[4] ________ - ________ Local differentiable quasigroups and connections that are associated with a three-web of multi-dimensional surfaces (Russian), Sib. Mat. Zh. 12(1971), 1181 – 1191.

[5] ________ - ________ Local analytic quasigroups and loops (Russian), Kalinin Gos Univ., Kalinin 1980.

[6] ALBERT, A.A. Quasigroups I, Trans. Amer. Math. Soc., 54(1943), 507 – 519.

[7] BLASCHKE, W. - G. BOL Geometrie der Gewebe, Springer Verlag, 1938.

[8] BOL, G. Über Dreigeweben im vierdimensionalen Raum, Math. Ann. 110(1935), 431 – 463.

[9] ________ Gewebe und Gruppen, Math. Ann. 114(1937), 414 – 431.

[10] BRUCK, R.H. A Survey of Binary Systems, Springer, 1971.

[11] CHERN, S.S. Abzählungen für Gewebe, Abh. Math. Sem. Hamburg 11 (1936), 163 – 170.

[12] ________ Eine Invariantentheorie der Dreigewebe aus r-dimensionalenMannigfaltigkeiten im R_{2r}, Abh. Math. Sem. Hamburg, 11(1936), 333 – 358.

[13] ________ Web geometry, Bull. Amer. Math. Soc.(N.S.) 6(1982), 1 – 8.

[14] GLAUBERMAN, G. On loops of odd order, J. Algebra 1(1964), 374—396

[15] HOFMANN, K.H. - K. STRAMBACH The Akivis algebra of a homogeneous loop, Mathematika 33(1986), 87 —95.

[16] HOLMES, J.P. - A.A. SAGLE Analytic H-spaces, Campbell-Hausdorff formula and alternative algebra, Pacific J. Math. 91(1980), 105 — 134.

[17] KIKKAWA, M. On local loops in affine manifolds, J. Sci. Hiroshima Univ. A-I 28(1964), 199 — 207.

[18] __________ On the decomposition of a linearly connectied manifold with torsion, J. Sci. Hiroshima Univ. A-I 33(1969), 1 — 9.

[19] __________ On locally reductive spaces and tangent algebras, Mem. Fac. Lit. & Sci., Shimane Univ., Nat. Sci. 5(1972), 1 — 13.

[20] __________ On some quasigroups of algebraic models of symmetric spaces, Mem. Fac. Lit. & Sci., Shimane Univ., Nat. Sci. 6(1973), 9 — 13.

[21] __________ On some quasigroups of algebraic models of symmetric spaces II, Mem. Fac. Lit. & Sci., Shimane Univ., Nat. Sci. 7(1974), 29 — 35.

[22] __________ Geometry of homogeneous Lie loops, Hiroshima Math. Soc. 5(1975), 141 — 179.

[23] __________ A note on subloops of a homogeneous Lie loop and subsystems of its Lie triple algebra, Hiroshima Math. J. 5 (1975), 439 — 446.

[24] __________ On some quasigroups of algebraic models of symmetric spaces III, Mem. Fac. Lit. & Sci., Shimane Univ., Nat. Sci. 9 (1975), 7 — 12.

[25] __________ On the left translations of homogeneous loops, Mem. Fac. Lit. & Sci., Shimane Univ., Nat. Sci. 10(1976), 19—25.

[26] __________ On homogeneous systems I, Mem. Fac. Lit. & Sci., Shimane Univ., Nat. Sci. 11(1977), 9 — 17.

[27] __________ On homogeneous systems II, Mem. Fac. Sci., Shimane Univ. 12(1978), 5 — 13.

[28] __________ Remarks on solvability of Lie triple algebras, Mem. Fac. Sci., Shimane Univ. 13(1979), 17 — 22.

[29] __________ On homogeneous systems III, Mem. Fac. Sci., Shimane Univ. 14(1980), 41 — 46.

[30] __________ On homogeneous systems IV, Mem. Fac. Sci., Shimane Univ. 15(1981), 1 — 7.

[31] __________ On Killing-Ricci forms of Lie triple algebras, Pacific J. Math. 96(1981), 153 — 166.

[32] __________ On the decomposition of homogeneous systems with nondegenerate Killing-Ricci tensor, Hiroshima Math. J.11(1981) 525 — 531.

[33] KIKKAWA, M. On the Killing radical of Lie triple algebras, Proc. Japan Acad. Ser.A 59(1982), 212 – 215.

[34] ________ Remarks on invariant forms of Lie triple algebras, Mem. Fac. Sci., Shimane Univ. 16(1982), 23 – 27.

[35] ________ On homogeneous systems V, Mem. Fac. Sci., Shimane Univ. 17(1983), 9 – 13.

[36] ________ Naturally reductive metrics on homogeneous systems, Proc. Kon. Nederland Akad. Wetensch. A-87(1984), 203 – 208.

[37] ________ Totally geodesic imbeddings of homogeneous systems into their enveloping Lie groups, Mem. Fac. Sci. , Shimane Univ. 18(1984), 1 – 8.

[38] ________ Canonical connections of homogeneous Lie loops and 3-webs, Mem. Fac. Sci., Shimane Univ. 19(1985), 37 – 55.

[39] ________ Remarks on canonical connections of loops with the left inverse property, Mem. Fac. Sci., Shimane Univ. 20 (1986), 9 – 19.

[40] ________ Affine homogeneous structures on analytic loops, Mem. Fac. Sci., Shimane Univ. 21(1987), 1 – 15.

[41] ________ Projectivity of left loops on R^n, Mem. Fac. Sci., Shimane Univ. 22(1988), 33 – 41.

[42] ________ Projectivity of homogeneous left loops on Lie groups I (Algebraic framework), Mem. Fac. Sci., Shimane Univ. 23 (1989), 17 – 22.

[43] ________ Projectivity of homogeneous left loops on Lie groups II (Local theory), Mem. Fac. Sci., Shimane Univ. 24(1990), to appear.

[44] KOBAYASHI, S. – K. NOMIZU Foundations of Differential Geometry Vol. I, Interscience, 1963.

[45] ________________ Foundations of Differential Geometry Vol. II, Interscience, 1969.

[46] KUZ'MIN, E.N. The connection between Mal'cev algebras and analytic Moufang loops (Russian), Alg. i Log. 10(1971), 3 –22.

[47] LICHNEROWICZ, A. Géométrie des Groupes de Transformations, Dunod, 1958.

[48] LOOS, O. Symmetric Spaces Vol. I, Benjamin, 1969.

[49] MAL'CEV, A.I. Analytic loops (Russian), Mat. Sb. (N.S.) 36(78) (1955), 569 – 576.

[50] MIKHEEV, P.O. On a G-property of a local analytic Bol loop (Russian), "The Principle of Inclusion and Invariant Tensors", Moscow 1986, 78 – 84.

[51] NAGY, P. On the canonical connection of a three-web, Pub. Math. Debrecen 32(1985), 93 – 99.

[52] NOMIZU,K. Invariant affine connections on homogeneous spaces, Amer. J. Math. 76(1954), 33 – 65.

[53] REIDEMEISTER, K. Gewebe und Gruppen, Math. Zeitschr. 29(1929), 427 – 439.

[54] SABININ, L.V. Geometry of loops (Russian), Mat. Zametki 12(1972), 605 – 616.

[55] ________ On the equivalence of the category of loops and the category of homogeneous spaces (Russian), Dok. Akad. Nauk SSSR 205(1972), 533 – 536.

[56] ________ Methods of nonassociative algebras in differential geometry (Russian), Supplement to the Russian translation of S. Kobayashi–K. Nomizu: Foundations of Differential Geometry Vol. 1, Nauka 1981, 293 – 339.

[57] ________ Geometric odules (Russian), Webs and Quasigroups, Kalinin Gos. Univ. 1987, 88 – 98.

[58] ________ – P.O. MIKHEEV Analytic Bol loops (Russian), Webs and Quasigroups, Kalinin Gos. Univ. 1982, 102 – 109.

[59] ________ – ________ A Symmetric connection in the space of analytic Moufang loop (Russian), Dokl. Akad. Nauk SSSR 262 (1982), 807 – 809.

[60] ________ – ________ The differential geometry of Bol loops (Russian), Dokl. Akad. Nauk SSSR 281(1985), 1055 – 1057.

[61] SAGLE, A.A. A note on anti-commutative algebras obtained from reductive homogeneous spaces, Nagoya Math. J. 31(1968), 105 – 124.

[62] ________ On anti-commutative algebras and homogeneous spaces, J. Math. Mech. 16(1976), 1381 – 1393.

[63] ________ – J. SCUMI Anti-commutative algebras and homogeneous spaces with multiplications, Pacific J. Math. 68(1977), 255 – 269.

[64] ________ – ________ Multiplications on homogeneous spaces, nonassociative algebras and connections, Pacific J. Math. 48(1973), 247 – 266.

[65] VARADARAJAN, V. Lie Groups, Lie Algebras and their Representations, Prentice-Hall, 1974.

[66] YAMAGUTI, K. On the Lie triple system and its generalization, J. Sci. Hiroshima Univ. A-I 21(1958), 155 – 160.

[67] ________ On cohomology groups of general Lie triple systems, Kumamoto J. Sci. A8(1969), 135 – 146.

Quadratic Dynamical Systems

Michael K. Kinyon
Dept. of Mathematics
University of Utah
SLC, Utah 84112, USA

and

Arthur A. Sagle
Dept. of Mathematics
University of Hawaii at Hilo
Hilo, Hawaii 96720–4091, USA

Abstract

Quadratic dynamical systems are described in terms of algebras with the theme that the structure of algebras helps determine the behavior of trajectories. Thus for the differential system $dX/dt = X^2$ or the discrete system $X(k+1) = X(k)^2$ occurring in an algebra A, semi–simplicity of A gives decoupling of the systems and solvability of A gives solutions by linear equations. Quadratic systems have a wide scope: higher degree polynomial systems may be embedded into quadratic systems in such a way that solutions to the quadratic system yield solutions to the polynomial system. Examples of quadratic differential systems given by algebras include smooth predator–prey models, the Lorenz system, and geodesic equations on homogeneous spaces. Examples of quadratic discrete systems given by algebras include the Henon map and discrete predator–prey models. Automorphisms of algebras preserve dynamical behavior such as periodic points, equilibria (of differential systems), and fixed points (of discrete systems). They also may be used to give explicit solutions such as periodic orbits.

1 Algebras and Examples

A <u>nonassociative algebra</u> [2] is a vector space A over a field (here usually the real numbers $\mathbb{R}$) with a bilinear multiplication $\beta : A \times A \to A$; denote this structure by (A, β) or just A when β is understood. For example, if A is an associative algebra, let $A^+ = (A, \beta)$ be the commutative Jordan algebra with multiplication $\beta(X,Y) = XY + YX$. Similarly one may form the anticommutative Lie algebra $A^- = (A, \beta)$ with $\beta(X,Y) = [X,Y] = XY - YX$. Generalizations of these algebras appear in many applications as noted below. Hereafter we shall often write $XY = \beta(X,Y)$.

A subalgebra B of an algebra A is a subspace of A such that $B^2 \subseteq B$. For $X \in A$, $\mathbb{R}[X]$ denotes the subalgebra generated by X and A is power–associative if $\mathbb{R}[X]$ is associative for every $X \in A$. An ideal I of an algebra A is a subspace of A such that $IA \subseteq I$ and $AI \subseteq I$. As in associative algebras, the quotient algebra can be formed and the map $A \to A/I : X \to X + I$ is an algebra homomorphism. A is a simple algebra if $A^2 \neq 0$ and A has no proper ideals; i.e. no proper homomorphisms. An algebra is semi–simple if it the direct sum of ideals which are simple algebras. The radical, $Rad\ A$, of an algebra is the smallest ideal of A such that $A/Rad\ A$ is semi–simple or the zero algebra. The radical is given by $Rad\ A = (Rad\ M)A$ where M is the associative algebra generated by the right and left multiplication functions $R(Z) : X \to \beta(X,Z)$ and $L(W) : X \to \beta(W,X)$. An algebra A is nilpotent if there exists an integer N such that all products with N factors are zero. A is solvable if for $A^{(1)} = A, A^{(2)} = AA, \ldots, A^{(k+1)} = A^{(k)}A^{(k)}$, there is an integer N such that $A^{(N)} = 0$; i.e., $A \supseteq A^{(2)} \supseteq A^{(3)} \supseteq \cdots \supseteq A^{(N)} = 0$. The radical is usually associated with a nilpotent or solvable ideal.

Definition. Let A be an algebra over $\mathbb{R}$. A *quadratic* mapping in A is one of the form

$$E(X) \equiv C + TX + X^2$$

where $C \in A$, $T : A \to A$ is linear. If we view $E(X)$ as defining a vector field on A, then we have an associated quadratic differential system [5, 6] of the form

$$\dot{X} = E(X).$$

On the other hand, if we view $E : A \to A$ as being a smooth mapping that we can iterate, then we have an associated quadratic discrete dynamical system of the form

$$X(k+1) = E(X(k))$$

where $E(X)$ is a quadratic vector field.

Remark: Any mapping on $\mathbb{R}^n$ of the form $F(X) = C + TX + \beta(X)$ where $\beta : \mathbb{R}^n \to \mathbb{R}^n$ is a quadratic function (homogeneous of degree 2) can be considered as a quadratic mapping occuring in an algebra. Indeed the quadratic term defines a commutative algebra multiplication (which we also denote by β) by the bilinearization process: $\beta(X,Y) \equiv \beta(X+Y) - \beta(X) - \beta(Y)$.

Homogenization processes may be used to remove the constant and linear terms of quadratic mappings at the expense of increasing the dimension of the system by 1. These allow the solutions of $\dot{X} = C + TX + X^2$ or $X(k+1) = C + TX(k) + X(k)^2$ to be obtained from systems of the form $\dot{\widetilde{X}} = \widetilde{X}^2$ and $\widetilde{X}(k+1) = \widetilde{X}(k)^2$.

For the differentiable case with E a vector field [6], set $E_u(X) = u^2C + uTX + X^2$, so that $\dot{X} = E_1(X)$ is the original system. Let $F_t^u(X)$ denote the flow for E_u. Let $\widetilde{A} = A \times \mathbb{R}$ and define a multiplication $\widetilde{\beta}$ on $\widetilde{A}$ as follows: set $\widetilde{X} = \begin{pmatrix} X \\ u \end{pmatrix}$, and define

$$\widetilde{\beta}(\widetilde{X}, \widetilde{X}) \equiv \begin{pmatrix} E_u(X) \\ 0 \end{pmatrix}.$$

It is easily checked that $\widetilde{\beta}$ is quadratic and thus gives a commutative algebra multiplication on $\widetilde{A}$. Letting $F_t(\widetilde{X})$ be the flow for $\dot{\widetilde{X}} = \widetilde{X}^2$, a straightforward calculation shows that

$$F_t(\widetilde{X}) = \begin{pmatrix} F_t^u(X) \\ u \end{pmatrix}.$$

To recover the solution to the original system, restrict the flow to the hyperplane given by $u \equiv 1$.

For the discrete case with E an iteratable mapping, set $\widetilde{X} = \begin{pmatrix} X \\ u \end{pmatrix}$ as before, and define a multiplication on $\widetilde{A} = A \times \mathbb{R}$ by

$$\widehat{\beta}(\widetilde{X}, \widetilde{X}) \equiv \begin{pmatrix} E_u(X) \\ u^2 \end{pmatrix}.$$

Then $\widehat{\beta}$ is quadratic and also gives a commutative algebra multiplication on $\widetilde{A}$. The iterates of the original quadratic vector field E in A are given by powers in $\widetilde{A}$. This follows from a calculation showing that for $n > 0$,

$$\widetilde{X}^{(n)} = \widehat{\beta}(\widetilde{X}^{(n-1)}, \widetilde{X}^{(n-1)}) = \begin{pmatrix} E(X(n-1)) \\ 1 \end{pmatrix} = \begin{pmatrix} X(n) \\ 1 \end{pmatrix}.$$

Example: (1) By Taylor's theorem, quadratic mappings occur as the quadratic approximation to smooth mappings

$$F(X) \sim F(0) + F^{(1)}(0)X + (1/2!)F^{(2)}(0)(X, X)$$

where the algebra multiplication on $\mathbb{R}^n$ satisfies $\beta(X, X) = F^{(2)}(0)X^2/2!$.

(2) The Lorenz model of thermal convection is given by the following quadratic differential system in $\mathbb{R}^3$:

$$\begin{bmatrix} \dot{x_1} \\ \dot{x_2} \\ \dot{x_3} \end{bmatrix} = \begin{bmatrix} -a & a & 0 \\ c & -1 & 0 \\ 0 & 0 & -b \end{bmatrix} \begin{bmatrix} x_1 \\ x_2 \\ x_3 \end{bmatrix} + \begin{bmatrix} 0 \\ -x_1x_3 \\ x_1x_2 \end{bmatrix}$$

for suitable parameters $a, b, c \in \mathbb{R}$. The commutative algebra $A = (\mathbb{R}^3, \beta)$ obtained from the quadratic part of this system is solvable as can be seen from its multiplication table relative to the natural basis $\{e_1, e_2, e_3\}$:

β^+	e_1	e_2	e_3
e_1	0	$\frac{1}{2}e_3$	$-\frac{1}{2}e_2$
e_2	$\frac{1}{2}e_3$	0	0
e_3	$-\frac{1}{2}e_2$	0	0

(3) Quadratic differential systems occur in linear control systems with a quadratic cost function. For $X \in \mathbb{R}^n$ (the states) and $U \in \mathbb{R}^q$ (the inputs), let the linear system be given by $dX/dt = FX + GU$ for suitable constant matrices F and G, and $t \in [a, b]$. Let the quadratic cost function be given by $J(U) = \int_a^b L(X(t), U(t))dt$ where $L(X, U) = 1/2(X^tQX + U^tRU)$ for suitable symmetric matrices Q and R. The system which is optimal over $[a, b]$ relative to the cost $J(U)$ is given by the feedback law $U(t) = -R^{-1}G^tP(t)X(t)$ where the nxn symmetric matrix $P(t)$ satisfies the quadratic Riccati matrix equation $dP/dt = -Q - PF - F^tP + P(GR^{-1}G^t)P$. The multiplication $\beta(P, P) = P(GR^{-1}G^t)P$ makes the vector space of nxn matrices into a Jordan algebra.

(4) The Euler equation for the motion of a rotating rigid body with no external forces is given by the quadratic system in $\mathbb{R}^3$

$$\begin{bmatrix} \dot{x}_1 \\ \dot{x}_2 \\ \dot{x}_3 \end{bmatrix} = \begin{bmatrix} (I_3 - I_2)/I_1 x_2 x_3 \\ (I_1 - I_3)/I_2 x_1 x_3 \\ (I_2 - I_1)/I_3 x_1 x_2 \end{bmatrix}$$

where the nonzero moments of inertia I_j satisfy $I_1 \neq I_2 \neq I_3 \neq I_1$. It is straightforward to show the commutative algebra $A = (\mathbb{R}^3, \beta^+)$ is simple [7].

(5) The differential geometry of invariant Lagrangian systems [3, 7, 9, 12] is given in terms of quadratic systems and extends the preceding example. Let G be a connected Lie group and let H be a closed (Lie) subgroup with Lie algebras g and h respectively. The homogeneous space G/H is <u>reductive</u> if there is a subspace m of g such that $g = m + h$ (direct sum) and $(AdH)(m) \subseteq m$; i.e., $[h, m] \subseteq m$. For example, let g and h be semi–simple and $m = h^\perp$ relative the Killing form of g. For a reductive space, there is a bijective correspondence between the set of G–invariant connections ∇ on G/H and the set of algebras (m, α) with $AdH \subseteq Aut(m, \alpha)$; i.e., $adh \subseteq Der(m, \alpha)$. In particular, a curve $\sigma(t)$ in G/H is a geodesic if its tangent field $X(t) = \dot{\sigma}(t)$ satisfies the quadratic equation

$$\dot{X} + \alpha(X, X) = 0.$$

Next let G/H be a configuration space for an invariant system with nondegenerate Lagrangian. Then a solution $\sigma(t)$ to the corresponding Euler–Lagrange equation satisfies an extended Euler field equation which reduces to the above geodesic equation when the Lagrangian is given by kinetic energy [7]. More general quadratic equations occur when the Lagrangian is not given by kinetic energy. The algebras occurring in this example are usually noncommutative.

(6) As an example of a discrete quadratic system occuring in an algebra, the Henon map in $\mathbb{R}^2$ given by

$$\begin{aligned} E(X) &= \begin{pmatrix} a + bx_2 + cx_1^2 \\ dx_1 \end{pmatrix} \\ &= \begin{pmatrix} a \\ 0 \end{pmatrix} + \begin{pmatrix} 0 & b \\ d & 0 \end{pmatrix} \begin{pmatrix} x_1 \\ x_2 \end{pmatrix} + \begin{pmatrix} cx_1^2 \\ 0 \end{pmatrix} \\ &= C + TX + X^2 \end{aligned}$$

occurs in the algebra $A = (\mathbb{R}^2, \beta)$, where

$$\beta(X, Y) = \begin{pmatrix} cx_1y_1 \\ 0 \end{pmatrix}$$

gives the algebra multiplication.

(7) Relative growth rate problems and related Lotka–Volterra predator–prey models are given by quadratic differential systems [8, 13]. Let the relative growth rate of n species be given by

$$\dot{x}_i/x_i = g_i(x_1, \ldots, x_n) \sim c_i + \sum_j b_{ij}x_j$$

for $i = 1, \ldots, n$ with the growth rate functions g_i having the indicated linear approximation. For $X \in \mathbb{R}^n$, this gives the quadratic equation $\dot{X} = TX + \beta(X, X)$:

$$\begin{bmatrix} \dot{x}_1 \\ \vdots \\ \dot{x}_n \end{bmatrix} = \begin{bmatrix} c_1 & & 0 \\ & \vdots & \\ 0 & & c_n \end{bmatrix} \begin{bmatrix} x_1 \\ \vdots \\ x_n \end{bmatrix} + \begin{bmatrix} x_1 \sum b_{1j}x_j \\ \vdots \\ x_n \sum b_{nj}x_j \end{bmatrix}.$$

The noncommutative growth rate algebra multiplication β on $\mathbb{R}^n$ is given by $\beta(X, Y) \equiv (diagX)BY$ where $diagX$ is the diagonal matrix formed from the vector X and $B = (b_{ij})$, and the algebra $A = (\mathbb{R}^n, \beta)$ has the following properties:
a. A satisfies the identity $\beta(\beta(Z, X), Y) = \beta(\beta(Z, Y), X)$.
b. If B^{-1} exists and all entries in B are nonzero, then A is simple and has right identity $e = B^{-1}l$ where $l = (1, \ldots, 1)^t \in \mathbb{R}^n$.

c. The multiplication functions are $R(X) = diag(BX)$ and $L(X) = (diagX)B$.
d. The vector field $E(X)$ has equilibrium point N given by the solution to $BN + c = 0$, where $c = (c_1, \ldots, c_n)^t \in \mathbb{R}^n$. Furthermore, analogous to the logistics equation, we have

$$E(X) = \beta(X, X - N) \text{ and } E'(N) = L(N).$$

(8) The discrete version of the previous example is the discrete predator–prey model given by

$$\begin{aligned} x_1(k+1) &= c_1 x_1(k) + \sum b_{1j} x_1(k) x_j(k) \\ x_2(k+1) &= c_2 x_2(k) + \sum b_{2j} x_2(k) x_j(k) \\ &\vdots \\ x_n(k+1) &= c_n x_n(k) + \sum b_{nj} x_n(k) x_j(k) \end{aligned}$$

Thus $X(k+1) = TX(k) + X(k)^2$ in the algebra A where

$$\beta(X, X) = \begin{pmatrix} \sum b_{1j} x_1 x_j \\ \vdots \\ \sum b_{nj} x_n x_j \end{pmatrix}$$

If the matrix $B = (b_{ij})$ is nonsingular and all $b_{ii} \neq 0$, then A is simple.

Many more equations may be regarded as quadratic systems as indicated by the following [1].

Theorem 1.1 *(1) Let $x^{(n)} = P(x, x^{(1)}, \ldots, x^{(n)})$ be an n-th order differential equation where $P(z_1, \ldots, z_n)$ is a polynomial in the z's. Then this equation may be imbedded into a quadratic system $\dot{X} = X^2$ whose solution gives the solution to the polynomial differential equation.*
(2) Let $x(n+1) = P(x(n), x(n-1), \ldots, x(0))$ be an n-th order discrete equation where $P(z_1, \ldots, z_n)$ is a polynomial in the z's. Then this equation may be embedded into a quadratic system $X(n+1) = X(n)^2$ whose solution gives the solution to the polynomial discrete equation.

The following example indicates the method of proof in the differentiable case.

Example: For the Van der Pol equation $\ddot{u} = (3cu^2 + d)\dot{u} + abu = P(u, \dot{u})$, set $x_1 = u$, $x_2 = \dot{u}$, $x_3 = u^2$. Then the polynomial equation becomes a quadratic system

$$\dot{X} = \begin{bmatrix} \dot{x_1} \\ \dot{x_2} \\ \dot{x_3} \end{bmatrix} = \begin{bmatrix} \dot{u} \\ \ddot{u} \\ 2u\dot{u} \end{bmatrix} = \begin{bmatrix} x_2 \\ (3cx_3 + d)x_2 + abx_1 \\ 2x_1x_2 \end{bmatrix}$$

$$= \begin{bmatrix} 0 & 1 & 0 \\ ab & d & 0 \\ 0 & 0 & 0 \end{bmatrix} \begin{bmatrix} x_1 \\ x_2 \\ x_3 \end{bmatrix} + \begin{bmatrix} 0 \\ 3cx_2x_3 \\ 2x_1x_2 \end{bmatrix}$$

$$(*) \qquad = TX + \beta(X, X).$$

Finally, homogenize this as above to obtain a purely quadratic system (without a linear term).

Remark: (1) The algebras $A = (\mathbb{R}^3, \beta)$ and $\tilde{A} = (\mathbb{R}^4, \tilde{\beta})$ given by the Van der Pol equation are both solvable.
(2) The idea behind the proof in the discrete case is similar to that in the differentiable case, but the algebra obtained may be infinite dimensional; for example, consider $X(k+1) = X(k)^3$ in $\mathbb{R}$. Let V be the vector space of all real sequences $(v_1, v_2, \ldots, v_j, \ldots)$. Let A be the following commutative infinite dimensional algebra (V, β): for $Z = \sum z_i e_i$, let $\beta(Z, Z) = \sum z_i z_{i+1} e_i$, here $e_i = (0, \ldots, 0, 1, 0, \ldots)$ with 1 in the i–position and 0 elsewhere. With the substitution $u_1(k) = X(k)$ and $u_{i+1}(k) = u_i(k)^2$ for $i = 1, 2, \ldots$, the system $X(k+1) = X(k)^3$ in $\mathbb{R}$ is equivalent to the quadratic system $U(k+1) = \beta(U(k), U(k))$ in A.

2 Solutions and Structure

The structure properties of an algebra are related to the behavior of solutions; for example, semi–simple algebras and decoupling, nilpotent algebras and unbounded solutions (in the differentiable case) or vanishing solutions (in the discrete case), and solvable algebras and linearly solvable equations. Because of the homogenization process, we shall usually focus on the purely quadratic case $T = 0$. First we shall examine solution forms to quadratic dynamical systems.

For quadratic differential systems $\dot{X} = X^2$ with $X(0) = X$, one has, at least locally, a power series solution that can be written in terms of the algebra as

$$X(t) = X + tX^2 + \frac{1}{2!}t^2(2XX^2) + \frac{1}{3!}t^3(4XXX^2 + 2X^2X^2) + \cdots,$$

using the fact that A is commutative. In general, if $X^{[k]}, k = 1, 2, \ldots$ denotes the k–linear function of X in the k–th term of the solution, then $X^{[k+1]} = \sum \binom{k-1}{j} X^{[j+1]} X^{[k-j]}$. The series solution shows that the trajectory through X

stays in the subalgebra $\mathbb{R}[X]$ generated by X. In particular, this subalgebra is invariant under the flow F_t of the quadratic vector field.

Setting $X^3 = XX^2, \ldots, X^k = XX^{k-1}, \ldots$, we can write

$$X(t) = X + X^2 t + X^3 t^2 + (X^2 X^2 + 2X^4)\frac{t^3}{3} + \cdots .$$

Using the series solution as a guide, one can obtain several results for differentiable systems.

Proposition 2.1 *[6] Let A be power–associative, i.e. the subalgebra $\mathbb{R}[X]$ generated by any element X is associative. Then the solution to $\dot{X} = X^2$, $X(0) = X$ is given by $X(t) = (I - tL(X))^{-1}X$ where $L(X)Y \equiv XY$ is the left multiplication operator. This solution blows up in finite time.*

Proof: Using the power series solution, we have.

$$\begin{aligned} X(t) &= X + X^2 t + X^3 t^2 + X^4 t^3 + \cdots \\ &= (I + tL(X) + t^2 L(X)^2 + t^3 L(X)^3 + \cdots)X \\ &= (I - tL(X))^{-1}X, \end{aligned}$$

which is valid for t sufficiently small. More directly, one can simply differentiate the expression $(I - tL(X))^{-1}X$ and notice that it satisfies the differential equation.

Proposition 2.2 *[5, 6, 14] (1) If X is nilpotent of index $N > 2$, then the solution to $\dot{X} = X^2$, $X(0) = X$ is unbounded. The index $N = 2$ if and only if the solution is an equilibrium point.*
(2) If X is an idempotent, i.e. $X^2 = X$, then the solution is unbounded with finite escape time along the ray through X.

Proof: From the series solution, if X is nilpotent of index $N > 2$, then the solution $X(t)$ is a polynomial in t of degree $N - 1$, and hence unbounded. If the index $N = 2$, then $X^2 = 0$ and the series solution shows that $X(t) \equiv X$ is an equilibrium , and conversely. This establishes (1). For (2), notice that the series solution reduces to $X(t) = (1 + t + t^2 + \cdots)X = (1/1 - t)X$. More formally, set $X(t) = x(t)X$, then $\dot{X} = X^2$ if and only if $\dot{x}X = x^2 X^2 = x^2 X$ if and only if $\dot{x} = x^2$. The solution to this equation is $1/1 - t$ which is unbounded and has finite escape time, and consequently gives the unbounded solution $X(t)$.

Letting $\mathcal{E}$ denote the set of equilibrium points for $\dot{X} = X^2$, we have the following.

Proposition 2.3 *Let $N \in \mathcal{E}$ be a nonzero equilibrium. Then N is not hyperbolic and the line $\{sN : s \in \mathbb{R}\} \subseteq \mathcal{E}$.*

Proof: The linearization of E at N is $E'(N) = 2L(N)$. But since $0 = E(N) = N^2$, we see that $E'(N)N = 0$, so that N is not hyperbolic. The rest follows from $(sN)^2 = s^2N^2 = 0$; see [5].

For discrete quadratic systems $X(n+1) = E(X(n)) = X(n)^2$ with $X(0) = X$, the orbit $\{E^{(n)}(X)\}$ through X is given by powers of X: $E^{(0)}(X) = X$, $E^{(1)}(X) = X^2$, $E^{(2)}(X) = X^2X^2$, ..., $E^{(n)}(X) = (\cdots((X^2)^2)^{\cdots})^2$. This yields the following proposition that should be contrasted with the preceding one.

Proposition 2.4 *(1) If X is nilpotent of index N, the orbit through X vanishes after, at most, $\lceil \log_2 N \rceil$ iterations, where $\lceil \cdot \rceil$ denotes the ceiling function (least integer greater than its argument).*
(2) X is an idempotent if and only if X is fixed point of $X(n+1) = X(n)^2$.

Letting $\mathcal{F}$ denote the set of fixed points for $X(n+1) = X(n)^2$, we have the following.

Proposition 2.5 *Let $P \in \mathcal{F}$ be a fixed point. Then P is unstable.*

Proof: Note that $E'(P)P = 2L(P)P = 2P^2 = 2P \neq 0$. Since $E'(N)$ has an eigenvalue greater than one, P is unstable.

Next we consider how the structure of the algebra affects the quadratic vector field and the associated differentiable and discrete dynamical systems.

Proposition 2.6 *If $A = A_1 \oplus \cdots \oplus A_k$ is semi-simple, then the vector field $E(X) = X^2$ decouples into vector fields $E_j(X) = X_j^2$ occurring in the simple algebras A_j. Hence the systems $\dot{X} = X^2$ and $X(n+1) = X(n)^2$ with $X(0) = X$ decouple into systems $\dot{X}_j = X_j^2$ and $X_j(n+1) = X_j(n)^2$ with $X_j(0) = X_j$.*

Proof: For $X \in A$, $X = X_1 + \cdots + X_k$ where $X_j \in A_j$, and the semi-simplicity of A implies that $X_iX_j = 0$ if $i \neq j$. Thus $E(X) = X^2 = (X_1 + \cdots + X_k)^2 = X_1^2 + \cdots + X_k^2$.

Propositions 2.2 and 2.3 imply the following global counterpart.

Proposition 2.7 *If A is a nilpotent algebra, then the flow $F_t(X)$ through X for $E(X) = X^2$ is unbounded, while the orbit $\{E^{(n)}(X)\}$ eventually vanishes.*

In the solvable case, we have the following result for differentiable systems [1, 6]. We state it in the case of a nonzero linear term.

Theorem 2.8 *Let A be a solvable algebra with $A \supseteq A^{(2)} \supseteq A^{(3)} \supseteq \cdots \supseteq A^{(N)} = 0$ where $A^{(k+1)} = A^{(k)}A^{(k)}$ are ideals of A and $TA^{(k)} \subseteq A^{(k)}$ where $T : A \to A$ is a linear map. Then the solution to the quadratic equation $\dot{X} = TX + X^2$ can be solved by solving finitely many linear equations.*

3 Automorphisms

Automorphisms measure structure and symmetries. The automorphism group of the quadratic mapping $E(X) = X^2$ is the same as the automorphism group of the associated commutative algebra. This can help one to locate interesting dynamical behavior of smooth and discrete systems algebraically. Most of the results in this section we state without proof; details can be found in [1].

Definition: An automorphism of an algebra $A = (\mathbb{R}^n, \beta)$ is an invertible linear transformation $\phi \in GL(\mathbb{R}^n)$, the general linear group, such that $\phi\beta(X,Y) = \beta(\phi X, \phi Y)$ for all $X, Y \in A$. A derivation of an algebra A is a linear transformation $D : A \to A$ satisfying the product rule $D\beta(X,Y) = \beta(DX, Y) + \beta(X, DY)$ for all $X, Y \in A$.

A similar concept is the following.

Definition: An automorphism of a smooth map $E : \mathbb{R}^n \to \mathbb{R}^n$ is an invertible linear transformation $\phi \in GL(\mathbb{R}^n)$ such that $E(\phi X) = \phi E(X)$ for all $X \in \mathbb{R}^n$. A derivation of a vector field is a linear transformation $D : \mathbb{R}^n \to \mathbb{R}^n$ satisfying $DE(X) = E'(X)DX$ for all $X \in \mathbb{R}^n$.

Remark: The set $Aut\ A$ of all automorphisms of A and the set $Aut\ E$ of all automorphisms of E are closed (Lie) subgroups of $GL(\mathbb{R}^n)$. The set $Der\ A$ of all derivations of A and the set $Der\ E$ of all derivations of E are Lie subalgebras of $gl(\mathbb{R}^n)$, the Lie algebra of $GL(\mathbb{R}^n)$. For any $D \in Der\ A$ (resp. $Der\ E$), $\exp\ D = I + D + D^2/2! + \cdots$ is in $Aut\ A$ (resp. $Aut\ E$); that is, the Lie algebra of $Aut\ A$ is $Der\ A$ and the Lie algebra of $Aut\ E$ is $Der\ E$. The two notions of automorphism are related by the following.

Theorem 3.1 *Let $\dot{X} = E(X) \equiv TX + X^2$ occur in an algebra A. Then*

$$\begin{aligned} Aut\ E &= \{\phi \in Aut\ A : T\phi = \phi T\} \text{ and} \\ Der\ E &= \{D \in Der\ A : TD = DT\}. \end{aligned}$$

In particular, when $E(X) = X^2$, $Aut\ E = Aut\ A$ and $Der\ E = Der\ A$.

Proof: Suppose $\phi \in Aut\ E$. Then from $E(X) = TX + X^2$, we obtain $T\phi X + (\phi X)^2 = E(\phi X) = \phi E(X) = \phi TX + \phi(X^2)$ which gives $T\phi = \phi T$ and $\phi(X^2) = (\phi X)^2$. Since A is commutative, the latter implies $\phi\beta(X,Y) = \beta(\phi X, \phi Y)$. Thus $\phi \in Aut\ A$. The reverse inclusion is obvious. A similar argument gives the other equality.

Proposition 3.2 *Let $E : \mathbb{R}^n \to \mathbb{R}^n$ be smooth. Thinking of E as a vector field, let $F_t(X)$ denote its flow through X, and thinking of E as a mapping, let $\{E^{(n)}(X)\}$ denote the iterates through X. Fix $\phi \in GL(\mathbb{R}^n)$. Then*
(1) $\phi \in Aut\ E$ if and only if $\phi \circ F_t = F_t \circ \phi$, and
(2) $\phi \in Aut\ E$ if and only if $\{\phi E^{(n)}(X)\}$ is the orbit through ϕX.

For a given vector field E on $\mathbb{R}^n$, *Aut* E and $(Aut\ E)_0$, the connected component of the identity, can be used to locate equilibria and periodic trajectories of differentiable dynamical systems and fixed points and periodic points of discrete dynamical systems. Of course, we shall be particularly interested in the quadratic case $E(X) = X^2$.

Recall that $\mathcal{E}$ denotes the set of equilibria of a given vector field E on $\mathbb{R}^n$. For $N \in \mathcal{E}$, let $Att(N)$ denote the domain of attraction of N.

Theorem 3.3 *For any vector field E on $\mathbb{R}^n$,*
(1) $(Aut\ E)\mathcal{E} = \mathcal{E}$, and
(2) if N is an isolated in $\mathcal{E}$, then $(Aut\ E)_0 N = N$ and $(Aut\ E)_0 Att(N) = Att(N)$.
For $E(X) = X^2$,
(3) if $\mathcal{L}(N) = \{sN : s \in \mathbb{R}\}$ is an isolated line in $\mathcal{E}$, then $(Aut\ E)_0\mathcal{L}(N) = \mathcal{L}(N)$.

Let $\mathcal{P}_\tau$ denote the set of all periodic points of period τ for a differential system $\dot{X} = E(X)$ associated with a given vector field E. For γ a trajectory in $\mathcal{P}_\tau$, let $Att(\gamma)$ denote the domain of attraction of γ. We say that $\gamma \in \mathcal{P}_\tau$ (a trajectory) is *isolated* if there is a tubular neighborhood $\mathcal{U}$ of γ so that no $\delta \in \mathcal{P}_\tau$ intersects $\mathcal{U}$

Theorem 3.4 *For any vector field E on $\mathbb{R}^n$,*
(1) $(Aut\ E)\mathcal{P}_\tau = \mathcal{P}_\tau$ (automorphisms preserve periods), and
(2) if $\gamma \subseteq \mathcal{P}_\tau$ is an isolated trajectory in $\mathcal{P}_\tau$, then $(Aut\ E)_0\gamma = \gamma$ (as sets), and $(Aut\ E)_0 Att(\gamma) = Att(\gamma)$.

Proposition 3.5 *Let $\dot{X} = X^2$ occur in A with A n-dimensional.*
(1) If $n = 2$, then there are no periodic trajectories.
(2) If $n = 3$, then the periodic trajectories lie on cones.
(3) If A is power-associative, then there are no periodic solutions.
(4) No periodic attractors exist; however, periodic ω or α limit sets may exist.

In the discrete case, recall that $\mathcal{F}$ denotes the set of fixed points of a given smooth mapping $E : \mathbb{R}^n \to \mathbb{R}^n$.

Theorem 3.6 *Let $E : \mathbb{R}^n \to \mathbb{R}^n$ be smooth. Then*
(1) $(Aut\ E)\mathcal{F} = \mathcal{F}$, and
(2) if $P \in \mathcal{F}$ is an isolated fixed point, then $(Aut\ E)_0 P = P$.

Let $\mathcal{P}_N$ denote the set of all periodic points of $X(n+1) = E(X(n))$ of period N.

Theorem 3.7 *Let $E : \mathbb{R}^n \to \mathbb{R}^n$ be smooth. Then*
(1) $(Aut\ E)\mathcal{P}_N = \mathcal{P}_N$, and
(2) if $P \in \mathcal{P}_N$ is an isolated periodic point, then $\phi E^{(n)}(P) = P$ for all $\phi \in (Aut\ E)_0$ and all nonnegative integers n.

Proposition 3.8 *If A is power-associative and the discrete system $X(n+1) = X(n)^2$ has a periodic point, then A has an idempotent.*

We close by describing situations in which automorphisms give solutions to differentiable or discrete systems, and give quadratic examples. We begin with the differentiable case where E is a vector field on $\mathbb{R}^n$.

Theorem 3.9 *Let $G \in Der\ E$ and $P \in \mathbb{R}^n$. Then $X(t) = (\exp tG)P$ is a solution to $\dot{X} = E(X)$ if and only if $GP = E(P)$. In this case,*
(1) $P \in \mathcal{E}$ if and only if $GP = 0$, and
(2) $(\exp tG)P \in \mathcal{P}_\tau$ if and only if $(\exp \tau G)P = P$; i.e., P is an eigenvector of $\exp \tau G$ with eigenvalue 1.

The following gives a criterion for when a periodic trajectory is given by an automorphism.

Theorem 3.10 *Let $\gamma \in \mathcal{P}_\tau$ be an isolated orbit in $\mathcal{P}_\tau$. If there exist $P \in \gamma$ and $D \in Der\ E$ such that $DP \neq 0$, then there exists $G \in Der\ E$ so that $\gamma(t) = (\exp tG)P$.*

In the quadratic case, we have the following

Example: Let A be a 3–dimensional commutative algebra which supports a periodic solution $X(t) = (\exp tG)P$ of $\dot{X} = X^2$ where $G \in Der\ A$ and $GP = P^2$. Then A is simple and has a basis $\{X_0, X_1, X_2\}$ with multiplication table

	X_0	X_1	X_2
X_0	λX_0	X_2	$-X_1$
X_1	X_2	μX_0	0
X_2	X_1	0	μX_0

where λ, $\mu \in \mathbb{R}$ and $\lambda\mu < 0$. Relative to the above basis, the matrix of G can be any scalar multiple of $\begin{bmatrix} 0 & 0 & 0 \\ 0 & 0 & -1 \\ 0 & 1 & 0 \end{bmatrix}$. Letting x_0, x_1, and x_2 denote coordinates in the basis, the initial point P can be any point on the cone $\mathcal{C}$ given by $\lambda x_0^2 + \mu(x_1^2 + x_2^2) = 0$. The trajectory on a fixed $P \in \mathcal{C}$, $P = \sum p_i X_i$, can be described by $x_0(t) \equiv p_0$, $x_1(t)^2 + x_2(t)^2 \equiv p_1^2 + p_2^2$. This trajectory is a ($\alpha$ or ω) limit set for every point on the cylinder with axis X_0 and radius $(p_1^2 + p_2^2)^{\frac{1}{2}}$.

Finally we consider the discrete case in which $E : \mathbb{R}^n \to \mathbb{R}^n$ is a smooth mapping.

Theorem 3.11 *Let E be smooth on $\mathbb{R}^n$. Let $P \in \mathbb{R}^n$ and $\phi \in Aut\ E$ be such that $\phi P = E(P)$. Then the iterates $E^{(n)}(P) = \phi^{(n)}P$ where $\phi^{(n)} = \phi\phi\cdots\phi$ (composition n times).*

Corollary 3.12 *Let $G \in Der\ E$ and $P \in A$ be such that $e^G P = E(P)$, then the iterates $E^{(n)}(P) = e^{nG}P$.*

The analysis of examples is a bit more delicate than in the differentiable case; see [1] for details. For a quadratic map $E(X) = X^2$, we have the following.

Example: Let A be a 3–dimensional commutative algebra which supports a periodic orbit $X(n) = (\exp nG)P$ of $X(n+1) = X(n)^2$ where $G \in Der\ A$ and $e^G P = P^2$. Then A is simple and has a basis $\{X_0, X_1, X_2\}$ with multiplication table

	X_0	X_1	X_2
X_0	λX_1	$\frac{b^2}{2}(\cos bX_2 + \sin bX_3)$	$\frac{b^2}{2}(-\sin bX_2 + \cos bX_3)$
X_1	$\frac{b^2}{2}(\cos bX_2 + \sin bX_3)$	μX_1	0
X_2	$\frac{b^2}{2}(-\sin bX_2 + \cos bX_3)$	0	μX_1

where $\lambda + \mu = b^2$, and b is the frequency. The orbits lie on cones as in the differentiable example.

References

[1] M.K. Kinyon and A.A. Sagle, Quadratic Dynamical Systems and Algebras, manuscript in preparation.

[2] R. Schafer, **Introduction to Nonassociative Algebras**, Academic Press, 1966.

[3] A.A. Sagle and R. Walde, **Introduction to Lie Groups and Lie Algebras**, Academic Press, 1973.

[4] A.A. Albert, The radical of a nonassociative algebra, Bull. A.M.S., **48** (1942), 891–897.

[5] L. Markus, Quadratic differential equations and nonassociative algebras, Ann. Math. Studies, **45** (1960), Princeton Univ. Press, 185–213.

[6] S. Walcher, **Algebras and Differential Equations**, Hadronic Press, 1989.

[7] A.A. Sagle, Invariant Lagrangian mechanics, connections and nonassociative algebras, Algebras, Groups and Geo., **3** (1986), 199–263.

[8] M. Peschel and W. Mende, **The Predator–Prey Model**, Springer–Verlag, 1986.

[9] A.A. Sagle, Jordan algebras and connections on homogeneous spaces, Trans. A.M.S., **187** (1974).

[10] J. Sotomayor and R. Paterlini, Quadratic vector fields with finitely many periodic orbits, **Geometric Dynamics**, Springer Lecture Notes 1007 (1983).

[11] H. Röhrl, Algebras and differential equations, Nagoya Math. J., **68** (1977), 59–122.

[12] V. Arnold, **Mathematical Methods of Classical Mechanics**, Springer–Verlag, 1978.

[13] H.R. Van der Vaart, Conditions for periodic solutions of Volterra differential systems, Bull. Math. Biology, **40** (1978), 133–160.

[14] J. Kaplan and J. Yorke, Nonassociative real algebras and quadratic differential equations, Nonlinear Analysis TMA, **3** (1979), 49–51.

SUR LA COHOMOLOGIE DES ALGEBRES DE MALCEV*

Akry KOULIBALY
Institut de Mathématiques et de Physique
Université de Ouagadougou
03 B.P. 7021
Ouagadougou 03, Burkina Faso

et

Kalifa TRAORE
Institut de Mathématiques et de Physique
Université de Ouagadougou
03 B.P. 7021
Ouagadougou 03, Burkina Faso

ABSTRACT

The cohomology theory of Malcev algebras has been developped by K.Yamaguti as a generalisation of the cohomology of Lie algebras. Little use has been made of this theory. In this paper, we give some conditions for the cohomology of a solvable Malcev algebra to be trivial.

Introduction.

La cohomologie d'une algèbre de Malcev a été introduite par K. YAMAGUTI comme une généralisation de celles des algèbres de Lie. Dans cet article, nous donnons les conditions nécessaires et suffisantes pour que le premier groupe de cohomologie d'une algèbre de Malcev résoluble soit nul et nous établissons un théorème de structure. Dans cet article, K désigne un corps commutatif. Toutes les algèbres de Malcev, de Lie sont supposées de dimension finie sur K.

1. Préléminaires.

Soit M une algèbre non associative sur un corps commutatif K. On dira que M est une K-algèbre de Malcev si $x^2 = 0$ pour tout x dans M et $J(x,y,xz) = J(x,y,z)x$ quels que soient x, y, z dans M où $J(x,y,z) = (xy)z + (yz)x + (zx)y$.

Définition 1.1. Soit M une algèbre de Malcev sur un corps commutatif K. Une structure d'espace vectoriel à opérateurs sur K de domaine M est la donnée d'une structure constituée par un K-espace vectoriel V et d'une application ρ de M dans la K-al-

---

* à paraître dans *Algebras, Groups and Geometries*, volume 7 (1990)
AMS 1980 Subject classification (1985 Revision) Primary: 17D10

gèbre des endomorphismes de V. Les endomorphismes $\rho(x)$, notés ρ_x, pour x dans M, sont appelés les opérateurs de la structure. On dit que ρ est une représentation faible de M dans V si :

1) ρ est une application K-linéaire,

2) $[D(x,y), \rho_z] = \rho_{[xyz]}$ quels que soient x, y, z dans M où $D(x, y) = [\rho_x, \rho_y] + \rho_{xy}$ et $[x\ y\ z] = x(yz) - y(xz) + (xy)z$. On dit que V est l'espace de ρ.

Pour x dans M, soit L_x l'endomorphisme de M défini par $L_x(y) = xy$. L'application $x \mapsto L_x$ est une représentation faible de M dite représentation régulière de M.

Soit $x \mapsto \rho_x$ une application K-linéaire de M dans $\mathrm{End}_K(V)$. Posons :

$\Delta(x,y) = [\rho_x, \rho_y] - \rho_{xy}$. Si $\Delta(x,y) = 0$ pour tous x,y dans M, alors ρ est une représentation faible de M dans V appelée représentation spéciale de M dans V.

Définition1.2. Soit ρ une représentation faible d'une K-algèbre de Malcev M dans V. Un sous espace U de V est appelé introverti pour ρ si U est stable par tous les opérateurs de V.

Soit U un sous espace de V introverti pour ρ; alors :

1) l'application $x \mapsto \tilde{\rho}_x = \rho_x|_U$ est une représentation faible de M dans U, dite sous représentation faible de ρ.

2) Si l'on note $\overline{\rho}_x$ l'endomorphisme de V/U déduit de ρ_x par passage au quotient, l'application $x \mapsto \overline{\rho}_x$ est une représentation faible de M dans V/U, dite représentation faible quotient de ρ.

On dira qu'une représentation faible ρ de M dans V est finie si V est de dimension finie sur K.

Lemme 1.3. *Soient ρ, ρ' deux représentations faibles finies de M dans V et V' respectivement et θ une application linéaire de V dans V' telle que :*

$\theta(\rho_x(v)) = \rho'_x(\theta(v))$ pour touts x dans M, v dans V. S'il existe une application linéaire θ' de V' dans V telle que $\theta' o \theta = 1_V$ (endomorphisme identique de V) et Kerθ' est un sous espace de V' introverti pour ρ', alors $\theta'(\rho'_x(v')) = \rho_x(\theta'(v'))$ quels que soient x dans M, v' dans V'.

En effet soit $\{v_1, \ldots, v_p\}$ une base de V. Comme θ est injective, alors $\{\theta(v_1), \ldots, \theta(v_p)\}$ est une famille libre de V'. Elle peut donc être prolongée en une base de V' soit $\{\theta(v_1), \ldots, \theta(v_p), v'_{p+1}, \ldots, v'_n\}$. Définissons donc $\theta':V' \to V$ par les conditions $\theta'(\theta(v_i)) = v_i$ pour $i = 1, \ldots, p$ et $\theta'(v'_j) = 0$ pour $j = p+1, \ldots, n$. Si $v' \notin \mathrm{Im}\theta$, alors $v' \in \mathrm{Ker}\theta'$. Puisque Ker θ' est introverti pour ρ', il vient que pour tout x dans M, $\rho'_x(v') \in \mathrm{Ker}\theta'$. D'où $\theta'(\rho'_x(v')) = 0 = \rho_x(\theta'(v'))$ quel que soit x dans M. Si $v' \in \mathrm{Im}\theta$, il existe $v \in V$ tel que $v' = \theta(v)$. Par suite $\theta'(\rho'_x(v')) = \theta'(\rho'_x(\theta(v))) = \theta'(\theta(\rho_x(v))) = \rho_x(v) = \rho_x(\theta'(v'))$ pour tout x dans M .

Lemme 1.4. *Soient ρ, ρ' deux représentations faibles finies de M dans V et V' respectivement et θ une application linéaire de V dans V' telle que $\theta(\rho_x(v)) = \rho'_x(\theta(v))$ pour tout x dans M, v dans V. S'il existe une application linéaire θ' de V' dans V telle que $\theta \, o \theta' = 1_{V'}$, et Imθ' est un sous espace de V introverti pour ρ, alors $\theta'(\rho'_x(v')) = \rho_x(\theta'(v'))$ quels que soient x dans M, v' dans V'.*

En effet, pour tout x dans M et tout v' dans V' on a $\theta'(\rho'_x(v')) = \theta'(\rho'_x(\theta(\theta'(v')))) = \theta'(\theta(\rho_x(\theta'(v'))))$. Comme Imθ' est introverti pour ρ, alors $\rho_x(\theta'(v')) \in \mathrm{Im}\theta'$. Il existe donc ω dans V' tel que $\rho_x(\theta(v')) = \theta'(\omega)$. D'où $\theta'(\rho'_x(v')) = \theta'(\theta(\theta'(\omega))) = \theta'(\omega) = \rho_x(\theta'(v'))$.

2. Cohomologie associée à une représentation faible.

Soit ρ une représentation faible d'une algèbre de Malcev M sur un corps commutatif K dans un K-espace vectoriel V. Désignons par $C^{2p-1}(M,V)$ le K-espace vectoriel des applications (2p-1)-linéaires f de $M \times M \times ... \times M$ ((2p-1) fois) dans V telles que : $f(x_1, ..., x_{2k-1}, x_{2k}, ..., x_{2p-1}) = 0$ dès que $x_{2k-1} = x_{2k}$. On pose $C^o(M,V) = V$. L'opérateur cobord δ_{2p-1} de $C^{2p-1}(M,V)$ dans $C^{2p+1}(M,V)$ est une application K-linéaire définie par les formules (cf[8]) : $\delta_o(v)(x) = \rho_x(v)$ pour tout v dans V et

$$(\delta_{2p-1}(f))(x_1, ..., x_{2p+1}) = (-1)^p \rho_{x_{2p+1}}[\rho_{x_{2p-1}}(f(x_1,...,x_{2p-2}, x_{2p}))$$

$$- \rho_{x_{2p}}(f(x_1, ..., x_{2p-1})) + f(x_1, ..., x_{2p-2}, x_{2p-1}x_{2p})]$$

$$+ \sum_{k=1}^{p}(-1)^{k+1}D(x_{2k-1}, x_{2k})(f(x_1, x_2,..., \hat{x}_{2k-1}, \hat{x}_{2k}, ..., x_{2p+1}))$$

$$+ \sum_{k=1}^{p} \sum_{j=2k+1}^{2p+1} (-1)^k \; f(x_1, x_2,..., \hat{x}_{2k-1}, \hat{x}_{2k},..., [x_{2k-1}\, x_{2k}\, x_j],...,x_{2p+1})$$

pour tous les éléments $x_1, ..., x_{2p+1}$ de M et toute cochaine f de $C^{2p-1}(M,V)$ où le symbole ^ sur une lettre indique que cette lettre doit être omise et $[x_{2k-1}\, x_{2k}\, x_j] = x_{2k-1}(x_{2k}x_j) - x_{2k}(x_{2k-1}x_j) + (x_{2k-1}x_{2k})x_j$. Dans [8], il est montré que $\delta_1 \circ \delta_0 = 0$ et $\delta_{2p+1} \circ \delta_{2p-1} = 0$ pour p = 1, 2, 3, ...

L'espace $C^{2p-1}(M,V)$ est appelé espace des (2p-1)-cochaines. On appelle cocycle une cochaine f telle que $\delta_{2p-1}(f) = 0$ et cobord une cochaine f de la forme $\delta_{2p-1}(g)$.

Le (2p-1)-groupe de cohomologie de YAMAGUTI de M dans V est le K-espace vectoriel $H^{2p-1}(M,V) = \text{Ker } \delta_{2p+1}/\text{Im } \delta_{2p-1}$. Par définition [8], $H^0(M,V) = \{v \in V / \rho_x(v) = 0, \forall x \in M\}$; posons $H^0(M,V) = V^M$. Il est évident que V^M est un sous-espace de V introverti pour ρ.

Lemme 2.1. *Soient ρ,ρ' deux représentations faibles d'une algèbre de Malcev M dans V et V' respectivement et θ une application linéaire de V dans V'. Si θ est injective, alors l'application linéaire $f \mapsto \overline{\theta}(f) = \theta o f$ de $C^{2p+1}(M,V)$ dans $C^{2p+1}(M,V')$ est injective pour p = 0, 1, 2, ...*

Lemme 2.2. *Soient ρ,ρ' deux représentations faibles d'une algèbre de Malcev M dans V et V' respectivement et θ une application linéaire de V dans V' telle que*

$\theta \circ \rho_x = \rho'_x \circ \theta$ pour tout x dans M. Alors il existe une application linéaire et une seule ψ_{2p+1} de $H^{2p+1}(M,V)$ dans $H^{2p+1}(M,V')$ rendant commutatif le diagramme

$$\begin{array}{ccc} \mathrm{Ker}\,\delta_{2p+1} & \xrightarrow{\theta^*} & \mathrm{Ker}\,\delta'_{2p+1} \\ \downarrow q & & \downarrow q' \\ H^{2p+1}(M,V) & \xrightarrow{\psi_{2p+1}} & H^{2p+1}(M,V') \end{array}$$

où δ_{2p+1}, δ'_{2p+1} désignent respectivement les opérateurs cobords de $C^{2p+1}(M,V)$ dans $C^{2p+3}(M,V)$ et $C^{2p+1}(M,V')$ dans $C^{2p+3}(M,V')$.

En effet comme pour tout x dans M et tout v dans V, on a $\theta(\rho_x(v)) = \rho'_x(\theta(v))$, alors pour tout f dans $C^{2p+1}(M,V')$ on a : $\overline{\theta}(\delta_{2p+1}(f))(x_1,\ldots, x_{2p+3}) = (\theta \circ \delta_{2p+1}(f))(x_1,\ldots, x_{2p+3}) = \delta'_{2p+1}(\theta o f)(x_1,\ldots, x_{2p+3})$ quels que soient les éléments $x_1, \ldots, x_{2p+3}$ de M. Par suite, il vient que $\overline{\theta}(\mathrm{Ker}\, \delta_{2p+1}) \subseteq \mathrm{Ker}\, \delta'_{2p+1}$ et ceci nous dit qu'il existe une application linéaire et une seule θ^* de Ker δ_{2p+1} dans Ker δ'_{2p+1} telle que $\theta^*(f) = \overline{\theta}(f)$ pour tout f dans Ker δ_{2p+1}. Désignons par q la surjection canonique de Ker δ_{2p+1} sur $H^{2p+1}(M,V)$ et par q' la surjection canonique de Ker δ'_{2p+1} sur $H^{2p+1}(M,V')$. Soit $f \in \mathrm{Im}\delta_{2p-1}$, alors il existe $\overline{f} \in C^{2p-1}(M,V)$ telle que $f = \delta_{2p-1}(\overline{f})$. D'où $(q' \circ \theta^*)(f) = q'(\overline{\theta}(f)) = q'(\theta \circ f) = q'(\theta \circ \delta_{2p-1}(\overline{f})) = q'(\delta'_{2p-1}(\theta \circ \overline{f})) = 0$. Il existe alors donc et une seule application linéaire ψ_{2p+1}de $H^{2p+1}(M,V)$ dans $H^{2p+1}(M,V')$ telle que $q' \circ \theta^* = \psi_{2p+1} \circ q$.

Soient ρ, ρ' des représentations faibles d'une algèbre de Malcev M dans V et V' respectivement. Désignons par $\mathbf{C}$ la catégorie dont les objets sont les espaces de représentations faibles et les morphismes, les applications linéaires θ de V dans V' vérifiant $\theta \circ \rho_x = \rho'_x \circ \theta$ pour tout x dans M. Les applications $V \mapsto H^{2p+1}(M,V)$ et $\theta \mapsto \psi_{2p+1}$ définissent un foncteur covariant $H^{2p+1}(M,-)$ de $\mathbf{C}$ dans la catégorie EV_K des K-espaces vectoriels. Posons $\psi_{2p+1} = H^{2p+1}(M,\theta)$.

Théorème 2.3. *Soient ρ, ρ', ρ'' des représentations faibles finies d'une algèbre de Malcev M dans V, V' et V'' respectivement. Si la suite*

$$O \longrightarrow V \xrightarrow{\theta} V' \xrightarrow{\theta'} V'' \longrightarrow 0$$ *est exacte dans* $\mathbf{C}$ *et le noyau de la rétraction r associée à θ est un sous-espace de V' introverti pour ρ', alors la suite*

$$0 \rightarrow V^M \rightarrow V'^M \rightarrow V''^M \xrightarrow{\phi_o} H^1(M,V) \rightarrow H^1(M,V') \rightarrow H^1(M,V'')$$

où ϕ_o est l'application linéaire nulle, est exacte dans EV_K.

En effet, comme $\theta \circ \rho_x = \rho'_x \circ \theta$ et $\theta' \circ \rho'_x = \rho''_x \circ \theta'$, il vient que

$\overline{\theta}(\mathrm{Ker}\delta_1) \subseteq \mathrm{Ker}\delta'_1$ et $\overline{\theta'}(\mathrm{Ker}\delta'_1) \subseteq \mathrm{Ker}\delta''_1$ où $\overline{\theta}$ est l'application linéaire définie dans le lemme 2.1 et $\overline{\theta'}$ l'application linéaire de $C^1(M,V')$ dans $C^1(M,V'')$ définie par $\overline{\theta'}(f) = \theta' \circ f$ pour tout f dans $C^1M,V')$. Par suite $\overline{\theta}$ (respectivement $\overline{\theta'}$) induit par restriction à $\mathrm{Ker}\delta_1$ et $\mathrm{Ker}\delta'_1$ (respectivement à $\mathrm{Ker}\delta'_1$ et $\mathrm{Ker}\delta''_1$) une application linéaire θ^* (respectivement θ'^*). De plus, puisque $\mathrm{Im}\,\delta_o \subseteq \mathrm{Ker}\delta_1$ il existe une et une seule application $\overline{\delta}_o$ de V dans $\mathrm{Ker}\delta_1$ telle que $\overline{\delta}_o(v) = \delta_o(v)$ pour tout v dans V. De façon analogue, on montre qu'il existe une application linéaire $\overline{\delta'}_o$ (respectivement $\overline{\delta''}_o$) de V' dans $\mathrm{Ker}\,\delta'_1$ (respectivement de V'' dans $\mathrm{Ker}\,\delta''_1$) telle que $\overline{\delta'}_o(v') = \delta'_o(v')$ pour tout v' dans V' (respectivement $\overline{\delta''}_o(v'') = \delta''_o(v'')$ pour tout v'' dans V''). Ainsi, nous pouvons considérer le diagramme:

$$\begin{array}{ccccccccc}
C & \longrightarrow & V & \xrightarrow{\theta} & V' & \xrightarrow{\theta'} & V'' & \longrightarrow & O \\
 & & \downarrow{\scriptstyle \overline{\delta_o}} & & \downarrow{\scriptstyle \overline{\delta_o}} & & \downarrow{\scriptstyle \overline{\delta''_o}} & & \\
C & \longrightarrow & \mathrm{Ker}\delta_1 & \xrightarrow{\theta^*} & \mathrm{Ker}\delta'_1 & \xrightarrow{\theta'^*} & \mathrm{Ker}\delta''_1 & &
\end{array}$$

Quels que soient x dans M,v dans V on a $[(\theta^* \circ \overline{\delta}_o(v)](x) = \overline{\theta}(\delta_o(v))(x) =$

$\theta(\rho_x(v)) = \rho'_x(\theta(v)) = [(\overline{\delta'}_o \circ \theta)(v)](x)$; d'où $\theta^* \circ \overline{\delta}_o = \overline{\delta'}_o \circ \theta$. De façon analogue,

on montre que $\theta'^* \circ \overline{\delta'}_o = \overline{\delta''}_o \circ \theta'$. Il en résulte que le diagramme ci-dessus est commutatif.

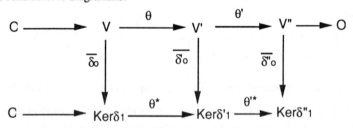

Montrons maintenant que la dernière ligne du diagramme ci-dessus est exacte. Soit $f \in \mathrm{Ker}\theta^*$, alors $0 = \overline{\theta}(f) = \theta \circ f$. θ étant injective on a $f = 0$. D'où l'exactitude en $\mathrm{Ker}\delta_1$. Pour tout f dans $\mathrm{Ker}\delta_1$ on a $(\theta'^* \circ \theta)(f) = \theta' \circ \theta \circ f = 0$; par suite $\theta'^* \circ \theta = 0$ et ceci nous dit que $\mathrm{Im}\,\theta^* \subseteq \mathrm{Ker}\,\theta'^*$. Soit $g \in \mathrm{Ker}\,\theta'^*$, alors $0 = \theta'^*(g) = \theta' \circ g$. Par suite, pour tout x dans M, $g(x) \in \mathrm{Ker}\,\theta' = \mathrm{Im}\,\theta$; il existe donc $v \in V$ tel que $g(x) = \theta(v)$. Posons $h = r \circ g$. Comme pour tous x, y, z dans M, $\delta_1(h)(x,y,z) = r(\delta'_1(g)(x,y,z)) = 0$ d'après le lemme 1.3, alors $h \in \mathrm{Ker}\delta_1$. Il s'en suit que pour tout x dans M, on a $\theta^*(h)(x) = (\theta \circ r)(g(x)) = \theta(r(\theta(v))) = \theta(v) = g(x)$ c'est-à-dire $g = \theta^*(h) \in \mathrm{Im}\,\theta^*$. D'où l'exactitude en $\mathrm{Ker}\,\delta'_1$. De l'injectivité de θ et de tout ce qui précède, il vient d'après le lemme du serpent qu'il existe une application linéaire ϕ_o de V''^M dans $H^1(M,V)$ telle que le diagramme

$$0 \to V^M \to V'^M \to V''^M \xrightarrow{\phi_o} H^1(M,V) \xrightarrow{H^1(M,\theta)} H^1(M,V') \xrightarrow{H^1(M,\theta')} H^1(M,V'')$$

soit exact dans EV_K.

Soit $\tau \in \mathrm{Ker}\,H^1(M,\theta)$, alors il existe f dans $\mathrm{Ker}\,\delta_1$ tel que $\tau = q(f)$ (cf. Lemme 2.2). Puisque $0 = H^1(M,\theta)(\tau) = (H^1(M,\theta) \circ q)(f) = q'(\theta \circ f)$ on a $\theta \circ f \in \mathrm{Ker} q' = \mathrm{Im}\delta'_o$. Il existe donc $v' \in V'$ tel que $\theta \circ f = \delta'_o(v')$. Par suite, pour tout x dans M, on a $\theta(f(x)) = \rho'_x(v')$. Ker(r) étant un sous-espace de V' introverti pour ρ' alors $f(x) = r(\rho'_x(v')) = \rho_x(r(v'))$ d'après le lemme 1.3. D'où $f(x) = \delta_o(r(v'))(x)$ et ceci nous dit que $f \in \mathrm{Im}\,\delta_o$. Il en résulte que $\tau = q(f) = 0$. Il s'en suit alors que $H^1(M,\theta)$ est injective. De l'exactitude du diagramme ci-dessus en $H^1(M,V)$ on a $\mathrm{Im}\,\phi_o = \mathrm{Ker}(H^1(M,\theta)) = 0$ d'où ϕ_o est le morphisme nul de V''^M dans $H^1(M,V)$. Ceci nous dit donc que $\mathrm{Im}\,(V'^M \mapsto V''^M) = \mathrm{Ker}\,\phi_o = V''^M$, c'est-à-dire que l'application linéaire de V'^M dans V''^M est surjective.

Soit ρ une représentation faible d'une algèbre de Malcev M sur un corps commutatif K dans V, N(M) le J-noyau de M c'est-à-dire $N(M) = \{x \in M \,/\, J(x,M,M) = 0\}$.

Dans tout ce qui suit, nous supposerons que pour tout choix de V, $V^{N(M)} = \{v \in V \,/\, \rho_x(v) = 0, \forall\, x \in N(M)\}$ est un sous-espace non nul de V introverti pour ρ et que pour tout idéal I de M/N(M), $(V^{N(M)})^I$ est un sous-espace de $V^{N(M)}$ introverti pour la représentation faible $\rho^*: \overline{x} \mapsto \rho^*_{\overline{x}} = \tilde{\rho}_x$ de M/N(M) dans $V^{N(M)}$.

Soit d_o l'opérateur cobord de $V^{N(M)} = C^o(M/N(M),V^{N(M)})$ dans $C^1(M/N(M),V^{N(M)})$ et d_1 l'opérateur cobord de $C^1(M/N(M),V^{N(M)})$ dans

$C^3(M/N(M),V^{N(M)})$ définis respectivement par $d_o(v)(\overline{x}) = \delta_o(v)(x)$ pour tout v dans $V^{N(M)}$ et tout x dans M et $d_1(f)(\overline{x}, \overline{y}, \overline{z}) = \delta_1(f \circ p)(x,y,z)$ pour tout f dans $C^1(M/N(M),V^{N(M)})$ et tous x, y, z dans M où p désigne la surjection canonique de M sur M/N(M) et $p(x) = \overline{x}$ pour tout x dans M.

Lemme 2.4. *Si ρ est une représentation faible d'une K-algèbre de Malcev M dans V, alors $(V^{N(M)})^{M/N(M)} = V^M$*

En effet comme pour tout v dans V^M on a $v \in V^{N(M)}$, il vient que pour tout x dans M, $\rho^*_{\overline{x}}(v) = \tilde{\rho}_x(v) = \rho_x(v) = 0$. D'où $v \in (V^{N(M)})^{M/N(M)}$ et par suite $V^M \subseteq (V^{N(M)})^{M/N(M)}$. Réciproquement si $v \in (V^{N(M)})^{M/N(M)}$ on a pour tout x dans M, $0 = \rho^*_{\overline{x}}(v) = \rho_x(v)$ et par suite $v \in V^M$. D'où le résultat .

Lemme 2.5. *Si ρ est une représentation faible d'une K-algèbre de Malcev M dans V, il existe une unique application linéaire injective ψ de $H^1(M/N(M),V^{(N(M)})$ dans $H^1(M,V)$ rendant commutatif le diagramme*

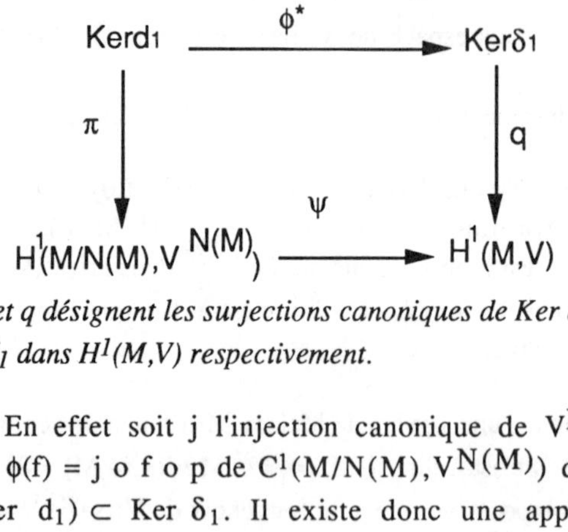

où π et q désignent les surjections canoniques de Ker d_1 sur $H^1(M/N(M),V^{N(M)})$ et de Ker δ_1 dans $H^1(M,V)$ respectivement.

En effet soit j l'injection canonique de $V^{N(M)}$ dans V. L'application $f \mapsto \phi(f) = j \circ f \circ p$ de $C^1(M/N(M),V^{N(M)})$ dans $C^1(M,V)$ est linéaire et ϕ (Ker d_1) $\subset$ Ker δ_1. Il existe donc une application ϕ^* et une seule de Ker d_1 dans Ker δ_1 telle que $\phi^*(f) = j \circ f \circ p$ quelque soit f dans Ker d_1. Soit $f \in$ Im d_o, alors il existe $v \in V^{N(M)}$ tel que $f = d_o(v)$. Par suite $q(\phi^*(f)) = q(j \circ d_o(v) \circ p)$. Comme pour tout x dans M, $(j \circ d_o(v) \circ p)(x) = \delta_o(v)(x)$, il vient que $j \circ d_o(v) \circ p \in$ Im δ_o. Il en résulte que $(q \circ \phi^*)(f) = 0$, c'est-à-dire $f \in$ Ker $(q \circ \phi^*)$ et ceci nous dit que Im$(d_o) \subset$ Ker$(q \circ \phi^*)$. Réciproquement: Soit $f \in$ Ker$(q \circ \phi^*)$, alors $j \circ f \circ p \in$ Ker q = Im δ_o. Il existe donc un v dans V tel que $j \circ f \circ p = \delta_o(v)$. D'où pour tout x dans N(M), on a $0 = f(p(x)) = \rho_x(v)$ et par suite $v \in V^{N(M)}$. Il en résulte que pour tout y dans M on a $(j \circ d_0(v) \circ p)(y) = \rho^*_{p(y)}(v) = \rho_y(v) = \delta_0(v)(y) = (j \circ f \circ p)(y)$ soit $j \circ d_0(v) \circ p = j \circ f \circ p$. Ceci nous dit que $f = d_0(v) \in \mathrm{Im}\, d_0$ c'est-à-dire $\mathrm{Ker}(q \circ \phi^*) \subseteq \mathrm{Im}\, d_0$. Il existe donc une et une seule application linéaire injective ψ de $H^1(M/N(M), V^{N(M)})$ dans $H^1(M,V)$ telle que $q \circ \phi^* = \psi \circ \pi$.

Lemme 2.6. *Soient M une algèbre de Malcev non de Lie de dimensions 4 sur un corps commutatif K de caractéristique différente de 2 et 3, ρ une représentation faible de M dans un K-espace vectoriel V. Si $H^1(M,V) = 0$, alors $V^M = 0$.*

En effet M étant à isomorphisme près, la K-algèbre de Malcev résoluble dont la table de multiplication relativement à une base $\{e_1, e_2, e_3, e_4\}$ s'écrit : $e_1e_2 = -e_2$, $e_1e_3 = -e_3$, $e_1e_4 = e_4$, $e_2e_3 = 2e_4$, $e_ie_j = -e_je_i$ pour i<j, tous les autres produits étant nuls, on a $N(M) = Ke_4$. Par suite M/N(M) est une K-algèbre de Lie résoluble. Si $H^1(M,V) = 0$, on a d'après le lemme 2.5. $H^1(M/N(M), V^{N(M)}) = 0$. Posons $I = K\overline{e_2} + K\overline{e_3}$ et $P = M/N(M)$. Comme $(V^{N(M)})^I$ est un sous-espace de V introverti pour la représentation faible ρ^*, alors l'application $\overline{y} + I \mapsto \rho^{**}_{\overline{y}+I} = \rho^*_{\overline{y}}$ est une représentation faible de P/I dans $(V^{N(M)})^I$. Comme dans le lemme 2.5., il existe une application linéaire injective de $H^1(P/I, (V^{N(M)})^I)$ dans $H^1(P,V^{N(M)})$. Ainsi $H^1(P,V^{N(M)}) = 0$ nous dit que $H^1(P/I, (V^{N(M)})^I) = 0$. La représentation faible quotient ρ^{**} de P/I dans $(V^{N(M)})^I$ est spéciale car $\dim_K P/I = 1$. Par suite d'après Barnes (cf[1]) on a: $((V^{N(M)})^I)^{P/I} = 0$; c'est-à-dire $0 = (V^{N(M)})^{N(M)} = V^M$.

Proposition 2.7. *Soient M une algèbre de Malcev résoluble sur un corps commutatif K de caractéristique différente de 2 et 3, ρ une représentation faible de M dans un K-espace vectoriel V. Si $H^1(M,V) = 0$, alors $V^M = 0$.*

En effet si M est de Lie, raisonnons par récurrence sur la dimension de M. Si $\dim_K M = 1$, alors ρ est nécessairement une représentation spéciale. Nous pouvons donc considérer V comme un M-module de Lie. Par suite, d'après Barnes (cf. [1], lemme 2) la relation $H^1(M,V) = 0$ nous dit que $V^M = 0$. Supposons la proposition vraie pour toute algèbre de Lie de dimension strictement inférieure à celle de M et $\dim_K M > 1$. Soit I un idéal propre de M. D'après le lemme 2.5, $H^1(M,V) = 0$ entraîne $H^1(M/I,V^I) = 0$. De l'hypothèse de récurrence il vient que $V^M = (V^I)^{M/I} = 0$. Si M n'est pas Lie, alors le lemme 2.5 nous dit que $H^1(M/N(M);V^{N(M)}) = 0$. Si M/N(M) est de Lie, il vient d'après ce qui précède que $V^M = (V^{N(M)})^{M/N(M)} = 0$. Si M/N(M) n'est pas de Lie, alors le lemme 2.6 nous dit que la proposition est vraie pour $\dim_K M/N(M) = 4$. Supposons donc la proposition vraie pour toute algèbre de Malcev résoluble de dimension strictement inférieure à celle de M. Posons I = N(M/N(M)) le J-noyau de M/N(M) et P = M/N(M) ; I est non nul car M/N(M) est résoluble. Du Lemme 2.5 on a $H^1(P/I,(V^{N(M)})^I) = 0$. D'après l'hypothèse de récurrence il vient que $((V^{N(M)})^I)^{P/I} = 0$ c'est-à-dire $V^M = (V^{N(M)})^{M/N(M)} = 0$.

Exemple 2.8. Soit M une algèbre de Malcev non de Lie de dimension 4 sur un corps commutatif K de caractéristique différente de 2 et 3. $\lambda: x \mapsto L_x$ la représentation régulière de M. Comme M est à isomorphisme près la K-algèbre de Malcev non de Lie résoluble dont la table de multiplication relativement à une base $\mathcal{U} = \{e_1,e_2,e_3,e_4\}$ s'écrit : $e_1e_2 = - e_2$, $e_1e_3 = - e_3$, $e_1e_4 = e_4$, $e_2e_3 = 2\, e_4$, $e_ie_j = - e_je_i$ pour $i< j$, tous les autres produits étant nuls, il vient que $N(M)= Ke_4$. Par suite $M^{N(M)} = Ke_2 + Ke_3 + Ke_4 = M^2$. Il est immédiat que M^2 est un sous-espace de M introverti pour λ. Soit $f\in \operatorname{Ker} \delta_1$ où δ_1 est l'opérateur cobord de $C^1(M,N)$ dans $C^3(M,N)$. Posons

$f(e_1) = \sum_{i=1}^{4} a_ie_i$, $f(e_2) = \sum_{i=1}^{4} b_ie_i$, $f(e_3) = \sum_{i=1}^{4} c_ie_i$, $f(e_4) = \sum_{i=1}^{4} \alpha_ie_i$. Des équations $0 = \delta_1(f)(e_i,e_j,e_k)$, i, j, k = 1,2,3,4 il vient que la matrice $\mathcal{M}_{\mathcal{U}}(f)$ s'écrit

$$\mathcal{M}_{\mathcal{U}}(f) = \begin{pmatrix} 0 & 0 & 0 & 0 \\ a_2 & 0 & 0 & 0 \\ 0 & 0 & 0 & 0 \\ a_4 & 0 & -2a_2 & 0 \end{pmatrix}$$

En posant $t = - a_2e_2 + a_4e_4$, on a $f = \delta_o(t)$. Par suite $f \in \operatorname{Im}\delta_o$. Il en résulte que $H^1(M,M) = 0$. D'où $M^M = Z(M) = 0$; Z(M) désigne le centre de M.

Théorème 2.9. *Soient M une algèbre de Malcev résoluble sur un corps commutatif K de caractéristique zéro, ρ une représentation spéciale finie de M dans un K espace vectoriel V. S'il existe x_o et y_o dans M tels que $\rho_{x_oy_o}$ soit un endomorphisme injectif de V, alors les assertions suivantes sont équivalentes : 1) $V^M = 0$; 2) $H^1(M,V) = 0$.*

En effet d'après la proposition 2.7 on a 2) $\Rightarrow$ 1). Supposons maintenant que $V^M = 0$. Si M est de Lie, il vient de [1] que $V^M = 0$ équivaut à $H^1(M,V) = 0$. Si M n'est pas de Lie, alors d'après le Lemme 2.5, il existe une application injective ψ de $H^1(M/N(M),V^{N(M)})$ dans $H^1(M,V)$ telle que $q\circ \phi^* = \psi \circ \pi$. Soit donc $f \in \operatorname{Ker} \delta_1$; alors quels que soient x et y dans M et z dans N(M) on a $0 = \delta_1(f)(x,y,z) = \rho_z(\rho_y(f(x))) - \rho_z(\rho_x(f(y))) - \rho_z(f(xy)) + D(x,y)(f(z)) - f([xyz])$. Ainsi si $V^{N(M)} = V$, il vient que $D(x,y)(f(z)) = f([xyz])$. Puisque $[xyz] = x(yz) - y(xz) + (xy)z$ et $J(x,y,z) = 0$,

alors $[xyz] = [xyz] + J(x,y,z) = 2\,(xy)z = 2\,L_{xy}(z)$. Par suite $\rho_{xy}(f(z)) = f(L_{xy}(z))$. Il en résulte que pour tout $n \geq 1$ on a : $\rho^n_{xy}(f(z))=f(L^n_{xy}(z))$. M étant résoluble sur un corps de caractéristique zéro, M^2 est alors nilpotente (cf [7]). D'où pour n assez grand on a $\rho^n_{xy}(f(z)) = 0$ pour tout x et tout y dans M. Ainsi s'il existe x_o et y_o dans M tels que $\rho_{x_oy_o}$ soit injectif, on a $f(z) = 0$ et ceci nous dit que $z \in \operatorname{Ker} f$. Il existe alors une et une seule application linéaire $\overline{f}$ de $M/N(M)$ dans $V^{N(M)} = V$ telle que $\overline{f} \circ p = f$ où p désigne la surjection canonique de M sur $M/N(M)$. De plus $\phi^*(\overline{f}) = f$. Il en résulte que ϕ^* est surjective. D'où ψ est un isomorphisme de $H^1(M/N(M), V^{N(M)})$ dans $H^1(M,V)$.
Raisonnons maintenant par récurrence sur la dimension de M.
Si $\dim_K M = 4$, $M/N(M)$ est une K-algèbre de Lie résoluble et par suite $0 = V^M = (V^{N(M)})^{M/N(M)}$ nous dit que $H^1(M,V) = 0$. Supposons donc que 2) $\Rightarrow$ 1) pour toute K-algèbre de Malcev résoluble de dimension strictement inférieure à celle de M. Comme $0 = V^M = (V^{N(M)})^{M/N(M)}$ et $\dim_K M/N(M) < \dim_K M$ (car M étant résoluble on a $N(M) \neq 0$; cf. [4]) il vient de l'hypothèse de récurrence de $H^1(M/N(M),V) = 0$. Par suite $H^1(M,V) = 0$.

Supposons maintenant que $V^{N(M)} \subset V$. Comme $(V^{N(M)})^{N(M)} = V^{N(M)}$ on a d'après ce qui précède $H^1(M/N(M),V^{N(M)})$ est isomorphe à $H^1(M,V^{N(M)})$. Puisque $0 = V^M = (V^{N(M)})^{M/N(M)}$, alors toujours d'après ce qui précède on a $H^1(M/N(M),V^{N(M)}) = 0$ et par suite $H^1(M,V^{N(M)}) = 0$. De l'exactitude de la suite (cf.théorème 2.3)

$$0 \longrightarrow (V^{N(M)})^M \longrightarrow V^M \longrightarrow (V/V^{N(M)})^M \longrightarrow H^1(M,V^{N(M)})$$

on a $(V/V^{N(M)})^M = 0$. Raisonnons maintenant par récurrence sur la dimension de V. Si $\dim_K V = 2$, alors $\dim_K V^{N(M)} = 1$. Il en résulte que $(V/V^{N(M)})^{N(M)} = V/V^{N(M)}$. Par suite d'après tout ce qui précède on a : $H^1(M/N(M),V/V^{N(M)})$ isomorphe à $H^1(M,V/V^{N(M)})$. Comme $0 = (V/V^{N(M)})^M = ((V/V^{N(M)})^{N(M)})^{M/N(M)}$, alors toujours d'après tout ce qui précède on a $H^1(M/N(M),V/V^{N(M)}) = 0$. Par suite $H^1(M,V/V^{N(M)}) = 0$. De l'exactitude de la suite (cf théorème 2.3)
$H^1(M,V^{N(M)}) \longrightarrow H^1(M,V) \longrightarrow H^1(M,V/V^{N(M)})$ il vient que $H^1(M,V) = 0$.
Supposons donc que 2) $\Rightarrow$ 1) pour tout espace de représentation spéciale de dimension inférieure à celle de V. De l'exactitude de la suite

$$H^1(M,V^{N(M)}) \longrightarrow H^1(M,V) \longrightarrow H^1(M,V/V^{N(M)})$$

et de l'hypothèse de récurrence, il vient que $H^1(M,V) = 0$.

3. Un théorème de structure.

Soit M une algèbre de Malcev sur un corps commutatif K et f un endomorphisme de M. On dira que f est un opérateur symétrique pour la multiplication de M si pour tout x et tout y dans M, on a, $f(x)y + xf(y) = 0$. L'ensemble des opérateurs symétriques pour la multiplication de M est noté $\mathcal{S}_M$. Le radical résoluble de M est noté R(M).

Lemme 3.1. *Si M est une algèbre de Malcev semi-simple sans centre sur un corps commutatif K de caractéristique zéro, alors* $\mathcal{S}_M = 0$.

En effet, M étant semi-simple, sa forme de Killing β est non dégénérée. Quels que soient x, y, z dans M, f dans $\mathcal{S}_M$ on a :$\beta(f(x)y,z) = \beta(f(y)x,z) = \beta(f(y),xz) = \beta(xz,f(y)) = \beta(x,zf(y)) = -\beta(yf(z),x) = \beta(yf(z),x) = \beta(y,f(z)x) = -\beta(y,zf(x)) = -\beta(yz,f(x)) = -\beta(f(x),yz) = -\beta(f(x)y,z)$. Ceci nous dit alors que $\beta(f(x)y,z) = 0$. Par suite, $f(x)y = 0$ pour tout y dans M. Il en résulte que $f(x) \in Z(M)$ (centre de M). M étant sans centre, il vient que $f(x) = 0$ quel que soit x dans M, c'est-à-dire $f = 0$.

Théorème 3.2. *Soient M une algèbre de Malcev sur un corps commutatif K de caractéristique zéro et* $M = R(M) \oplus S$ *une décomposition de Levi de M. Si S est sans centre et M possède un opérateur symétrique inversible, alors M est résoluble.*

En effet soit $f \in \mathcal{S}_M$ avec f inversible. Désignons pour tout t dans M la projection de f(t) sur R(M) et S par $f(t)_1$ et $f(t)_2$ respectivement. Soit g l'endomorphisme de S défini par $g(x) = f(x)_2$ pour tout x dans S. Quels que soient x et y dans S on a $f(x)y + xf(y) = 0$. Par suite $f(x)_1 y + f(x)_2 y + xf(y)_1 + xf(y)_2 = 0$. D'où $f(x)_2y + xf(y)_2 = - f(x)_1y - xf(y)_1 \in R(M) \cap S = 0$ et par suite $f(x)_2 y + xf(y)_2 = 0$ c'est-à-dire $g(x)y + xg(y) = 0$. Il en résulte que $g \in \mathcal{S}_S$. Comme S est semi-simple sans centre, alors $g = 0$ d'après le Lemme3.1 Il s'en suit que $f(S) \subseteq R(M)$. Comme pour tout y dans M, z dans R(M) on a $f(y)z+yf(z) = 0$, alors $y_2f(z)_2 = 0$ pour tout y dans M. Par suite $yf(z)_2 = 0$ pour tout y dans S. D'où $f(z)_2 \in Z(S)$ (centre de S). S étant sans centre on a alors $f(z)_2 = 0$. Il en résulte que $f(R(M)) \subseteq R(M)$. Comme f est inversible, il vient des relations $f(S) \subseteq R(M)$ et $f(R(M)) \subseteq R(M)$ que $S = 0$. Par suite $M = R(M)$ et ceci montre que M est résoluble.

Bibliographie

1. D.W. Barnes, On the cohomology of soluble Lie algebras, *Math. Zeitoch.*, **101** (1967), 343-349.
2. C. Chevalley, *Théorie des groupes de Lie*, Hermann (1968).

3. J. Dixmier, Cohomologie des algèbres de Lie nilpotentes, *Acta. Sci. Math. szeged* **16** (1955), 246-250.
4. A. Elduque et A. A. El Malek , On the J-nucleus of a Malcev algebra, *Algebras, Groups and Geometries* **3** (1986), 493-503.
5. A. Koulibaly, Contribution à la théorie des algèbres de Malcev, *Cahiers Mathémathiques*, n° **33**, USTL, Montpellier (1985).
6. A. A. Sagle, Malcev algebras,*Trans amer. Math. Soc.* **101** (1961), 426-458.
7. E. L. Stitzinger, Supersolvable Malcev algebras, *J. Algebra* **103** (1986) 69-79.
8. K. Yamaguti, On the theory of Malcev algebra, *Kumainoto J. Sci.* Ser. A, **6**, n°1 (1963), 9-45.

AN IDENTITY ON SYMMETRIC ALGEBRAS OF LIE ALGEBRAS

FUJIO KUBO

Department of Mathematics
Kyushu Institute of Technology, Tobata, Kitakyushu, 804, Japan

ABSTRACT

We study the Lie algebras whose symmetric algebras satisfy an identity. Then we describe the structures of such Lie algebras.

Introduction.

Let L be a finite-dimensional Lie algebra with a Lie product [,] over an algebraically closed field of characteristic zero and $S(L)$ the symmetric algebra of L. The Lie product of L is canonically extended to that of $S(L)$ and $S(L)$ turns a Poisson algebra, which is the algebraic concept of the classical Poisson Lie algebra and in general an infinite-dimensional Lie algebra with an additional associative algebra structure related by the derivation property to the Lie bracket (K.H. Bhaskara and K. Viswanath [3]).

Kubo and Mimura [6] gave a bracket (,) on the Poisson algebra $S(L)$ with a Lie product [,] relative to an associative derivation D of $S(L)$ by $(a,b) = [a,b] + D(a)b - aD(b)$ $(a, b \in S(L))$. It is shown in [6] that $S(L)$ is a Lie algebra under the bracket (,) if and only if D is a Lie derivation of $S(L)$ and satisfies the identity $(*)$ $[a,b]D(c) + [b,c]D(a) + [c,a]D(b) = 0$ for any $a, b, c \in S(L)$.

In this paper we deal with the derivations D of $S(L)$ extended from derivations of L (hence the first condition in the above paragraph is automatically satisfied) satisfying the identity $(*)$ given above. We will show that if there exists a nonzero derivation of L satisfying the identity $(*)$ on L then L is the direct sum of its center and the 3-dimensional split simple Lie algebra, or L has an abelian ideal U of L such that L/U is the Heisenberg algebra of dimension 3 or the 2-dimensional nonabelian Lie algebra [Theorems 2.3, 3.3]. We will then give the descriptions of these Lie algebras in §4, 5.

1. Notations and preliminary results.

An associative commutative algebra A over a field of characteristic zero containing an identity, equipped with a Lie bracket $[\ ,\ ]$ such that $[ab,c]=a[b,c]+b[a,c]$ $(a,b,c\in A)$, is called a *Poisson algebra* (K.H.Bhaskara and K.Viswanath [3,p2]). Some examples of such algebras, which are not the integral domains, are found in Kubo and Mimura [5]. Let D be an associative derivation of A, and define a bracket $(\ ,\ )$ on A by $(a,b)=[a,b]+D(a)b-aD(b)$ $(a,b\in A)$. We call the algebra A with this bracket $(\ ,\ )$ the *D–extension* of the Poisson algebra $(A,[\ ,\ ])$ ([6]). Then the bracket $(\ ,\ )$ satisfies the Jacobi identity if and only if D satisfies the conditions: $D([a,b])=[D(a),b]+[a,D(b)]$ and $[a,b]D(c)+[b,c]D(a)+[c,a]D(b)=0$ for any $a,b,c\in A$ ([6,Theorem 8]).

Throughout this paper let $\mathbf{k}$ be an algebraically closed field of chracteristic zero, L a finite-dimensional Lie algebra over $\mathbf{k}$ with a Lie product $[\ ,\ ]$ and $S(L)$ the symmetric algebra of L. Let $\{x_1,\cdots,x_n\}$ be a basis of L. We regard $S(L)$ a Poisson algebra with a Lie product

$$[a,b]=\sum_{i,j}[x_i,x_j]\frac{\partial a}{\partial x_i}\frac{\partial b}{\partial x_j}.$$

Let $\mathrm{Der}(L)$ be the set of all derivations of L. For $D\in\mathrm{Der}(L)$ we write

$$J_D(a,b,c)=[a,b]D(c)+[b,c]D(a)+[c,a]D(b)$$

$(a,b,c\in S(L))$, where we extend D to the associative derivation of $S(L)$. Then we denote by

$$D^*(L)=\{D\in\mathrm{Der}(L)|J_D=0\text{ on }L\}.$$

As we see in the above paragragh the bracket $(\ ,\ )$ on $S(L)$, defined by $(a,b)=[a,b]+D(a)b-aD(b)$ $(a,b\in S(L))$, satisfies the Jacobi identity if and only if $J_D=0$ on L.

Let I be a Lie algebra over $\mathbf{k}$, K a Lie subalgebra of $\mathrm{Der}(I)$. We put $L=I\dot{+}K$ the direct sum of vectorspaces I and K. We define a Lie product on L making it into a Lie algebra by

$$[i+k,j+l]=[i,j]+k(j)-l(i)+[k,l]\quad(i,j\in I,k,l\in K).$$

This Lie algebra is called the *split extension* of I by K ([2,p22]).

We employ the following notations: $\sigma(L)$ and $\zeta(L)$ are the solvable radical and the center of L respectively. Let U be an ideal of L. For $x\in L$ we write $\mathrm{ad}_U x(u)=[u,x]$ $(u\in U)$ and $C_L(U)=\{x\in L|[x,U]=0\}$. For subspaces M,N of L, MN means the subspace $\{\sum_i m_in_i|m_i\in M,n_i\in N\}$ of $S(L)$. Triangular brakets $<\ ,\ >$ denote the Lie subalgebra generated by their contents. We write $L^2=[L,L]$.

We begin with the following

LEMMA 1.1. *Let $D \in D^*(L)$,M,N be subspaces of L and U an ideal of L Then*
(1) *if $[M, M] \neq 0$ and $[M, N] = 0$, then $D(N) = 0$.*
(2) *if $[M, M] \neq 0$ and $D(M) = 0$, then $D(L) = 0$.*
(3) *if L/U is nonabelian then $D(U) \subseteq U$.*
(4) *if U and L/U are nonabelian then $D(L) = 0$.*

PROOF. (1),(2): Take $x, y \in M$ with $[x, y] \neq 0$. By $J_D(x, y, z) = [x, y]D(z) = 0$ for any $z \in N$ (resp. $z \in L$) we have (1) (resp. (2)).

(3): Let M be a subspace of L such that $L = U + M, U \cap M = \{0\}$. Take $x, y \in M$ with $[x, y] \notin U$. Then for any $u \in U$, $J_D(x, y, u) \equiv [x, y]D(u) \pmod{UL}$. Thus $D(u) \in U$.

(4): Take $u, v \in U$ with $[u, v] \neq 0$. Then for any $x \in L$, $J_D(u, v, x) \equiv [u, v]D(x) \pmod{UU}$ by (3). Hence $D(L) \subseteq U$. Take $x, y \in L$ with $[x, y] \notin$ U. Then for any $w \in U$, $J_D(x, y, w) \equiv [x, y]D(w) \pmod{UU}$ and so $D(w) = 0$. Therefore $J_D(u, v, x) = [u, v]D(x) = 0$ for any $x \in L$, and we have $D(L) = 0$. ∎

2. Nonsolvable case.

In this section we shall obtain the structures of L such that $D^*(L) \neq 0$ for the case that L is nonsolvable.

LEMMA 2.1. *Let $D \in D^*(L)$. If S is a Levi factor of L, then $D(S) \subseteq S$.*

PROOF. It is nothing to prove for the case $S = 0$. If $S \neq 0$, then $D(\sigma(L)) \subseteq \sigma(L)$ by Lemma 1.1 (3). Putting $V = \sigma(L)$, we write $D(x) = x_V + x_S$ ($x_V \in V, x_S \in S$) for $x \in L$. Take any triple (h, e, f) of elements of S such that $[e, f] = h, [e, h] = 2e, [f, h] = -2f$. Then $J_D(e, f, h) = h(h_V + h_S) - 2f(e_V + e_S) - 2e(f_V + f_S) \equiv hh_V - 2fe_V - 2ef_V \pmod{SS}$. Hence $h_V = e_V = f_V = 0$, and $D(\{e, f, h\}) \subseteq S$. Since such triples generate S (Jacobson [4,Chapter IV]), $D(S) \subseteq S$. ∎

LEMMA 2.2. *Let $D \in D^*(L)$. If L is simple of dimension > 3, then $D(L) = 0$.*

PROOF. Let Φ be the root system for L relative to a Cartan subalgebra H of L, $L = H + \sum_{\alpha \in \Phi} \mathrm{k}e_\alpha$ the decomposition of L into the rootspaces and $\{\alpha_1, \ldots, \alpha_n\}$ a simple system of Φ. We choose the generators $h_i \in H, e_i \in \mathrm{k}e_{\alpha_i}, f_i \in \mathrm{k}e_{-\alpha_i}$ ($i = 1, \ldots, n$) for L such that $[h_i, h_j] = 0, [e_i, f_j] = \delta_{ij}h_i$ (δ_{ij} : Kronecker's δ),

$[e_i, h_j] = A_{ji}e_i$, $[f_i, h_j] = -A_{ji}f_i$ where $A_{ii} = 2, 0 \le A_{ij}A_{ji} < 4$ if $i \neq j$ ([4,Chapter IV]). Since $D \in \mathrm{Der}(L) = \mathrm{ad}_L L$, we write $D = \mathrm{ad}_L x, x = h + \sum_{\beta\in\Phi} b_\beta e_\beta$ ($b_\beta \in \mathbf{k}, h \in H$). Then for any $\alpha \in \Phi$

$$\begin{aligned} 0 &= J_D(h_i, h_j, e_\alpha) \\ &= [h_i, h_j][e_\alpha, x] + [h_j, e_\alpha][h_i, x] + [e_\alpha, h_i][h_j, x] \\ &= \sum_{\beta\in\Phi} b_\beta(\alpha(h_j)\beta(h_i) - \alpha(h_i)\beta(h_j))e_\alpha e_\beta. \end{aligned}$$

By our assumption, $n \ge 2$. Choose i, j with $A_{ij}A_{ji} < 4$. Then by $J_D(h_i, h_j, e_i) = \sum_{\beta\in\Phi} b_\beta(A_{ji}\beta(h_i) - 2\beta(h_j))e_i e_\beta = 0$ and $J_D(h_i, h_j, e_j) = \sum_{\beta\in\Phi} b_\beta(2\beta(h_i) - A_{ij}\beta(h_j))$ $e_j e_\beta = 0$, $b_\beta(4 - A_{ij}A_{ji})\beta(h_i) = 0$ for $i = 1, \ldots, n$. Hence if $b_\beta \neq 0$, then $\beta = 0$. This implies $x \in H$. Similarly by $J_D(h_i, e_i, e_j) = J_D(h_j, e_i, e_j) = 0$, we have $(4 - A_{ij}A_{ji})\alpha_i(x) = 0$ for $i = 1, \ldots, n$. Hence $x = 0$. Therefore $D = 0$. ∎

THEOREM 2.3. *Let L be a nonsolvable Lie algebra over* $\mathbf{k}$. *If $D^*(L) \neq 0$, then $L = \zeta(L) \oplus S$, where S is simple of dimension* 3.

PROOF. Let D be a nonzero element of $D^*(L)$. Take a Levi factor S of L and write $S = S_1 \oplus \cdots \oplus S_r$ (S_i: a simple ideal of S). If $r \ge 2, D(L) = 0$ by Lemma 1.1 (4), a contradiction. Therefore S is simple. Assume that $\dim S > 3$. Since $D(S) \subseteq S$ by Lemma 2.1, $D(S) = 0$ by Lemma 2.2. Whence $D(L) = 0$ by Lemma 1.1 (2), a contradiction. These shows that S is simple of dim 3, while $\sigma(L)$ is abelian by Lemma 1.1 (4).

Now write $S = \mathbf{k}h + \mathbf{k}e + \mathbf{k}f$ with $[e, f] = h, [e, h] = 2e, [f, h] = -2f$. Assume that $[\sigma(L), S] \neq 0$. Considered $\sigma(L)$ an S-module, there exists an irreducible S-submodule $V = \{v_0, \ldots, v_m\}$ of dimension ≥ 2 such that $[v_i, h] = (m - 2i)v_i, [v_i, e] = (-mi + i(i-1))v_{i-1}, [v_i, f] = v_{i+1}$ where $i = 0, \ldots, m, [v_0, e] = [v_m, f] = 0$ ([4,p85]). As $J_D(h, e, v_0) = -2eD(v_0) + mv_0D(e) = 0$, we can write $D(e) = 2pe, D(v_0) = pmv_0$ for some $p \in \mathbf{k}$ (Lemmas 1.1(3),2.1). Then $J_D(v_0, v_1, e)$ $= -pm^2v_0^2 = 0$ and so $p = 0$. Hence $D(e) = D(v_0) = 0$. On the other hand $J_D(h, f, v_0) = -v_1D(h) + mv_0D(f) = 0$ implies $D(h) = D(f) = 0$, because $D(h), D(f) \in S$ by Lemma 2.1. Therefore we have $D(S) = 0$ and $D(L) = 0$ (Lemma 1.1 (2)), a contradiction. ∎

3. Solvable case.

In this section we study the structures of L such that $D^*(L) \neq 0$ for the case that L is solvable.

LEMMA 3.1. *Let U be an abelian ideal of L. Then*
(1) *If $D \in D^*(L)$ then $[u,x]D(v) = [v,x]D(u)$ for $u,v \in U, x \in L$.*
(2) *$D^*(L) \neq 0$ then*

$$[v,x][u,y] = [v,y][u,x]$$

for $u,v \in U, x,y \in L$.

PROOF. (1): This follows from $J_D(u,v,x) = 0$ for $u,v \in U, x \in L$.

(2): Let D be a nonzero element of $D^*(L)$. Put $s = [v,x][u,y] - [v,y][u,x]$. We only consider the case that $[v,x] \neq 0$. For the case that $[v,y] \neq 0$ the proof will be done similarly. If $D(v) = 0$ then $D(U) = 0$ by (1). Hence by $J_D(v,x,y) = J_D(u,x,y) = 0$ we have $sD(x) = 0$. If $D(x) = 0$ then $D(L) = 0$ by Lemma 1.1 (2), a contradiction. Thus $D(x) \neq 0$ and so $s = 0$. Assume that $D(v) \neq 0$. If $D(v) = p[v,x]$ for some nonzero $p \in \mathbf{k}$ then $D(u) = p[u,x]$ by (1). Hence $J_D(v,u,y) = [u,y]D(v) + [y,v]D(u) = ps = 0$ and so $s = 0$. If $D(v)$ and $[v,x]$ are linearly independent then by (1) we write $[u,x] = q[v,x], D(u) = qD(v)$ for some $q \in \mathbf{k}$. Then by $J_D(u,v,y) = 0, [u,y] = q[v,y]$. This shows $s = 0$. ∎

The following lemma is an immediate consequence of the lemma 1.2 in Amayo [1].

LEMMA 3.2. *Let L be a nonabelian solvable Lie algebra over $\mathbf{k}$ and U an ideal of L maximal with respect to L/U being nonabelian. Then L/U is the 2-dimensional nonabelian Lie algebra or the Heisenberg algebra, with a basis $\{x_i, y_i, z | i = 1, \ldots, n\}$ and multiplications $[x_i, y_j] = \delta_{ij}z, [x_i, x_j] = [y_i, y_j] = [x_i, z] = [y_j, z] = 0$.*

THEOREM 3.3. *Let L be a nonabelian solvable Lie algebra over $\mathbf{k}$ and suppose that $D^*(L) \neq 0$ then there exists an abelian ideal U of L such that L/U is the Heisenberg algebra of dimension 3 or the 2-dimensional nonabelian Lie algebra.*

PROOF. Let D be a nonzero element of $D^*(L)$. Take an abelian ideal U maximal with respect to L/U being nonabelian. By Lemma 1.1 (3),(4), $D(U) \subseteq U$ and U is abelian.

Observing Lemma 3.2, we assume contrary that L/U is the Heisenberg algebra of dimension > 3. Take elements x_1, x_2, y_1, y_2, z of $L \setminus U$ such that $[x_1, y_1] = z + u_1, [x_2, y_2] = z + u_2$, the other Lie products of them belong to U. Since $J_D(x_j, y_j, x_i) \equiv zD(x_i)$ and $J_D(x_j, y_j, y_i) \equiv zD(y_i)$ (mod UL) for $i \neq j$, $D(x_i), D(y_i)$ $\in U$ for $i = 1, 2$. Hence $J_D(x_j, y_j, x_i) \equiv zD(x_i)$ and $J_D(x_j, y_j, y_i) \equiv zD(y_i)$ (mod UU), and so $D(x_i) = D(y_i) = 0$ for $i = 1, 2$. This shows that $D(L) = 0$ by Lemma 1.1 (2), a contradiction. ∎

REMARK. We can find a nonnilpotent Lie algebra L which satisfies the followings : (1) there exists an abelian ideal U of L such that L/U is the Heisenberg algebra of dimension 3; (2) there is no ideal V of L such that L/V is the 2-dimensional nonabelian Lie algebra.

Let L be a Lie algebra over $\mathbf{k}$ described in terms of a basis $\{v_1, v_2, x, y, z\}$ by the following multiplication table for the basis:

$$\begin{array}{lll} [v_1, x] = v_1, & [v_2, x] = v_1 + v_2, & [x, y] = z, \\ [v_1, y] = -v_1, & [v_2, y] = v_1 - v_2, & \end{array}$$

the product is zero if it is not in the table. Then $U =< v_1, v_2 >$ is an abelian ideal of L satisfying (1) given above. Let V be an ideal of L. It is easy to see that if $V \not\subseteq < v_1, v_2, z >$ then $U \subseteq V$. Thus every ideal V of L is in $< v_1, v_2, z >$ or contains U. Therefore L/V is of dimension ≥ 3 or nilpotent. This shows (2). In the next section we will see that if $D^*(L) \neq 0$ and L has the properties (1) and (2) given above then L is nilpotent.

4. Descriptions of structures (Nilpotent case).

In this section we assume that $D^*(L) \neq 0$. Let U be an abelian ideal of L such that L/U is the Heisenberg algebra of dimension 3 and suppose that there is no ideal V of L with L/V being the 2-dimensional nonabelian Lie algebra. Take elements $x, y, z \in L \backslash U$ with $[x, y] = z, [y, z] \equiv [z, x] \equiv 0 \pmod U$.

Assume that $[u, z] \neq 0$ for some $u \in U$ and let D be a nonzero element of $D^*(L)$. By $J_D(x, y, z) \equiv zD(z) \pmod{UL}$, $D(z) \in U$. Then observing $D(U) \subseteq U$, $J_D(u, x, z) \equiv [z, u]D(x), J_D(u, y, z) \equiv [z, u]D(y) \pmod{UU}$. Hence $D(\{x, y, z\}) \subseteq U$. Therefore $J_D(x, y, v) \equiv zD(v)$ for any $v \in U$ and $J_D(x, y, z) \equiv zD(z) \pmod{UU}$. This says that $D(U) = 0, D(z) = 0$. This contradicts Lemma 1.1 (2). Thus we have

$$[U, z] = 0.$$

4.1. *We study the structure of L for the case that $C_L(U) =< U, z >$.*

Take an element $w \in U$ with $[w, x] \neq 0$. Assume that $[w, y] = p[w, x]$ for some $p \in \mathbf{k}$. Then by Lemma 3.1 (2) we have $[u, y] = p[u, x]$ for any $u \in U$. Hence $y - px \in C_L(U)$, a contradiction. Thus

(4.1.1) $\quad [w, x]$ and $[w, y]$ are linearly independent.

Let u be an any element in U. Then by (4.1.1) and Lemma 3.1 (2), we can write

$$[u, x] = \alpha(u)[w, x], [u, y] = \alpha(u)[w, y]$$

for some linear function α on U. Hence $u - \alpha(u)w \in \zeta(L)$. Thus we have

(4.1.2) $\qquad U = (\zeta(L) \cap U) + < w >.$

Since $< U, z >$ is an abelian ideal of L, we have $[w, x][z, y] = [z, x][w, y]$ by Lemma 3.1 (2). Hence by (4.1.1) we can write

(4.1.3) $\qquad [y, z] = r[w, y], [z, x] = -r[w, x]$ for some $r \in \mathbf{k}$.

Let u be any element in U. Since $[u, x, y] = [u, y, x] \in < [w, x] > \cap < [w, y] > = 0, \alpha([u, x]) = 0$. Hence $[u, x, x] = \alpha([u, x])[w, x] = 0$. Similarly $[u, y, y] = 0$. Thus $[U, L] \subseteq \zeta(L)$. By (4.1.3) we have $L^2 \subseteq \zeta(L) + < z >$ and then

(4.1.4) $\qquad L^3 \subseteq \zeta(L)$.

This shows that L is nilpotent.

4.2. *We study the structure of L for the case that $L > C_L(U) >< U, z >$.*

Suppose $x \notin C_L(U)$. Take a nonzero $s \in C_L(U)$ and write $s = y - px \quad (p \in \mathbf{k})$. Choose $w \in U$ such that $[w, x] \neq 0$. Applying Lemma 3.1 (2) for the abelian ideal $< U, z >$, we have $[w, x][z, s] = [w, s][z, x] = 0$. Then we have

(4.2.1) $\qquad [z, s] = 0$.

Hence $C_L(U)$ is an abelian ideal of codimension 1.

Assume that ad$x$ is not nilpotent. Write $M = C_L(U)$ and let $M = \oplus_\alpha M_\alpha$ be a decomposion of M into the eigenspaces M_α relative to ad$x$. Choose a nonzero eigenvalue α. We consider the ascending chain of ad$x$-invariant subspaces $0 = M_\alpha^0 \subsetneq M_\alpha^1 \subsetneq \cdots \subsetneq M_\alpha^k = M_\alpha$ where $M_\alpha^i = \{v \in M_\alpha | (\text{ad}x - \alpha)^i v = 0\}$ $(i = 1, \cdots, k)$. Let $\{\overline{v_1}, \cdots, \overline{v_s}\}$ be a basis of $M_\alpha^k / M_\alpha^{k-1}$ and write $V = \sum_{i=2}^s \mathbf{k} v_i + \oplus_{\beta \neq \alpha} M_\beta$. Then V is an abelian ideal of L such that L/V is the 2-dimensional nonabelian Lie algebra, a contradiction. Therefore

(4.2.2) $\qquad$ ad$x$ is nilpotent.

By Engel's theorem L is nilpotent.

It is easy to see the structure of L for the case that $C_L(U) = L$. Gathering the above results togethere we have

THEOREM 4.1. *Let L be a finite-dimensional solvable Lie algebra over* $\mathbf{k}$. *Suppose that $D^*(L) \neq 0$ and that there is no ideal V of L such that L/V is the 2-dimensional Lie algebra. Then L is nilpotent and there exists an abelian ideal U of L such that*

(1) $L = U \oplus H$ *where* H *is the Lie algebra generated by* x, y, z *with multiplications* $[x, y] = z$, $H^2 \subseteq \zeta(H)$,

(2)$L = U \oplus H$, *where* H *is the Lie algebra with a basis* $\{v_1, v_2, v_3, x, y, z\}$ *and a multiplication table for the basis*

$$[x, y] = z, \quad [y, z] = rv_2, \quad [z, x] = -rv_3$$
$$[v_1, x] = v_3, \quad [v_1, y] = v_2$$

for some $r \in \mathbf{k}$, *the product is zero if it is not in the table, or*

(3) L *is the split extension* $U \dot{+} < \delta >$ *of* U *by* $< \delta >$ *where* δ *is a nilpotent linear transformation of* U.

5. Descriptions of structures (Nonnilpotent solvable case).

In this section we assume that $D^*(L) \neq 0$. Let U be an abelian ideal of L such that L/U is the 2-dimensional nonabelian Lie algebra. Take elements $x, y \in L \backslash U$ with $[x, y] \equiv x \pmod{U}$.

Assume that $[U, x] \neq 0$ and choose $w \in U$ with $[w, x] \neq 0$. If $[w, y] = p[w, x]$ for some $p \in \mathbf{k}$. By Lemma 3.1 (2) we have $[u, y] = p[u, x]$ for any $u \in U$. Hence $[w, x] = [w, x, y] - [w, y, x] = 0$, a contradiction. Thus

(5.1) $\qquad [w, x]$ and $[w, y]$ are linearly independent.

Similarly to the proof of (4.1.2) and (4.1.4), we have

(5.2) $\qquad U = (\zeta(L) \cap U) + < w >, \qquad [U, L] \subseteq \zeta(L).$

Write $[x, y] = x + u_0$ ($u_0 \in U$). Since $[u_0, y] \in \zeta(L)$ by (5.2) we have

(5.3) $\qquad [[x, y, y], y] = [x, y, y].$

It is easy to see the structures of L for the cases that $[U, x] = 0, [U, y] \neq 0$ and that $[U, L] = 0$. Therefore we have

THEOREM 5.1. *Let* L *be a finite-dimensional nonnilpotent Lie algebra over* $\mathbf{k}$. *Suppose that* $D^*(L) \neq 0$. *Then there exists an abelian ideal* U *of* L *such that*

(1) $L = U \oplus H$ *where* H *is the 2-dimensional nonabelian Lie algebra,*

(2) $L = U \oplus H$, *where* H *is the Lie algebra with a basis* $\{v_1, v_2, v_3, x, y, z\}$ *and a multiplication table for the basis*

$$[x, y] = x, \qquad [w, x] = v_2, \qquad [w, y] = v_3,$$

the product is zero if it is not in the table or

(3) *L is the split extension* $U \dot{+} < \delta >$ *of* U *by* $< \delta >$ *where* δ *is a nonnilpotent linear transformation of* U.

Acknowledgement.

We wish to express our gratitude to Professor Kiyoshi Yamaguch for hospitality in Hiroshima.

REFERENCES

1. R.K. Amayo, *On a class of finite-dimensional Lie algebras*, Proc. Amer. Math. Soc. **78** (1980),193-197

2. R.K. Amayo and I.N. Stewart, *Infinite-Dimensional Lie Algebras*, Noordhoff, Leyden,1974

3. K.H. Bhaskara and K. Vismanath, *Poisson Algebras and Poisson Manifolds*, Pitman, London,1988

4. N. Jacobson, *Lie Algebras*, Interscience, New York,1962

5. F. Kubo and F. Mimura, *Lie structures on differential algebras*, Hiroshima Math.J. **18** (1988),479-484

6. F. Kubo and F. Mimura, *Extensions of Poisson algebras by derivations*, Hiroshima Math.J., **20** (1990),37-46

Second fundamental form of a real hypersurface in a complex projective space

Sadahiro Maeda

Department of Mathematics
Nagoya Institute of Techonology
Gokiso, Shôwa, Nagoya, 466, JAPAN

ABSTRACT

The purpose of this paper is to study real hypersurfaces M in a complex projective space by using some conditions on the derivative of the second fundamental form of M.

0. Introduction

Let $P_n(C)$ be an n - dimensional complex projective space with Fubini-Study metric of constant holomorphic sectional curvature 4, and let M be a real hypersurface of $P_n(C)$. M has an almost contact metric structure (ϕ, ξ, η, g) induced from the complex structure of $P_n(C)$ (see, § 1). Many differential geometers have studied M (cf. [2], [7], [9], [11] and [14]) by using the structure (ϕ, ξ, η, g). Typical examples of real hypersurfaces in $P_n(C)$ are homogeneous ones. R. Takagi ([12]) showed that all homogeneous real hypersurfaces in $P_n(C)$ are realized as the tubes of constant radius over compact Hermitian symmetric spaces of rank 1 or rank 2. Namely, he showed the following:

Theorem T([12]). Let M be a homogeneous real hypersurface of $P_n(C)$. Then M is a tube of radius r over one of the following Kaehler submanifolds:

(A_1) hyperplane $P_{n-1}(C)$, where $0 < r < \pi/2$,
(A_2) totally geodesic $P_k(C)$ $(1 \leqq k \leqq n-2)$, where $0 < r < \pi/2$,
(B) complex quadric Q_{n-1}, where $0 < r < \pi/4$,
(C) $P_1(C) \times P_{(n-1)/2}(C)$, where $0 < r < \pi/4$ and $n(\geqq 5)$ is odd,
(D) complex Grassmann $G_{2,5}(C)$, where $0 < r < \pi/4$ and $n = 9$,
(E) Hermitian symmetric space $SO(10)/U(5)$, where $0 < r < \pi/4$ and $n = 15$.

Due to his classification, we find the number of distinct constant principal curvatures of a homogeneous real hypersurface is 2, 3 or 5. Here note that the vector ξ of any homogeneous real hypersurface M (which is a tube of radius r) is a principal curvature vector with principal curvature $\alpha = 2\cot 2r$ with multiplicity 1 (for further details, see [13]).

Now it is well-known that there does not exist a real hypersurface M with parallel second fundamental form A (, that is, $g((\nabla_X A)Y, Z) = 0$ for any vector fields X, Y and Z, where g and ∇ denote the induced

Riemannian metric and the induced Riemannian connection, respectively). So it is natural to investigate real hypersurfaces M by using some conditions (on the derivative of A) which are weaker than $\nabla A = 0$. Recently, from this point of view Kimura, Udagawa and the present author have studied real hypersurfaces M in $P_n(C)$ in terms of the derivative of the second fundamental form A (cf. [5], [6], [8]).

The purpose of this paper is to survey these three papers [5], [6] and [8].

1. Preliminaries

Let M be an orientable real hypersurface (with unit normal vector field N) of $P_n(C)$. The Riemannian connections $\widetilde{\nabla}$ in $P_n(C)$ and ∇ in M are related by the following formulas for arbitrary vector fields X and Y on M:

$$(1.\ 1)\quad \widetilde{\nabla}_X Y = \nabla_X Y + g(AX, Y)N,$$

$$(1.\ 2)\quad \widetilde{\nabla}_X N = -AX,$$

where g denotes the Riemannian metric of M induced from the Fubini-Study metric G of $P_n(C)$ and A is the shape operator of M in $P_n(C)$. An eigenvector X of the shape operator A is called a <u>principal curvature vector</u>. Also an eigenvalue λ of A is called a <u>principal curvature</u>. In what follows, we denote by V_λ the eigenspace of A associated with eigenvalue λ. It is known that M has an almost contact metric structure induced from the complex structure J on $P_n(C)$, that is, we define a tensor field ϕ of type (1, 1), a vector field ξ and a 1-form η on M by $g(\phi X, Y) = G(JX, Y)$ and $g(\xi, X) = \eta(X) = G(JX, N)$. Then we have

$$(1.\ 3)\quad \phi^2 X = -X + \eta(X)\xi,\ \ g(\xi, \xi) = 1,\ \ \phi\xi = 0.$$

It follows from (1. 1) that

$$(1.\ 4)\quad (\nabla_X \phi)Y = \eta(Y)AX - g(AX, Y)\xi,$$

$$(1.\ 5)\quad \nabla_X \xi = \phi AX.$$

Let $\widetilde{R}$ and R be the curvature tensors of $P_n(C)$ and M, respectively. Since the curvature tensor $\widetilde{R}$ has a nice form, we have the following Gauss and Codazzi equations:

$$(1.\ 6)\quad g(R(X, Y)Z, W) = g(Y, Z)g(X, W) - g(X, Z)g(Y, W) + g(\phi Y, Z)g(\phi X, W)$$
$$- g(\phi X, Z)g(\phi Y, W) - 2g(\phi X, Y)g(\phi Z, W)$$

$$+ g(AY, Z)g(AX, W) - g(AX, Z)g(AY, W),$$

(1. 7) $(\nabla_X A)Y - (\nabla_Y A)X = \eta(X)\phi X - \eta(Y)\phi X - 2g(\phi X, Y)\xi.$

From (1. 3) and (1. 6) we get

(1. 8) $SX = (2n + 1)X - 3\eta(X)\xi + hAX - A^2X,$

where h = trace A, S is the Ricci tensor of type (1, 1) on M and I is the identity map.

In [5] we introduced the following notion: The second fundamental form is *η-parallel* if $g((\nabla_X A)Y, Z) = 0$ for any X, Y and Z which are orthogonal to ξ. We here note that there exist real hypersurfaces with η-parallel second fundamental form in $P_n(C)$.

In §2 we study real hypersurfaces M of $P_n(C)$ by using the notion "A is η-parallel" (cf. Theorem 1 and Theorem 2). In §3 we investigate M by using a condition similar to that of the η-parallel of A (cf. Theorem 3). In §4 we classify real hypersurfaces M with $\nabla_\xi A = 0$, that is, the second fundamental form A is parallel in the direction of ξ (cf. Theorem 4). In the following, we use the same terminology and notations as above unless otherwise stated. Now we prepare without proof the following in order to prove our results:

Theorem K ([3]). *Let M be a real hypersurface of $P_n(C)$. Then M has constant principal curvatures and ξ is a principal curvature vector if and only if M is locally congruent to a homogeneous real hypersurface.*

Proposition 1 ([9]). *If ξ is a principal curvature vector, then the corresponding principal curvature α is locally constant.*

Proposition 2 ([9]). *Assume that ξ is a principal curvature vector and the corresponding principal curvature is α. If $AX = rX$ for $X \perp \xi$, then we have $A\phi X = ((\alpha r + 2)/(2r - \alpha))\phi X$.*

Proposition 3 ([9]). *Let M be a real hypersurface of $P_n(C)$. Then the following are equivalent:*

(i) *M is locally congruent to one of homogeneous ones of type A_1 and A_2.*

(ii) $g((\nabla_X A)Y, Z) = -\eta(Y)g(\phi X, Z) - \eta(Z)g(\phi X, Y)$ *for any vector fields X, Y and Z on M.*

Proposition 4 ([5]). *Let M be a real hypersurface of $P_n(C)$. Then the following are equivalent:*

(i) *The holomorphic distribution* $T^0M = \{ X \in TM \mid \eta(X) = 0 \}$ *is integrable.*

(ii) $g((\phi A + A\phi)X, Y) = 0$ for any X, $Y \in T^0M$

Proposition 5 ([8]). Let M be a real hypersurface of $P_n(C)$. Suppose that ξ is a principal curvature vector and the corresponding principal curvature is nonzero. If $\nabla_\xi A = 0$, then M is a tube of radius r over one of the following Kaehler submanifolds:
(A_1) hyperplane $P_{n-1}(C)$, where $0 < r < \pi/2$ and $r \neq \pi/4$,
(A_2) totally geodesic $P_k(C)$ $(1 \leq k \leq n-2)$, where $0 < r < \pi/2$ and $r \neq \pi/4$.

Proposition 6 ([8]). Let M be a real hypersurface of $P_n(C)$. Then "$A\xi = 0$" implies "$\nabla_\xi A = 0$".

Proposition 7 ([1]). Let M be a connected orientable real hypersurface (with unit normal vector N) in $P_n(C)$ on which ξ is a principal curvature vector with principal curvature $\alpha = 2 \cot 2r$. Then the following hold:
(i) M lies on a tube (in the direction of $\eta = \gamma'(r)$, where $\gamma(r) = \exp_x(rN)$ and x is a base point of the normal vector N) of radius r over a certain Kaehler submanifold $\widetilde{N}$ in $P_n(C)$.
(ii) Let $\cot\theta$ be a principal curvature of the shape operator A_η at $y = \gamma(r)$ of the Kaehler submanifold $\widetilde{N}$. Then the real hypersurface M has a principal curvature $\cot(\theta - r)$ at $x = \gamma(0)$.

Proposition 8 ([6]). Let M be a real hypersurface of $P_n(C)$. Then "$\nabla_\xi A = 0$" implies "ξ is a principal curvature vector of M".

2. Real hypersurfaces M with η-parallel second fundamental form A

Our aim here is to prove the following

Theorem 1 ([5]). Let M be a real hypersurface of $P_n(C)$. Then the second fundamental form of M is η-parallel and ξ is a principal curvature vector if and only if M is locally congruent to one of homogeneous real hypersurfaces of type A_1, A_2 and B.

Outline of Proof. Suppose that ξ is a principal curvature vector. Now we shall show that the real hypersurface M with η-parallel second fundamental form must be homogeneous. Let X be a principal curvature (unit) vector orthogonal to ξ with principal curvature r. For any Y $(\perp \xi)$, we get

$$(\nabla_Y A)X = \nabla_Y(AX) - A(\nabla_Y X)$$

$$= (Yr)X + (rI - A)\nabla_Y X.$$

So we have

$$g((\nabla_Y A)X, X) = Yr + g((rI - A)\nabla_Y X, X)$$

$$= Yr + g(\nabla_Y X, (rI - A)X) = Yr.$$

This, together with the hypothesis that the second fundamental form is η-parallel, shows

(2. 1) $Yr = 0$ for any $Y(\perp \xi)$.

Now it follows from the Codazzi equation (1. 7) that

$$(\nabla_X A)\xi - (\nabla_\xi A)X = -\phi X \quad \text{for any } X(\perp \xi).$$

On the other hand, from Propositions 1 and 2 we find

$$(\nabla_X A)\xi - (\nabla_\xi A)X = \nabla_X(A\xi) - A\nabla_X \xi - \nabla_\xi(AX) + A(\nabla_\xi X)$$

$$= r\{\alpha - (\alpha r + 2)/(2r - \alpha)\}\phi X - (\xi r)X$$

$$- (rI - A)\nabla_\xi X$$

for any unit vector $X(\varepsilon V_r)$ which is orthogonal to ξ.
Since $g(\phi X, X) = 0$ and $g((rI - A)\nabla_\xi X, X) = 0$, the above calculation yields

(2. 2) $\xi r = 0$.

And hence the equations (2. 1) and (2. 2) show that r must be constant. Therefore by virtue of Proposition 1 and Theorem K we can see that our real hypersurface must be homogeneous. Next we shall show that the second fundamental form of a real hypersurface of type A_1, A_2 and B is η-parallel: As an immediate consequence of Proposition 3 we find that the second fundamental form of a real hypersurface of type A_1 and A_2 is η-parallel. Now let M be of type B. Then M has three distinct constant principal curvatures (say, r_1, r_2 and α), so that $T_pM = V_{r_1} \oplus V_{r_2} \oplus \{\xi\}_R$ at any point p of M. Here we choose a local field of orthonormal frames $e_1, \ldots, e_{n-1}$ (resp. $\phi e_1, \ldots, \phi e_{n-1}$) is an orthonormal basis of V_{r_1} (resp. V_{r_2})(for details, see [13]).
A direct computation shows the following:

(2. 3) $(\nabla_{e_i} A)\phi e_j = r_1(\alpha - r_2)\delta_{ij}\xi$,

(2. 4) $(\nabla_{\phi e_i} A)e_j = r_2(r_1 - \alpha)\delta_{ij}\xi$,

(2. 5) $(\nabla_{e_i} A)e_j = (\nabla_{\phi e_i} A)\phi e_j = 0$ for $1 \leq i, j \leq n-1$.

Therefore the equations (2. 3), (2. 4) and (2. 5) assert that the second fundamental form of M is η-parallel. So the rest of Proof is to show that the second fundamental form of M, which is congruent to a real hypersurface of type C, D and E, is not η-parallel. Suppose that the second fundamental form of M is η-parallel. We here review the following: Our real hypersurface M has five distinct constant principal curvatures (say, r_1, r_2, r_3, r_4 and α), so that $T_pM = V_{r_1} \oplus V_{r_2} \oplus V_{r_3} \oplus V_{r_4} \oplus \{\xi\}_R$ at any point p of M. Let $x = \cot\theta$ $(0 < \theta < \pi/4)$. Then we may write

(2. 6) $r_1 = x$, $r_2 = -1/x$, $r_3 = (1 + x)/(1 - x)$, $r_4 = (x - 1)/(x + 1)$ and $\alpha = x - 1/x$.

Now we choose $X(\varepsilon V_s)$ and $Y(\varepsilon V_t)$, where $s, t = r_1, r_2, r_3, r_4$.
From the hypothesis that the second fundamental form of M is η-parallel, we see

$$(\nabla_X A)Y = \nabla_X(tY) - A(\nabla_X Y) = (tI - A)\nabla_X Y \varepsilon \{\xi\}_R$$

so that $\nabla_X Y \varepsilon V_t \oplus \{\xi\}_R$. Here and in the following we denote by $(*)_\lambda$ the V_λ-component of $(*)$. Then we have

$$\nabla_X Y = (\nabla_X Y)_t + g(\nabla_X Y, \xi)\xi = (\nabla_X Y)_t - g(Y, \nabla_X \xi)\xi$$

$$= (\nabla_X Y)_t - g(Y, \phi AX)\xi \quad \text{(by virtue of (1. 5))} \quad \text{so that}$$

(2. 7) $\nabla_X Y = (\nabla_X Y)_t - s\cdot g(\phi X, Y)\xi$ for $X \varepsilon V_s$ and $Y \varepsilon V_t$.

For any $X(\varepsilon V_r$, $r = r_1, r_2, r_3, r_4)$, from (1. 5), Propositions 1 and 2 we get

$$(\nabla_X A)\xi - (\nabla_\xi A)X = \nabla_X(\alpha\xi) - A(\nabla_X \xi) - \nabla_\xi(rX) + A\nabla_\xi X$$

$$= r\{\alpha - (\alpha r + 2)/(2r - \alpha)\}\phi X - (rI - A)\nabla_\xi X.$$

On the other hand, from (1. 7) we see

$$(\nabla_X A)\xi - (\nabla_\xi A)X = -\phi X \quad \text{so that}$$

(2. 8) $(rI - A)\nabla_\xi X = \{r(\alpha - (\alpha r + 2)/(2r - \alpha)) + 1\}\phi X$ for $X(\varepsilon V_r$, $r = r_1, r_2, r_3, r_4)$.

And hence we find

$\nabla_\xi X \;\varepsilon\; V_r \oplus \{\phi X\}_R$ for $X(\varepsilon V_r,\; r = r_3, r_4)$.

Now it follows from (2. 8) that

$$\{r - (\alpha r + 2)/(2r - \alpha)\}g(\nabla_\xi X, \phi X) = \{r(\alpha - (\alpha r + 2)/(2r - \alpha)) + 1\} \cdot g(\phi X, \phi X).$$

Therefore a straightforward computation yields

(2. 9) $g(\nabla_\xi X, \phi X) = (\alpha/2)g(\phi X, \phi X)$ for $X(\varepsilon V_r,\; r = r_3, r_4)$.

Let $X \varepsilon V_{r_1}$ and $Z \varepsilon V_{r_3}$. Then $\phi X \varepsilon V_{r_1}$ and $\phi Z \varepsilon V_{r_4}$ from Proposition 2. So the Gauss equation (1. 6) gives

(2. 10) $g(R(X, \phi X)Z, \phi Z) = -2g(\phi X, \phi X) \cdot g(\phi Z, \phi Z)$.

On the other hand, the equation (2. 7), together with (2. 9), shows

$$g(R(X, \phi X)Z, \phi Z) = r_1 \alpha \cdot g(\phi X, \phi X) \cdot g(\phi Z, \phi Z).$$

This, combined with (2. 6) and (2. 10), tells us that $x\{x - (1/x)\} = -2$, that is, $x^2 = -1$. This is a contradiction. Q. E. D.

Remark 1. If we omit the hypothesis that ξ is a principal curvature vector, Theorem 1 is not true.

The following example is worth mentioning.

Example. We shall construct a ruled real hypersurface M in $P_n(C)$ (with complex structure J), that is, we shall construct M in which the distribution $T^0M = \{ X \varepsilon TM \mid \eta(X) = 0 \}$ is integrable and its integral manifold is a totally geodesic submanifold $P_{n-1}(C)$. Now let $\gamma(t)$ $(t \varepsilon I)$ be an arbitrary (regular) curve in $P_n(C)$. Then for every t (εI) there exists a totally geodesic submanifold $P_{n-1}(C)$ (in $P_n(C)$) which is orthogonal to the plane τ_t spanned by $\{\gamma'(t), J\gamma'(t)\}$. We here denote by $P^{(t)}_{n-1}(C)$ such a totally geodesic submanifold $P_{n-1}(C)$. Let $M = \{x \varepsilon P^{(t)}_{n-1}(C) \mid t \varepsilon I\}$. Then the construction of M asserts that M is a ruled real hypersurface. Moreover, the construction of M tells us that there are many ruled real hypersurfaces. Here we point out that the mean curvature of any ruled real hypersurface M except the minimal ruled real hypersurface is not constant. Now, for later use we shall write down the shape operator A of a ruled real hypersurface M in $P_n(C)$ (for details, see [4]):

$$A\xi = \mu\xi + \nu U \quad (\nu \neq 0),$$
(2. 11) $$AU = \nu\xi,$$

$$AX = 0 \qquad\qquad (\text{for any } X \perp \xi, U),$$

where U is a unit vector orthogonal to ξ, μ and ν are differentiable functions on M.

So the equation (2. 11) shows that the vector field ξ of any ruled real hypersurface is not principal. Nevertheless the second fundamental form of any ruled real hypersurface M is η-parallel (for details, see Proposition 4 in [5]).

Now we shall provide a characterization of a ruled real hypersurface M in $P_n(C)$.

Theorem 2 ([5]). Let M be a real hypersurface of $P_n(C)$. Then the second fundamental of M is η-parallel and the holomorphic distribution $T^0M = \{ X \varepsilon TM \mid \eta(X) = 0 \}$ is integrable if and only if M is locally congruent to a ruled real hypersurface of $P_n(C)$.

Proof. Suppose that the holomorphic distribution T^0M is integrable and the second fundamental form of M is η-parallel. Now it follows from Proposition 4 that

$$(2.\ 12)\quad g(AX, \phi Y) = g(\phi X, AY) \quad \text{for any } X,\ Y \varepsilon T^0M.$$

And hence, in particular, for any $Z(\varepsilon T^0M)$ we get

$$Z(g(AX, \phi Y)) = Z(g(\phi X, AY)).$$

Thus, for any X, Y and $Z \varepsilon T^0M$, we have

$$(2.\ 13)\quad g((\nabla_Z A)X + A\nabla_Z X, \phi Y) + g((\nabla_Z \phi)Y + \phi(\nabla_Z Y), AX)$$

$$= g((\nabla_Z \phi)X + \phi\nabla_Z X, AY) + g((\nabla_Z A)Y + A\nabla_Z Y, \phi X).$$

Here it follows from the hypothesis and (1. 4) that

$$(2.\ 14)\quad g((\nabla_Z A)X, \phi Y) = g((\nabla_Z A)Y, \phi X) = 0,$$

$$(2.\ 15)\quad (\nabla_Z \phi)X = - g(AZ, X)\xi$$

and

$$(2.\ 16)\quad (\nabla_Z \phi)Y = - g(AZ, Y)\xi \quad \text{for any } X,\ Y \text{ and } Z \varepsilon T^0M.$$

Now we put $\nabla_Z X = (\nabla_Z X)_0 + g(\nabla_Z X, \xi)\xi$, where $(*)_0$ denotes the T^0M-component of $(*)$. Then, from (2. 12) we see

$$(2.17)\ g(A(\nabla_Z X)_0, \phi Y) = g(\phi(\nabla_Z X)_0, AY)$$

and

$$(2.18)\ g(A(\nabla_Z Y)_0, \phi X) = g(\phi(\nabla_Z Y)_0, AX) \quad \text{for any } X, Y, Z \in T^0M.$$

Substituting (2. 14)~(2. 18) into (2. 13), we find

$$(2.19)\ g(\nabla_Z X, \xi)g(A\xi, \phi Y) - g(AZ, Y)g(\xi, AX) + g(\nabla_Z Y, \xi)g(\phi\xi, AX)$$
$$= -g(AZ, X)g(\xi, AY) + g(\nabla_Z X, \xi)g(A\xi, \phi Y) + g(\nabla_Z Y, \xi)g(A\xi, \phi X).$$

Thus, from (1. 3), (1. 5) and (2. 19) we obtain

$$(2.20)\ g(X, \phi AZ)g(A\xi, \phi Y) + g(AZ, Y)g(AX, \xi)$$
$$= g(AZ, X)g(\xi, AY) + g(Y, \phi AZ)g(A\xi, \phi X) \quad \text{for any } X, Y, Z \in T^0M.$$

We put

$$(2.21)\ A\xi = \mu\xi + \nu U, \text{ where } \xi \text{ and } U \text{ are orthonormal.}$$

Here we may suppose that $\nu \neq 0$ (in local).
(Suppose that $\nu = 0$, that is, ξ is a principal curvature vector. This, together with the assertion (ii) in Proposition 4, implies that $\phi A + A\phi = 0$, which is a contradiction (cf. Lemma 2. 3 of [9])). Then, from (1. 3), (2. 20) and (2. 21) we find

$$(2.22)\ g(X, \phi AZ)g(U, \phi Y) + g(AZ, Y)g(U, X)$$
$$= g(AZ, X)g(U, Y) + g(Y, \phi AZ)g(U, \phi X).$$

Putting $Y = U$ in (2. 22), we get

$$(2.23)\ g(AZ, U)g(U, X) = g(AZ, X) + g(U, \phi AZ)g(U, \phi X).$$

And we put $X = \phi U$ in (2. 23) so that

$$0 = g(AZ, \phi U) - g(U, \phi AZ) = 2g(A\phi U, Z),$$

that is, $A\phi U$ is proportional to ξ. Moreover, from (2. 21) we see

$g(A\phi U, \xi) = g(\phi U, \mu\xi + \nu U) = 0$ so that

$$(2.24)\ A\phi U = 0.$$

It follows from (2. 23) and (2. 24) that

(2. 25) $g(AZ, U)g(U, X) = g(AZ, X)$ for any $X, Z \varepsilon T^0M$.

Then the equation (2. 25) implies that

$g(AX, Z) = 0$ for any $X(\perp U)$ and $Z \varepsilon T^0M$.

On the other hand, from (2. 21) we have

$g(AX, \xi) = g(\mu\xi + \nu U, X) = 0$ for any $X(\perp U) \varepsilon T^0M$.

And hence

(2. 26) $AX = 0$ for any $X(\perp U) \varepsilon T^0M$.

Now, putting $X = U$ and $Y = \phi U$ in (2. 12), from (2. 24) we get

$g(AU, U) = 0$.

Moreover, from (2. 26) we see

$g(AU, Z) = 0$ for any $Z(\perp U) \varepsilon T^0M$.

These, combined with (2. 21), imply that

(2. 27) $AU = \nu\xi$.

Thus, from (1. 6), (2. 26) and (2. 27) we find that

$g(R(X, \phi X)\phi X, X) \equiv 4$ for any $X \varepsilon T^0M$.

Therefore we conclude that our real hypersurface M must be a ruled real hypersurface (cf. Theorem 1 of [4]). Q. E. D.

Remark 2. Any ruled real hypersurface in $P_n(C)$ is not complete.

Remark 3. The holomorphic distribution T^0M of any homogeneous real hypersurface in $P_n(C)$ is not integrable (cf. Proposition 4).

3. Real hypersurfaces M with $\nabla^0 A^0 = 0$

Let M be a real hypersurface of $P_n(C)$. Then TM is a subbundle of $TP_n(C)$ over M and $T^0M = \{ X \varepsilon TM \mid \eta(X) = 0 \}$ is a subbundle of TM. Thus each of TM and T^0M has a metric connection induced from $TP_n(C)$. The orthogonal complement of T^0M in $TP_n(C)$ with respect to the metric on $TP_n(C)$ is denoted by N^0M, which is also a subbundle of $TP_n(C)$ with the induced metric connection.

Denote by ∇^0 and $\nabla^\perp$ the connections of T^0M and N^0M, respectively. Now we have

$$(3.1)\quad \nabla_X Y = \overset{\circ}{\nabla}_X Y + A_1(X)(Y),$$

$$(3.2)\quad \widetilde{\nabla}_X Y = \overset{\circ}{\nabla}_X Y + A_2(X)(Y) \quad \text{for any } Y \in C^\infty(T^0M) \text{ and } X \in TM,$$

where A_1 and A_2 are the second fundamental forms of the subbundle T^0M in TM and $TP_n(C)$, respectively. Note that the second fundamental form of TM in $TP_n(C)$ coincides with the ordinary second fundamental form of the immersion $M \to P_n(C)$. A_2 is interpreted as a smooth section of $\mathrm{Hom}(TM, \mathrm{Hom}(T^0M, N^0M))$. Set $A^0 = A_2|_{T^0M}$, which is a smooth section of $\mathrm{Hom}(T^0M, \mathrm{Hom}(T^0M, N^0M))$. Note that any ruled real hypersurfaces in $P_n(C)$ may be characterized by the condition $A^0 \equiv 0$. We here consider the covariant derivative of A^0 with respect to the connection on $\mathrm{Hom}(T^0M, \mathrm{Hom}(T^0M, N^0M))$ induced from $TP_n(C)$.

First of all we show the following fundamental relations.

Proposition 9 ([8]).
(i) $A_1(X)(Y) = -g(\phi AX, Y)\xi$,
(ii) $A_2(X)(Y) = g(AX, Y)N - g(\phi AX, Y)\xi$,
(iii) $\nabla^0\phi = 0$,
(iv) $\nabla^\perp_X \xi = g(AX, \xi)N$
(v) $\nabla^\perp_X N = -g(AX, \xi)\xi$,
<u>where</u> $X \in TM$ <u>and</u> $Y \in C^\infty(T^0M)$.

The connection on $\mathrm{Hom}(T^0M, \mathrm{Hom}(T^0M, N^0M))$ is also denoted by ∇^0. The covariant derivative of A^0 is defined by

$$(\overset{\circ}{\nabla}_X A^0)(Y)(Z) = \nabla^\perp_X A^0(Y)(Z) - A^0(\overset{\circ}{\nabla}_X Y)(Z) - A^0(Y)(\overset{\circ}{\nabla}_X Z)$$
for any $X \in TM$ and $Y, Z \in C^\infty(T^0M)$.

Now we state the following in order to prove Theorem 3.

Proposition 10 ([8]). <u>For any</u> $X \in TM$ <u>and</u> $Y, Z \in C^\infty(T^0M)$,
$$(3.3)\quad (\overset{\circ}{\nabla}_X A^0)(Y)(Z) = \Psi(X,Y,Z)N + \Psi(X,Y,\phi Z)\xi,$$
<u>where Ψ is the trilinear tensor defined by</u>
$$(3.4)\quad \Psi(X,Y,Z) = g((\nabla_X A)Y, Z) - \eta(AX)g(\phi AY, Z) - \eta(AY)g(\phi AX, Z) - \eta(AZ)g(\phi AX, Y).$$

Recall the definition of η-parallel of A. We say that A^0 is <u>η-parallel</u> if $\overset{\circ}{\nabla}_X A^0 = 0$ for any $X \in C^\infty(T^0M)$.

The main purpose of this section is to prove the following

Theorem 3 ([8]). Let M be a real hypersurface of $P_n(C)$. Assume that A^0 is η-parallel. Then M is locally congruent to one of the following:
(1) a homogeneous real hypersurface of type A_1,
(2) a homogeneous real hypersurface of type A_2,
(3) a homogeneous real hypersurface of type B,
(4) a real hypersurface in which T^0M is integrable and its integral manifold is a totally geodesic $P_{n-1}(C)$ (, that is, M is a ruled real hypersurface),
(5) a real hypersurface in which T^0M is integrable and its integral manifold is a complex quadric Q_{n-1}.

Proof. By Proposition 10, A^0 is η-parallel if and only if $\Psi(X,Y,Z) = 0$ for any X, Y, $Z \in C^\infty(T^0M)$, that is,

$$(3.\ 5)\quad g((\nabla_X A)Y, Z) = \eta(AX)g(\phi AY, Z) + \eta(AY)g(\phi AX, Z) + \eta(AZ)g(\phi AX, Y)$$

for any X, Y, $Z \in C^\infty(T^0M)$.

Therefore we must study real hypersurfaces (in $P_n(C)$) which satisfy (3. 5). Since the Codazzi equation (1. 7) tells us that $g((\nabla_X A)Y, Z)$ is symmetric for any X, Y and $Z(\in T^0M)$, exchanging X and Y in (3. 5), we obtain

$$g(Y, \phi AX)\eta(AZ) = g(X, \phi AY)\eta(AZ) \quad \text{so that}$$

$$(3.\ 6)\quad \eta(AZ)g((A\phi + \phi A)X, Y) = 0 \quad \text{for any } X,\ Y,\ Z(\in T^0M).$$

Now we assume that $\eta(AZ) = 0$ for any $Z(\in T^0M)$, that is, ξ is a principal curvature vector. Then the equation (3. 6) shows that $g((\nabla_X A)Y, Z) = 0$ for any $X, Y, Z(\in T^0M)$, that is, the second fundamental form A of M is η-parallel. And hence our real hypersurface M is locally congruent to one of homogeneous ones of type A_1, A_2 and B (cf. Theorem 1). Next we assume that ξ is not a principal curvature vector. Then the equation (3. 6) tells us that the holomorphic distribution T^0M is integrable (cf. Proposition 4). Of course the integral manifold M^0 of T^0M is a complex hypersurface (with complex structure ϕ) in $P_n(C)$. Moreover, the second fundamental form A^0 of M^0 is parallel (, which is equivalent to (3. 5)). Therefore we conclude that M^0 is locally congruent to $P_{n-1}(C)$ or Q_{n-1} (cf. [10]). Q. E. D.

Remark 4. Theorem 3 gives us the following:

Corollary 1. Let M be a real hypersurface of $P_n(C)$. Then A^0 is η-

parallel and ξ is a principal curvature vector if and only if M is locally congruent to one of homogeneous real hypersurfaces of type A_1, A_2 and B.

Corollary 2. Let M be a real hypersurface of $P_n(C)$. Then A^0 is η-parallel and the second fundamental form A of M is η-parallel if and only if M is locally congruent to one of homogeneous real hypersurfaces of type A_1, A_2 and B or a ruled real hypersurface.

Remark 5. The author does not know how to construct a real hypersurface M with $M^0 = Q_{n-1}$ (, that is, M is of case (5) in Theorem 3).

4. Real hypersurfaces M with $\nabla_\xi A = 0$

This section is devoted to the classification of real hypersurfaces M with $\nabla_\xi A = 0$ (, that is, the second fundamental form A is parallel in the direction of ξ) in $P_n(C)$. We have

Theorem 4 ([6]). Let M be a real hypersurface in $P_n(C)$. If $\nabla_\xi A = 0$, then M is locally congruent to one of the following:

(i) a non-homogeneous real hypersurface which lies on a tube of radius $\pi/4$ over a certain Kaehler submanifold $\tilde{N}$ in $P_n(C)$,

(ii) a homogeneous real hypersurface which lies on a tube of radius r over a totally geodesic $P_k(C)$ $(1 \leq k \leq n-1)$, where $0 < r < \pi/2$.

Proof. By virtue of Proposition 8 we find that ξ is a principal curvature vector (with principal curvature α). From Propositions 1 and 6, our discussion is divided into two cases: (i) $\alpha = 0$ and (ii) $\alpha \neq 0$.

Case of (i) $\alpha = 0$. Statement (i) of Proposition 7 asserts that our real hypersurface M lies on a tube of radius $\pi/4$ over a Kaehler submanifold $\tilde{N}$ in $P_n(C)$. But the converse is not true. Note that, in general a tube of radius $\pi/4$ over an arbitrary Kaehler submanifold $\tilde{N}$ is not a real hypersurface of $P_n(C)$. In fact, for example let $\tilde{N}$ be a complex quadric Q_{n-1}. Then a tube of radius $\pi/4$ over $\tilde{N}$ is $P^n(R)$ (which is the real part of $P_n(C)$) (cf. [3]). Statement (ii) of Proposition 7 shows the following:

Let $\tilde{N}$ be a Kaehler submanifold (with unit normal vector N) in $P_n(C)$. Suppose that the shape operator (with respect to N) A_N does not have the principal curvature 1. Then a tube (in the direction N) of radius $\pi/4$ over $\tilde{N}$ is a real hypersurface M. As a matter course the real hypersurface M admits the vector ξ as a principal curvature vector with principal curvature 0 (cf. [1]). Finally we remark that a homogeneous real hypersurface M with $A\xi = 0$ lies on a tube of radius $\pi/4$ over a totally geodesic $P_k(C)$ $(1 \leq k \leq n-1)$ (cf. [13]).

Case of (ii) $\alpha \neq 0$. See, Proposition 5. Q.E.D.

Remark 6. Proof of Theorem 4 shows that there exist many real hypersurfaces M's which are of case (i) in Theorem 4.

Remark 7. Theorem 4 gives us the following

Corollary 3. Let M be a real hypersurface of $P_n(C)$. Suppose that $A\xi \neq 0$. If $\nabla_\xi A = 0$, then M lies on a tube of radius r over a totally geodesic $P_k(C)$ $(1 \leq k \leq n-1)$, where $0 < r < \pi/2$ and $r \neq \pi/4$.

Acknowledgements

The author thanks Professor Kiyoshi YAMAGUTI for his devotion as a organizer of the Symposium.

References

1. T. E. Cecil and P. J. Ryan, Focal sets and real hypersurfaces in complex projective space, Trans. Amer. Math. Soc. 269(1982), 481-499.
2. U. H. Ki, H. Nakagawa and Y. J. Suh, Real hypersurfaces with harmonic Weyl tensor of a complex space form, Hiroshima Math. J. 20 (1990), 93-102.
3. M. Kimura, Real hypersurfaces and complex submanifolds in complex projective space, Trans. Amer. Math. Soc. 296(1986), 137-149.
4. M. Kimura, Sectional curvatures of holomorphic planes on a real hypersurface in $P_n(C)$, Math. Ann. 276(1987), 487-497.
5. M. Kimura and S. Maeda, On real hypersurfaces of a complex projective space, Math. Z. 202(1989), 299-311.
6. M. Kimura and S. Maeda, On real hypersurfaces of a complex projective space II, a preprint.
7. M. Kon, Real minimal hypersurfaces in a complex projective space, Proc. Amer. Math. Soc. 79(1980), 285-288.
8. S. Maeda and S. Udagawa, Real hypersurfaces of a complex projective space in terms of holomorphic distribution, Tsukuba J. Math. 14 (1990), 39-52.
9. Y. Maeda, On real hypersurfaces of a complex projective space, J. Math. Soc. Japan 28(1976), 529-540.
10. H. Nakagawa and R. Takagi, On locally symmetric Kaehler submanifolds in a complex projective space, J. Math. Soc. Japan 28(1976), 638-667.
11. M. Okumura, On some real hypersurfaces of a complex projective space, Trans. Amer. Math. Soc. 212(1975), 355-364.
12. R. Takagi, On homogeneous real hypersurfaces in a complex projective space, Osaka J. Math. 10(1973), 495-506.

13. R. Takagi, Real hypersurfaces in a complex projective space with constant principal curvatures I, II, J. Math. Soc. Japan 27(1975), 43-53, 507-516.
14. S. Udagawa, Bi-order real hypersurfaces in a complex projective space; Kodai Math. J. 10(1987), 182-196.

Key words. Second fundamental form, Real hypersurface, Complex projective space
1980 Mathematics subject classifications: (1985 Revision). 53B25, 53C40.

On Lie algebras in which the Frattini subalgebra is equal to the derived ideal

ATSUSHI MITSUKAWA
Department of Mathematics, Hiroshima University,*
Hiroshima, 730, Japan

Abstruct

Barnes has given the characterization of a finite dimensional nilpotent Lie algebra. In this paper, we shall generalize Barnes' result.

1. Introduction

The Frattini subalgebra $F(L)$ of a Lie algebra L is the intersection of the maximal subalgebras of L or is L if there are no maximal subalgebras. The Frattini ideal $\Phi(L)$ of L is the largest ideal of L contained in $F(L)$. Marshall[4] has investigated $F(L)$ of a finite dimensional Lie algebra L. Stitzinger[5–6] and Towers[8–10] have investigated finite dimensional Lie algebras with trivial Frattini subalgebras. On the other hand, Amayo[1], and Amayo and Stewart[2] have investigated not necessarily finite dimensional Lie algebras with good Frattini structure, which implies nilpotency of $\Phi(L)$. Barnes[3] has shown that when L is finite dimensional, $F(L) = L^2$ if and only if L is nilpotent.

In this paper we study a not necessarily finite dimensional Lie algebra L over a field $\mathfrak{k}$ of arbitrary characteristic. We shall generalize Barnes' result in order to give characterizations of nilpotency and local nilpotency in terms of the Frattini subalgebra. Let $\mathfrak{U}$ be the class of Lie algebras over $\mathfrak{k}$ in which the Frattini subalgebra is equal to the derived ideal. Then the followings are our main results: (1)

*Present address: Department of Mathematics, Fukuyama University, Higashimuramachi, Fukuyama, 729-02, Japan.

$\mathfrak{F}\mathfrak{N} \cap \mathfrak{U} = \mathfrak{N}$ (Theorem 4.3(1)). (2) $\mathrm{L(wser)}\mathfrak{F} \cap \mathfrak{U} = \mathrm{L}\mathfrak{N}$ (Theorem 4.3(2)). (3) $\mathrm{L}\mathfrak{N} \lneqq \mathrm{L}\mathfrak{F} \cap \mathfrak{U}$ (Theorem 5.1).

2. Notation and terminology

Throughout this paper $\mathfrak{k}$ is a field of arbitrary characteristic unless otherwise specified, and L is a not necessarily finite dimensional Lie algebra over $\mathfrak{k}$. Let $\bar{\mathfrak{k}}$ be an algebraic closure of $\mathfrak{k}$ and $L_{\bar{\mathfrak{k}}}$ be $L \otimes_{\mathfrak{k}} \bar{\mathfrak{k}}$. We mostly follow Amayo and Stewart[2] for the use of notation and terminology. In particular $H \leq L$ means that H is a subalgebra of L. H is a weakly serial subalgebra of L, denoted by H wser L, provided that for some totally ordered set Σ, there exists a collection $\{V_\sigma, \Lambda_\sigma \mid \sigma \in \Sigma\}$ of subspaces of L such that

(1) $H \subseteq \Lambda_\sigma$ and $H \subseteq V_\sigma$ for all $\sigma \in \Sigma$,

(2) $\Lambda_\tau \subseteq V_\sigma \subseteq \Lambda_\sigma$ if $\tau < \sigma$,

(3) $L \setminus H = \cup_{\sigma\in\Sigma}(\Lambda_\sigma \setminus V_\sigma)$,

(4) $[\Lambda_\sigma, H] \subseteq V_\sigma$ for all $\sigma \in \Sigma$.

We call $\{V_\sigma, \Lambda_\sigma \mid \sigma \in \Sigma\}$ the series from H to L. For a subset X of L, $\langle X \rangle$ denotes a subalgebra of L generated by X. For any $x, y \in L$ and $H \leq L$, we write $[y,_n x] = y(\mathrm{ad}\, x)^n$, $[H,_n x] = H(\mathrm{ad}\, x)^n$. For $\alpha \in \mathfrak{k}$, $L_\alpha(x) = \{y \in L \mid y(\mathrm{ad}\, x - \alpha 1_L)^n = 0$ for some $n \in \mathbb{N}\}$ and we write $L_*(x) = \bigoplus_{\alpha\neq 0} L_\alpha(x)$.

A class $\mathfrak{X}$ of Lie algebras over $\mathfrak{k}$ is a collection of Lie algebras over $\mathfrak{k}$ such that $\{0\} \in \mathfrak{X}$ and if $H \in \mathfrak{X}$ and $H \cong K$, then $K \in \mathfrak{X}$. $\mathrm{L}\mathfrak{X}$ (resp. $\mathrm{L(wser)}\mathfrak{X}$) is the class of Lie algebras such that any finite subset of L lies inside a subalgebra (resp. a weakly serial subalgebra) of L, belonging to $\mathfrak{X}$. Let $\mathfrak{X}$ and $\mathfrak{Y}$ be two classes of Lie algebras over $\mathfrak{k}$. Then $\mathfrak{X}\mathfrak{Y}$ is the class of Lie algebra L over $\mathfrak{k}$ having an ideal $I \in \mathfrak{X}$ such that $L/I \in \mathfrak{Y}$.

$\mathfrak{F}$, $\mathfrak{A}$, $\mathfrak{N}$ and $\mathfrak{E}$ denote respectively the classes of finite dimensional, abelian, nilpotent and Engel Lie algebras over $\mathfrak{k}$. $\mathrm{L}\mathfrak{F}$, $\mathrm{L}\mathfrak{N}$ and $\mathrm{L(wser)}\mathfrak{F}$ denote respectively the classes of locally finite, locally nilpotent and weakly serially finite Lie algebras over $\mathfrak{k}$. It is known that $\mathrm{L}\mathfrak{N} = \mathrm{L(wser)}\mathfrak{N} \lneqq \mathrm{L(wser)}\mathfrak{F}$ by Tôgô[7].

3. Preliminaries

We begin with quoting some facts of Amayo and Stewart[2]. An element x of L is a nongenerator if whenever $L = \langle X, x\rangle$ for a subset X of L it follows that $L = \langle X\rangle$.

LEMMA 3.1. *Let L be a Lie algebra. Then:*

(1) *The Frattini subalgebra is the set of nongenerators of L. Therefore $L = H$ if $L = V + H$ for some finite dimensional subspace V of $F(L)$ and for some subalgebra H of L.*

(2) *If L is locally nilpotent, then a maximal subalgebra of L is an ideal of L and has codimension one. Therefore $F(L) = L^2$ for a locally nilpotent Lie algebra L.*

We write $L^\omega = \bigcap_{n=1}^{\infty} L^n$.

LEMMA 3.2. *Let L be a locally finite Lie algebra over an algebraically closed field $\mathfrak{k}$ and $x \in L$, then*

$$L = L_0(x) \oplus L_*(x)$$

and $L_*(x)$ is contained in L^ω. Moreover if L is weakly serially finite, then $L_*(x)$ is finite dimensional.

PROOF. The first half is clear by using the Jordan decomposition of $\mathrm{ad}\, x$. If L is weakly serially finite, then there exists a finite dimensional weakly serial subalgebra H containing x and the series $\{V_\sigma, \Lambda_\sigma \mid \sigma \in \Sigma\}$ from H to L. Let α be a nonzero element of $\mathfrak{k}$ and V be a finite dimensional x-invariant indecomposable subspace of $L_\alpha(x)$. We can choose a basis $\{v_1, v_2, \dots, v_n\}$ of V such that $[v_i, x] = \alpha v_i + v_{i+1}$ for $1 \le i \le n-1$ and $[v_n, x] = \alpha v_n$.

If $v_n \notin H$, then there exists $\sigma \in \Sigma$ such that $v_n \in \Lambda_\sigma \setminus V_\sigma$, but $\alpha v_n = [v_n, x] \in [\Lambda_\sigma, x] \subseteq [\Lambda_\sigma, H] \subseteq V_\sigma$, which contradicts to the choice of σ. Therefore $v_n \in H$. If $v_i \notin H$ and $v_{i+1}, \dots, v_n \in H$, then there exists $\tau \in \Sigma$ such that $v_i \in \Lambda_\tau \setminus V_\tau$, but $\alpha v_i = [v_i, x] - v_{i+1} \in [\Lambda_\tau, H] + H \subseteq V_\tau + H = V_\tau$, which is a contradiction. Therefore $v_i \in H$. By induction on i we have $V \subseteq H$. Now $L_\alpha(x)$ is a sum of finite dimensional x-invariant indecomposable subspaces. Therefore $L_\alpha(x) \subseteq H$ and $L_*(x) \subseteq H$.

LEMMA 3.3. *Let L be a locally finite (resp. a weakly serially finite) Lie algebra over $\mathfrak{k}$. Then $L_{\bar{\mathfrak{k}}}$ is also a locally finite (resp. a weakly serially finite) Lie algebra over $\bar{\mathfrak{k}}$.*

PROOF. Let L be locally finite (resp. weakly serially finite) and let $x'_1, x'_2, \cdots, x'_n \in L_{\bar{\mathfrak{k}}}$. Then there exists a finite dimensional subspace V of L such that $x'_1, x'_2, \cdots, x'_n \in V_{\bar{\mathfrak{k}}}$. There exists a finite dimensional subalgebra (resp. weakly serial subalgebra) H of L containing V since L is locally finite (resp. weakly serially finite). Therefore $H_{\bar{\mathfrak{k}}}$ is a finite dimensional subalgebra (resp. weakly serial subalgebra) over $\bar{\mathfrak{k}}$ containing $x'_1, x'_2, \cdots, x'_n$.

4. Local nilpotency

We consider the following two conditions:
$(P1)$ There exists an integer m such that L^m is locally finite and for any $x \in L$ there exists an integer $n = n(x)$ such that $[L,_n x]$ is finite dimensional.
$(P2)$ L is locally finite and for any $x \in L$, $L_0(x)$ is finite codimensional.
When L satisfies $(P1)$ or $(P2)$, we say that L satisfies (P).

The following proposition gives a generalization of Barnes' result[3].

PROPOSITION 4.1. *Let L satisfy (P) and let A and B be ideals of L such that $B \subseteq A \cap F(L)$. Suppose A/B is nilpotent. Then A is locally nilpotent.*

PROOF. Let L satisfy $(P1)$. Then for any $x \in L$ there exists an integer $m \geq n$ such that $[L,_m x] = [L,_{m+1} x]$, therefore by Amayo and Stewart[2]

$$L = [L,_m x] + L_0(x).$$

But for any $x \in A$

$$[L,_m x] = \bigcap_{n=1}^{\infty} [L,_n x] \subseteq A^{\omega} \subseteq B \subseteq F(L),$$

and $[L,_m x]$ is finite dimensional. Therefore by Lemma 3.1(1)

$$L = L_0(x)$$

for any $x \in A$. Then $A \in \mathfrak{E}$ since $A = A_0(x)$ for any $x \in A$. Hence A is locally nilpotent since $A \in \mathfrak{E}$ and $A^m \in \mathrm{L}\mathfrak{N}$. Next let L satisfy $(P2)$ and let $x \in A$. Since $(L_{\bar{\mathfrak{k}}})_0(x) = (L_0(x))_{\bar{\mathfrak{k}}}$ and $L_{\bar{\mathfrak{k}}} = (L_{\bar{\mathfrak{k}}})_0(x) + (L_{\bar{\mathfrak{k}}})_*(x)$ by Lemma 3.3 and Lemma 3.2, $(L_{\bar{\mathfrak{k}}})_*(x)$ is finite dimensional. Also since $(L_{\bar{\mathfrak{k}}})_*(x) \subseteq (A_{\bar{\mathfrak{k}}})^\omega = (A^\omega)_{\bar{\mathfrak{k}}} \subseteq B_{\bar{\mathfrak{k}}} \subseteq F(L)_{\bar{\mathfrak{k}}}$, there exists a finite dimensional subspace H of $F(L)$ such that $(L_{\bar{\mathfrak{k}}})_*(x) \subseteq H_{\bar{\mathfrak{k}}}$. Then since $L_{\bar{\mathfrak{k}}} = (L_{\bar{\mathfrak{k}}})_0(x) + (L_{\bar{\mathfrak{k}}})_*(x) = (L_0(x))_{\bar{\mathfrak{k}}} + H_{\bar{\mathfrak{k}}}$,

$$L = L_0(x) + H = L_0(x)$$

for any $x \in A$. Hence $A = A_0(x)$ for any $x \in A$, which implies that A is locally nilpotent.

The following corollary gives a sufficient condition for the Frattini ideal to be locally nilpotent.

COROLLARY 4.2. *Let L satisfy (P). Then $\Phi(L)$ is locally nilpotent.*

Using this proposition we give characterizations of nilpotency and local nilpotency in terms of the Frattini subalgebra.

THEOREM 4.3. (1) $\mathfrak{F}\mathfrak{N} \cap \mathfrak{U} = \mathfrak{N}$.
(2) $\mathrm{L}(\mathrm{wser})\mathfrak{F} \cap \mathfrak{U} = \mathrm{L}\mathfrak{N}$.

PROOF. (1) Let $L \in \mathfrak{F}\mathfrak{N} \cap \mathfrak{U}$. Then L is locally finite since $L^n \in \mathfrak{F}$ for some n, and $L \in \mathrm{L}\mathfrak{N}$ by Proposition 3.1. Therefore for any $x \in L$, $x^* = \mathrm{ad}\, x\,|_{L^n}$ is a nilpotent derivation of L^n since $\langle L^n, x\rangle$ is nilpotent. By Engel's theorem there exists an integer m such that

$$x_1^* \cdots x_m^* = 0$$

for any $x_1, \cdots, x_m \in L$. Hence

$$\begin{aligned} L^{n+m} &= \sum_{x_1,\cdots,x_m \in L} [\cdots[L^n, x_1], \cdots, x_m] \\ &= \sum_{x_1,\cdots,x_m \in L} L^n x_1^* \cdots x_m^* \\ &= 0. \end{aligned}$$

Therefore L is nilpotent.

(2) Let $L \in \mathrm{L}(\mathrm{wser})\mathfrak{F} \cap \mathfrak{U}$. Then $L_{\bar{\mathfrak{k}}}$ is also weakly serially finite by Lemma 3.3. Hence $(L_{\bar{\mathfrak{k}}})_0(x)$ is a finite codimensional subspace of $L_{\bar{\mathfrak{k}}}$ for any $x \in L$. But since

$(L_0(x))_{\bar{\mathfrak{k}}} = (L_{\bar{\mathfrak{k}}})_0(x)$, $L_0(x)$ is a finite codimensional subspace of L. Therefore by Proposition 4.1 L is locally nilpotent. The converse is clear since $\mathrm{L}\mathfrak{N} \leq \mathrm{L}(\mathrm{wser})\mathfrak{F}$ by Tôgô[7].

REMARK. The following Theorem 5.1 shows that we cannot replace $\mathrm{L}(\mathrm{wser})\mathfrak{F}$ with $\mathrm{L}\mathfrak{F}$ in Theorem 4.3(2).

5. Example

In this section we an construct example which shows that $\mathrm{L}\mathfrak{N} \lneq \mathrm{L}\mathfrak{F} \cap \mathfrak{U}$.

Let V be a vector space over $\mathfrak{k}$ with basis $\{e_i \mid i \in \mathbf{N}\}$ and think of V as an abelian Lie algebra. Then there is a derivation x such that

$$e_1 x = e_1, \; e_{i+1} x = e_{i+1} + e_i \text{ for any } i \in \mathbf{N}.$$

Hence we can form the split extension

$$L = V + \langle x \rangle.$$

It is clear that $L \in \mathfrak{A}^2 \cap \mathrm{L}\mathfrak{F}$. We claim that V is the unique maximal subalgebra of L. Suppose that H is a maximal subalgebra of L which is different from V and that $v + x \in H$ for some $v \in V$. For any integer n there exists an integer $m \geq n$ such that $\sum_{i=1}^{m} a_i e_i \in H$ for some $a_i \in \mathfrak{k}$ and $a_m \neq 0$. Then

$$\sum_{i=1}^{m-k} a_{i+k} e_i = [\sum_{i=1}^{m-k+1} a_{i+k-1} e_i, v + x] - \sum_{i=1}^{m-k+1} a_{i+k-1} e_i$$

for $1 \leq k \leq m$. By induction on n and k we have $e_n \in H$ for any $n \in \mathbf{N}$. Hence $V \subseteq H$, a contradiction. Therefore $F(L) = V = L^2$, which implies $L \in \mathfrak{U}$. Since $[e_{1,n} x] = e_1 \neq 0$ for any $n \in \mathbf{N}$, $L \notin \mathrm{L}\mathfrak{N}$.

Now we have the following

THEOREM 5.1. $\mathrm{L}\mathfrak{N} \lneq \mathrm{L}\mathfrak{F} \cap \mathfrak{U}$.

Acknowledgements

The author would like to express his thanks to Professor N. Kawamoto for his valuable comments in preparing this paper.

References

1. R. K. Amayo, *Frattini subalgebras of finitely generated soluble Lie algebras,* *Trans. Amer. Math. Soc.* **236** (1978), 297-306.

2. R. K. Amayo and I. Stewart, *Infinite-dimensional Lie Algebras,* Noordhoff, Leyden, 1974.

3. D. W. Barnes, *On the cohomology of soluble Lie algebras,* *Math. Z.* **101** (1967), 343-349.

4. E. I. Marshall, *The Frattini subalgebra of a Lie algebra,* *J. London Math. Soc.* **42** (1967), 416-422.

5. E. L. Stitzinger, *On the Frattini subalgebra of a Lie algebra,* *J. London Math. Soc.* (2) **2** (1970), 429-438.

6. E. L. Stitzinger, *Frattini subalgebras of a class of solvable Lie algebras,* *Pacific J. Math.* **34** (1970), 177-182.

7. S. Tôgô, *Serially finite Lie algebras,* *Hiroshima Math. J.* **16** (1986), 443-448.

8. D. A. Towers, *Elementary Lie algebras,* *J. London Math. Soc.* (2), **7** (1973), 295-302.

9. D. A. Towers, *A Frattini theory for algebras,* *Proc. London Math. Soc.* (3), **27** (1973), 440-462.

10. D. A. Towers, *On complemented Lie algebras,* *J. London Math. Soc.* (2), **22** (1980), 63-65.

Coverings, Schur multipliers and K_2 of Kac-Moody groups

JUN MORITA
Institute of Mathematics
University of Tsukuba
Tsukuba, Ibaraki, 305, JAPAN

Abstract. Group coverings, Schur multipliers and K_2 of Kac-Moody groups are discussed, and a relation between K_2 and $K_2\mathrm{Sp}$ of Laurent polynomials is described. Also a new class of simply connected groups is presented.

0. Introduction.

This is a report to the Proceedings of the conference on "Non-associative algebras and related topics" held at Hiroshima in 1990 as a satellite meeting of ICM 90 in Kyoto. The purpose of this article is to give a survay of the results in [31],[32],[33] on Kac-Moody groups.

Contents

The groups associated with Kac-Moody algebras have been constructed and studied by many authors: (see, for examples,)

[5], [7], [12], [14], [15], [17], [18], [25], [36], [39], [40], [45], [46], [48].

Their BN-pair structures have been obtained by Moody-Teo [25] in the case of adjoint representations, by Marcuson [17] in the case of highest weight integrable representations, and by Peterson-Kac [36] in the case of general integrable representations. And also there are many approaches to construct "full" Kac-Moody groups: algebraic group theoretic [13], Lie group theoretic [44], axiomatic and presentation theoretic [47], Hopf algebra theoretic [1], K-theoretic [2], and so on. We will, here, choose and study the subgroup generated by 1-parameter subgroups corresponding to real roots. It might be better to call such a subgroup the "elementary" Kac-Moody group. Here we simply call it a Kac-Moody group. We will construct a universal covering group of a Kac-Moody group, and study the Schur multiplier of a Kac-Moody group and the associated K_2 (cf. Section 7). As an application of the discussion in the affine case, we will obtain some interesting exact sequence on K_2 and $K_2\mathrm{Sp}$ of Laurent polynomials (cf. Section 8). Also we can find a lot of Kac-Moody groups which are homologically simply connected. We will give some criterion for a Kac-Moody group to have such a property (cf. Section 9). In the finite dimensional case, the corresponding results are well-known, which are due to Matsumoto [19], Steinberg [42] and Suslin [43].

The main results stated in Section 7 were established in a joint work with Ulf Rehmann in University of Bielefeld, where the author visited during 1988-1989.

1. Coverings and Schur multipliers.

The Schur multiplier of a group is defined using the second homology group. Originally it was studied by Schur [37,38] for a finite group in terms of central extensions and linearizations of projective representations. That is, the Schur multiplier of a finite group is considered as the kernel of some extension (cf. [8]). He has established that the Schur multiplier of a finite group is finite. Here we shall review the theory of universal central extensions for any (finite or infinite) perfect groups (cf. [42]).

Let G and E be groups, and π a homomorphism of E to G. We call (π, E) an extension of G if π is an epimorphism. We call (π, E) a central extension of G if (π, E) is an extension of G and Ker π is contained in the center $Z(E)$ of E. We call (π, E) a universal central extension of G if (π, E) is a central extension of G and has the property that for every central extension (π', E') there uniquely exists a homomorphism ϕ of E to E' such that $\pi'\phi = \pi$. A group G has a universal central extension if and only if G is perfect, that is, $G = [G, G]$. By the definition, a universal central extension is uniquely determined, up to isomorphism, if it exists. We call G homologically simply connected if $(id., G)$ is a universal central extension of G (cf. [41],[42]).

If (π, E) is a universal central extension of G, then every projective representation of G can be uniquely lifted to a linear representation of E (cf. [42]). In this sence, here we call Ker π the Schur multiplier of G, and denote it by $M(G)$. A universal central extension may be called a universal covering in the sense of homology (cf. [41]).

EXAMPLE 1 ([21],[22],[30],[43]).

$$M(\mathrm{SL}(2,\mathbb{Q})) \simeq C_\infty \oplus (\bigoplus_{p:prime} C_{p-1})$$

Here C_m is the cyclic group of order m.

2. Kac-Moody algebras.

We will review the definitions of Kac-Moody algebras, Weyl groups and Chevalley involutions (cf. [5],[9],[11],[16],[23],[48]).

An $n \times n$ integral matrix $A = (a_{ij})_{1\le i,j\le n}$ is called a generalized Cartan matrix if $a_{ii} = 2$ $(1 \le i \le n)$, $a_{ij} \le 0$ $(1 \le i \ne j \le n)$, $a_{ij} = 0 \Longleftrightarrow a_{ji} = 0$ $(1 \le i,j \le n)$. A triplet $(\mathfrak{h}, \Pi, \Pi^\vee)$ is called a realization of A if $\mathfrak{h}$ is a vector space over $\mathbb{C}$ of dimension $2n - rank(A)$; $\Pi = \{\alpha_1, \cdots, \alpha_n\}$ is a set of n linearly independent elements of $\mathfrak{h}^* = \mathrm{Hom}_{\mathbb{C}}(\mathfrak{h}, \mathbb{C})$; $\Pi^\vee = \{h_1, \cdots, h_n\}$ is a set of n linearly independent elements of $\mathfrak{h}$; and $\alpha_i(h_j) = a_{ji}$ for all $1 \le i,j \le n$. Let $\mathfrak{g}$ be the Lie algebra over $\mathbb{C}$ generated by the so-called Cartan subalgebra $\mathfrak{h}$ and the so-called Chevalley generators $e_1, \cdots, e_n, f_1, \cdots, f_n$ with the following defining relations:

$$\begin{aligned}
&[h,h'] = 0 \quad (h,h' \in \mathfrak{h}), \quad [e_i, f_j] - \delta_{ij} h_i = 0 \quad (1 \le i,j \le n),\\
&[h,e_i] - \alpha_i(h)e_i = [h,f_i] + \alpha_i(h) f_i = 0 \quad (h \in \mathfrak{h},\ 1 \le i \le n),\\
&(\mathrm{ad}\ e_i)^{n(i,j)} e_j = (\mathrm{ad}\ f_i)^{n(i,j)} f_j = 0 \quad (1 \le i \ne j \le n),
\end{aligned}$$

where $n(i,j) = -a_{ij} + 1$. This Lie algebra $\mathfrak{g}$ is called a Kac-Moody algebra over $\mathbb{C}$ associated with A.

Let w_i be the involutive automorphism of $\mathfrak{h}^*$ defined by $w_i(\mu) = \mu - \mu(h_i)\alpha_i$ for all $\mu \in \mathfrak{h}^*$. The subgroup of $\mathrm{GL}(\mathfrak{h}^*)$ is generated by $w_1, \cdots, w_n$ is called the Weyl group and denoted by W. Let ω be the involutive automorphism of the Lie algebra $\mathfrak{g}$ defined by $\omega(e_i) = -f_i$, $\omega(f_i) = -e_i$ and $\omega(h) = -h$ for all $1 \le i \le n$ and $h \in \mathfrak{h}$. Then ω is called the Chevalley involution.

3. Root systems and fundamental subsets.

We will discuss here Kac-Moody root systems and some rank two fundamental subsets (cf. [3],[10],[11],[24],[26]).

Under the adjoint action, $\mathfrak{h}$ is diagonalizable on $\mathfrak{g}$, that is, $\mathfrak{g} = \oplus_{\alpha\in\mathfrak{h}^*} \mathfrak{g}^\alpha$, where $\mathfrak{g}^\alpha = \{x \in \mathfrak{g} \mid [h,x] = \alpha(h)x \text{ for all } h \in \mathfrak{h}\}$. Let $\Delta = \{\alpha \in \mathfrak{h}^* \mid \mathfrak{g}^\alpha \ne 0\}$. We call Δ the root system of $\mathfrak{g}$, and Π the set of simple roots of $\mathfrak{g}$. We see that $\mathfrak{g}^{\alpha_i} = \mathbb{C} e_i$, $\mathfrak{g}^{-\alpha_i} = \mathbb{C} f_i$ and $\mathfrak{g}^0 = \mathfrak{h}$, hence $\{0\} \cup \{\pm\alpha_i\}_{i=1}^n \subset \Delta$. Let

$\Delta_+ = \Delta \cap (\sum_{i=1}^n \mathbb{Z}_{\geq 0}\alpha_i \setminus \{0\})$, the set of positive roots, and $\Delta_- = -\Delta_+$, the set of negative roots. Then $\Delta = \Delta_+ \cup \{0\} \cup \Delta_-$.

One sees that Δ is W- stable. Therefore, we can define $\Delta^{re} = \cup_{i=1}^n W\alpha_i$, the set of real roots, as a subset of Δ. For each $\alpha \in \Delta$, we obtain

$$\alpha \in \Delta^{re} \iff \mathbb{Z}\alpha \cap \Delta = \{0, \ \pm\alpha\}.$$

Furthermore, dim $\mathfrak{g}^\alpha = 1$ if $\alpha \in \Delta^{re}$. For each $\alpha = w\alpha_i \in \Delta^{re}$ with $w \in W$ and $\alpha_i \in \Pi$, we put $h_\alpha = wh_i$. The h_α are well-defined.

A subset $\Pi' = \{\beta_1, \cdots, \beta_m\}$ of Δ is called a fundamental subset if $\Pi' \subset \Delta^{re}$; Π' is linearly independent; and $\beta_i - \beta_j \notin \Delta$ for distinct i, j. For a fundamental subset $\Pi' = \{\beta_1, \cdots, \beta_m\}$, we set $A' = (a'_{ij})$, where $a'_{ij} = \beta_j(h_{\beta_i})$. Then we write $A' < A$. A fundamental subset is corresponding to the set of simple roots of some "standard regular" subalgebra. Now we define $\Omega(A)$ by the set consisting of all $\mu(i_1, i_2, \cdots, i_s)$ satisfying $1 \leq i_1, i_2, \cdots, i_s \leq n$; $\mu(i_1, i_2, \cdots, i_s) > 1$ and $\mu(i_s, \cdots, i_2, i_1) = 1$, where $\mu(j_1, j_2, \cdots, i_r) = | a_{j_1 j_2} a_{j_2 j_3} \cdots a_{j_{r-1} j_r} |$. Then, as in the proof of the main theorem in [28], we can establish the following. (The statement of the theorem in [29] should be corrected as follows.)

THEOREM 2. (1) *Suppose that A is tree. Then*

$$\begin{pmatrix} 2 & -a \\ -1 & 2 \end{pmatrix} < A \iff a \in \Omega(A).$$

(2) *Suppose that A is symmetrizable (,that is, there are positive integers $d_1, \cdots, d_n$ satisfying $d_i a_{ij} = d_j a_{ji}$), then*

$$\begin{pmatrix} 2 & -a \\ -1 & 2 \end{pmatrix} < A \implies a \in \Omega(A).$$

In general, the converse of (2) is not true (cf. [34],[35]).

4. Chevalley bases for real roots.

We will choose a basis $e_\alpha \in \mathfrak{g}_\alpha$ for each $\alpha \in \Delta^{re}$, and study the relations among the e_α (cf. [27]).

For each $\alpha \in \Delta^{re}$, a pair $(e_\alpha, e_{-\alpha}) \in \mathfrak{g}^\alpha \times \mathfrak{g}^{-\alpha}$ is called a Chevalley pair for α if

$$[e_\alpha, e_{-\alpha}] = h_\alpha, \quad \omega(e_\alpha) + e_{-\alpha} = 0.$$

There are precisely two Chevalley pairs for each $\alpha \in \Delta^{re}$. If one is $(e_\alpha, e_{-\alpha})$, then $(-e_\alpha, -e_{-\alpha})$ is the other. We choose and fix a Chevalley pair for each $\alpha \in \Delta_+^{re} =$

$\Delta^{re} \cap \Delta_+$ with $e_{\alpha_i} = e_i$ and $e_{-\alpha_i} = f_i$ for $1 \le i \le n$. Then we obtain the set $\mathcal{C} = \{e_\alpha \mid \alpha \in \Delta^{re}\}$, which is called a Chevalley basis for Δ^{re}.

Now we define the numbers $\eta_{\alpha\beta}$ and $N_{\alpha\beta}$ by the following:

$$(\exp\ \text{ad}\ e_\alpha)(\exp\ -\text{ad}\ e_{-\alpha})(\exp\ \text{ad}\ e_\alpha)e_\beta = \eta_{\alpha\beta}e_{\beta'}$$

for all $\alpha, \beta \in \Delta^{re}$ and

$$[e_\alpha, e_\beta] = N_{\alpha\beta}e_{\alpha+\beta}$$

for all $\alpha, \beta \in \Delta^{re}$ with $\alpha + \beta \in \Delta^{re}$, where $\beta' = \beta - \beta(h_\alpha)\alpha \in \Delta^{re}$. Then:

THEOREM 3. (1) $\eta_{\alpha\beta} = \pm 1$.
(2) *If p is the largest integer satisfying $\beta - p\alpha \in \Delta^{re}$, then* $N_{\alpha\beta} = \pm(p+1)$.

5. Kac-Moody groups.

We will construct Kac-Moody groups and study several relations between typical elements (cf. [27]).

A $\mathfrak{g}$-module V is called integrable if $V = \oplus_{\mu \in \mathfrak{h}^*} V^\mu$ and each e_α is locally nilpotent on V, where $V^\mu = \{v \in V \mid hv = \mu(h)v \text{ for all } h \in \mathfrak{h}^*\}$ and $\alpha \in \Delta^{re}$. A basis of V is called standard if it consists of the bases of the V_μ and its $\mathbb{Z}$-span is stable under $e_\alpha^{(m)}$ for all $\alpha \in \Delta^{re}$ and $m \in \mathbb{Z}_{\ge 0}$, where

$$e_\alpha^{(m)} = \frac{e_\alpha^m}{m!}.$$

We take an integral $\mathfrak{g}$-module V with a standard basis. Let $V_\mathbb{Z}$ be the $\mathbb{Z}$-span of this standard basis in V. For a commutative ring, R, with 1, we put $V(R) = R \otimes V_\mathbb{Z}$. Then, for each $\alpha \in \Delta^{re}$ and $t \in R$, we can consider the R-linear operator $x_\alpha(t)$ on $V(R)$ defined by

$$x_\alpha(t)(r \otimes v) = \sum_{m=0}^{\infty} t^m r \otimes e_\alpha^{(m)} v.$$

Since $x_\alpha(-t)$ is the inverse of $x_\alpha(t)$, the operator $x_\alpha(t)$ lies in $\mathrm{GL}(V(R))$. Now we define the (elementary) Kac-Moody group $G_V(A, R)$ as the subgroup of $\mathrm{GL}(V(R))$ generated by $x_\alpha(t)$ for all $\alpha \in \Delta^{re}$ and $t \in R$. Using rank two Kac-Moody root systems and Theorem 3, we obtain the following relations in $G_V(A, R)$.

THEOREM 4.
(A) $\quad x_\alpha(s)x_\alpha(t) = x_\alpha(s+t)$,
(B1) $\quad [x_\alpha(s), x_\beta(t)] = 1$
if $Q_{\alpha\beta} = \emptyset$,
(B2) $\quad [x_\alpha(s), x_\beta(t)] = x_{\alpha+\beta}(N_{\alpha\beta}st)$

if $Q_{\alpha\beta} = \{\alpha+\beta\} \subset \Delta^{re}$,

(B3) $\quad [x_\alpha(s), x_\beta(t)] = x_{\alpha+\beta}(\pm st)x_{2\alpha+\beta}(\pm s^2 t)$

if $Q_{\alpha\beta} = \{\alpha+\beta,\ 2\alpha+\beta\} \subset \Delta^{re}$,

(B4) $\quad [x_\alpha(s), x_\beta(t)] = x_{\alpha+\beta}(\pm 2st)x_{2\alpha+\beta}(\pm 3s^2 t)x_{\alpha+2\beta}(\pm 3st^2)$

if $Q_{\alpha\beta} = \{\alpha+\beta,\ 2\alpha+\beta,\ \alpha+2\beta\} \subset \Delta^{re}$,

(B5) $\quad [x_\alpha(s), x_\beta(t)] = x_{\alpha+\beta}(\pm st)x_{2\alpha+\beta}(\pm s^2 t)x_{3\alpha+\beta}(\pm s^3 t)x_{3\alpha+2\beta}(\pm 2s^3 t^2)$

if $Q_{\alpha\beta} = \{\alpha+\beta,\ 2\alpha+\beta,\ 3\alpha+\beta,\ 3\alpha+2\beta\} \subset \Delta^{re}$,

(B′) $\quad w_\alpha(u)x_\beta(t)w_\alpha(-u) = x_{\beta'}(\eta_{\alpha\beta}u't)$,

(C) $\quad h_\alpha(u)h_\alpha(v) = h_\alpha(uv)$,

where $s, t \in R$, $\alpha \in \Delta^{re}$, $Q_{\alpha\beta} = (\mathbb{Z}_{>0}\alpha + \mathbb{Z}_{>0}\beta) \cap \Delta$, $\beta' = \beta - \beta(h_\alpha)\alpha$, $u' = u^{-\beta(h_\alpha)}$, *and* $w_\alpha(u) = x_\alpha(u)x_{-\alpha}(-u^{-1})x_\alpha(u)$ *and* $h_\alpha(u) = w_\alpha(u)w_\alpha(-1)$ *with* u *in the multiplicative group,* $R^\times$*, of* R.

The signatures in the above relations (B3)-(B5) are depending only on the choice of a Chevalley basis $\mathcal{C}$, and they can be calculated using the numbers $N_{\gamma\delta}$ (cf. [4],[42],[47]).

6. Steinberg-Tits presentations and associated K_2.

We will discuss a presentation of a universal Kac-Moody group, and define the associated Steinberg group and K_2 (cf. [47]).

Tits [47] shows that, for every field F, there exists a Kac-Moody group whose presentation is given by $x_\alpha(t)$ for all $\alpha \in \Delta^{re}$ and $t \in F$ as generators and by (A), (B1)-(B5), (B′), (C) as defining relations. Therefore, such a group dominates all other Kac-Moody groups. In this sense, it is called a universal Kac-Moody group, and denoted by $G_u(A, F)$. He has established that the standard Borel subgroup of a Kac-Moody group over a field has some amalgamated free product decomposition, which gives a universality as well as a presentation to $G_u(A, F)$.

Let $\mathrm{St}(A, F)$ be the group presented by the generators $\tilde{x}_\alpha(t)$ for all $\alpha \in \Delta^{re}$ and $t \in F$ and by the defining relations (A), (B1)-(B5), (B′), where $x_\alpha(t)$ and $w_\alpha(u)$ are replaced by $\tilde{x}_\alpha(t)$ and $\tilde{w}_\alpha(u)$ respectively. Then there is a canonical homomorphism, called ψ, of $\mathrm{St}(A, F)$ onto $G_u(A, F)$ given by $\psi(\tilde{x}_\alpha(t)) = x_\alpha(t)$ for all $\alpha \in \Delta^{re}$ and $t \in F$. Put $K_2(A, F) = \ker\psi$, then, we obtain the following exact sequence:

$$1 \longrightarrow K_2(A, F) \longrightarrow \mathrm{St}(A, F) \xrightarrow{\psi} G_u(A, F) \longrightarrow 1.$$

Then, $K_2(A, F)$ is generated by $\{u, v\}_\alpha$ for all $u, v \in F^\times$, the multiplicative group of F, and $\alpha \in \Delta^{re}$, where $\{u, v\}_\alpha = \tilde{h}_\alpha(u)\tilde{h}_\alpha(v)\tilde{h}_\alpha(uv)^{-1}$ and $\tilde{h}_\alpha(u) = \tilde{w}_\alpha(u)\tilde{w}_\alpha(-1)$.

7. Main results.

We will present three results.

For a generalized Cartan matrix A and a field F, we let $L(A, F)$ be the abelian group generated by the symbols $c_i(u, v)$ for all $1 \leq i \leq n$ and $u, v \in F^{\times}$ with the defining relations:

(L1) $$c_i(u, v)c_i(uv, w) = c_i(u, vw)c_i(v, w);$$

(L2) $$c_i(1, 1) = 1;$$

(L3) $$c_i(u, v) = c_i(u^{-1}, v^{-1});$$

(L4) $$c_i(u, v) = c_i(u, (1-u)v) \quad \text{if } u \neq 1;$$

(L5) $$c_i(u, v^{a_{ji}}) = c_j(u^{a_{ij}}, v);$$

(L6) $$c_{ij}(uv, w) = c_{ij}(u, w)c_{ij}(v, w);$$

(L7) $$c_{ij}(u, vw) = c_{ij}(u, v)c_{ij}(u, w);$$

for all $1 \leq i \neq j \leq n$ and $u, v, w \in F^{\times}$, where $c_{ij}(u, v) = c_i(u, v^{a_{ji}}) = c_j(u^{a_{ij}}, v)$. Then there is a natural homomorphism, called λ, of $L(A, F)$ to $K_2(A, F)$ defined by $\lambda(c_i(u, v)) = \{u, v\}_{\alpha_i}$ for all $1 \leq i \leq n$ and $u, v \in F^{\times}$. It follows from Tits [47] (cf. Section 6) that λ is surjective. Furthermore, using the same method as in Matsumoto [19], we can establish the following.

THEOREM 5. *λ is an isomorphism.*

This result is well-known if A is of finite type, which is due to Matsumoto [19].

Next, using the same argument as in Steinberg [42], we obtain the following.

THEOREM 6. *If F is an infinite field, then* St(A, F) *is a universal central extension of $G_u(A, F)$.*

This result is well-known if A is of finite type, in which case the condition $| F | = \infty$, for example, can be replaced by $| F | > 9$ (cf. [42]). Of course, we have already known a complete information in case of $| F | \leq 9$ according to circumstances, since $G_u(A, F)$ is a finite group of Lie type (cf. [6],[8],[42]).

Using Theorems 5 and 6, we can see the following.

THEOREM 7. *If F is an infinite field, then $M(G_u(A, F))$ is isomorphic to $L(A, F)$, where $M(G_u(A, F))$ is the Schur multiplier of $G_u(A, F)$.*

8. An application to the stable K_2 and $K_2\mathrm{Sp}$.

We will establish a polynomial version of some exact sequence in Suslin [43] (cf. [33]).

Let $W(F)$ be the Witt ring of a field F, and $I(F)$ the fundamental ideal of $W(F)$ consisting of the classes with even rank (cf. [22]). Then there is the following exact sequence:

$$1 \longrightarrow I^3(F) \longrightarrow K_2\mathrm{Sp}(F) \longrightarrow K_2(F) \longrightarrow 1$$

(cf. [20],[43]). Here we obtain an $F[z, z^{-1}]$–version of this sequence, using the structure of $K_2(C_n^{(1)}, F)$ associated with a universal Kac-Moody group over F of affine type $C_n^{(1)}$. The results described in Section 7 impliy

$$K_2(C_n^{(1)}, F) \simeq K_2\mathrm{Sp}(F) \oplus I^2(F).$$

Hence, we obtain the following.

THEOREM 8. (1) *$K_2\mathrm{Sp}_{2n}(F[z, z^{-1}])$ has the stability from $n = 1$.*
(2) *There is the following exact sequence:*

$$1 \longrightarrow I^3(F) \oplus I^2(F) \longrightarrow K_2\mathrm{Sp}(F[z, z^{-1}]) \longrightarrow K_2(F[z, z^{-1}]) \longrightarrow 1.$$

9. A new class of simply connected groups.

We will give a sufficient condition for a universal Kac-Moody group $G_u(A, F)$ to be homologically simply connected (cf. [31]).

Put $Q = \oplus_{i=1}^n \mathbb{Z}\,\alpha_i$, the root lattice of $\mathfrak{g}$. Let Q' be the additive subgroup of Q generated by the elements

$$a_{ji}\alpha_i - a_{ij}\alpha_j$$

for all $1 \le i, j \le n$. Set $P(A) = Q/Q'$. Then we obtain the following.

THEOREM 9. *Let A be a generalized Cartan matrix. Then $K_2(A, F) = 1$ for every field F if and only if $P(A) = 0$.*

As a direct consequence of this result, we can find a homologically simply connected Kac-Moody group.

THEOREM 10. *If $P(A) = 0$ and F is an infinite field, then $G_u(A, F)$ is homologically simply connected.*

The simplest example with this property will be given in the following example, which was observed at first. We can make lots of examples of generalized Cartan matrices A such that $K_2(A, F) = 1$ for every field F. Such generalized Cartan matrices must be non-symmetrizable.

EXAMPLE 11. *Let*

$$A = \begin{pmatrix} 2 & -2 & -1 \\ -1 & 2 & -1 \\ -1 & -1 & 2 \end{pmatrix}.$$

Then $K_2(A, F) = 1$ *for every field* F.

Finally, we must make a remark of misprint. In [31;Theorem 1(1)], the word "row" should be changed into "column".

REFERENCES

1. E. Abe and M. Takeuchi, *Groups associated with some types of infinite dimensional Lie algebras.* (to appear)

2. A. Bak, *Non-abelian K-theory and applications to Kac-Moody theory*, A talk in the informal seminor on K-theory during ICM90 in Kyoto, 1990.

3. N. Bourbaki, "Groupes et algèbres de Lie," Chap. 4-6, Hermann, Paris, 1968.

4. R. W. Carter, "Simple groups of Lie Type," J. Wiley - Sons, London, New York, 1972.

5. H. Garland, *The arithmetic theory of loop groups*, IHES Publ. Math. **52** (1980), 181 – 312.

6. R. L. Griess, Jr., *Schur multipliers of finite simple groups of Lie type*, Trans. Amer. Math. Soc. **183** (1973), 355 – 421.

7. E. Gutkin, *Schubert calculus on flag varieties of Kac-Moody groups*, Algebras, Groups and Geometries **3** (1986), 27 – 59.

8. B. Huppert, "Endliche Gruppen I," Springer, Berlin, 1967.

9. V. G. Kac, *Simple irreducible graded Lie algebras of finite growth*, Math. USSR-Izv. **2** (1968), 1271 – 1311.

10. V. G. Kac, *Infinite root systems, representations of graphs and invariant theory*, Invent. Math. **56** (1980), 57 – 92.

11. V. G. Kac, "Infinite dimensional Lie algebras," Progress in Math., Birkhäuser, Boston, 1983.

12. V. G. Kac, *Constructing groups associated to infinite dimensional Lie algebras*, in "Infinite dimensional groups with applications (ed. by V. G. Kac)," Reserch Inst. Publ. 4, Springer, New York, Berlin, 1985, pp. 167 – 232.

13. V. G. Kac - D. H. Peterson, *Regular functions on certain infinite dimensional groups*, in "Arithmetic and Geometry," Progress in Math. 36, Birkhäuser, Boston, 1983, pp. 143 – 166.

14. V. G. Kac and D. H. Peterson, *Unitary structure in representations of infinite dimensional groups and a convexity theorem*, Invent. Math. **76** (1984), 1 – 14.

15. S. Kummar, *Demazure character formula in arbitrary Kac-Moody setting*, Invent. Math. **89** (1987), 395 – 423.

16. J. Lepowsky, "Lectures on Kac-Moody Lie algebras," Paris University, Paris, 1978.

17. R. Marcuson, *Tits' systems in generalized nonadjoint Chevalley groups*, J. Algebra **34** (1975), 84 – 96.

18. O. Mathieu, *Formules de charactères pour les algèbres de Kac-Moody générales*, Astérisque **159 – 160** (1988), 1 – 267.

19. H. Matsumoto, *Sur les sous-groupes arithmétiques des groupes semi-simples déployés*, Ann. Scient. Ec. Norm. Sup. **(4) 2** (1969), 1 – 62.

20. A. S. Merkurjev and A. A. Suslin, *K–cohomology of Severi-Brauer varieties and the norm residue homomorphism*, Math. USSR-Izv. **21** (1983), 307 – 340.

21. J. Milnor, "Introduction to algebraic K–theory," Ann. Math. Studies 72, Princeton Univ. Press, Princeton, 1971.

22. J. Milnor and D. Husemoller, "Symmetric bilinear forms," Ergebnisse der Mathematik und Ihrer Grenzgebiete 73, Springer, Berlin, New York, 1973.

23. R. V. Moody, *A new class of Lie algebras*, J. Algebra **10** (1968), 211 – 230.

24. R. V. Moody, *Root systems of hyperbolic type*, Adv. in Math. **33** (1979), 144 – 160.

25. R. V. Moody and K. L. Teo, *Tits' systems with crystallographic Weyl groups*, J. Algebra **21** (1972), 178 – 190.

26. R. V. Moody and T. Yokonuma, *Root systems and Cartan matrices*, Canad. J. Math. **(1) 34** (1982), 63 – 79.

27. J. Morita, *Commutator relations in Kac-Moody groups*, Proc. Japan Acad. Ser. A **(1) 63** (1987), 21 – 22.

28. J. Morita, *Root strings with three or four real roots in Kac-Moody root systems*, Tôhoku Math. J. **(4) 40** (1988), 645 – 650.

29. J. Morita, *Certain rank two subsystems of Kac-Moody root systems*, in "Infinite dimensional Lie algebras and groups (ed. by V. G. Kac)," Adv. Ser. in Math. Phys. 7, World Scientific, Singapore, New Jersey, 1989, pp. 52 – 56.

30. J. Morita, *On the group structure of rank one K_2 of some $\mathbb{Z}_S$*, Bull. Soc. Math. Belgique **XLII** (1990), 561 – 575.

31. J. Morita, *Kac-Moody groups over fields with trivial K_2*, Comm. Algebra **(5)19** (1991), 1541 – 1544.

32. J. Morita and U. Rehmann, *A Matsumoto-type theorem for Kac-Moody groups*, Tôhoku Math. J. **(4)42** (1990), 537 – 560.

33. J. Morita and U. Rehmann, *Symplectic K_2 of Laurent polynomials, associated Kac-Moody groups and Witt rings*, Math. Z. **206** (1991), 57 – 66.

34. S. Naito, *Subalgebras of Kac-Moody Lie algebras and highest weight representations*, Master Thesis (in Japanese), Kyoto University, 1990.

35. S. Naito, *On regular subalgebras of Kac-Moody algebras and their invariant forms — Symmetrizable case —*, preprint, 1990.

36. D. H. Peterson and V. G. Kac, *Infinite flag varieties and conjugacy theorems*, Proc. Nat. Acad. Sci. USA **80** (1983), 1778 – 1782.

37. J. Schur, *Über die Darstellungen der Endlichen Gruppen durch gebrochene lineare Substitutionen*, J. Reine Angew. Math. **127** (1904), 20 – 50.

38. J. Schur, *Über die Darstellungen der Endlichen Gruppen durch gebrochene lineare Substitutionen*, J. Reine Angew. Math. **132** (1907), 85 – 137.

39. P. Slodowy, *Singularitäten, Kac-Moody Liealgebren, assozierte Gruppen und Veralgemeinerungen*, Habilitationsschrift, Universität Bonn, Bonn, 1984.

40. P. Slodowy, *An adjoint quotient for certain groups attached to Kac-Moody algebras*, in "Infinite dimensional groups with applications (ed. by V. G. Kac)," Reserch Inst. Publ. 4, Springer, New York, Berlin, 1985, pp. 307 - 333.

41. M. R. Stein, *Generators, relations and coverings of Chevalley groups over commutative rings*, Amer. J. Math. **93** (1971), 965 - 1004.

42. R. Steinberg, "Lectures on Chevalley groups," Yale University, New Haven, CT, 1968.

43. A. A. Suslin, *Torsion in K_2 of fields*, K-theory **1** (1987), 5 - 29.

44. K. Suto, *Groups associated with unitary forms of Kac-Moody algebras*, J. Math. Soc. Japan (1) **40** (1988), 85 - 104.

45. J. Tits, *Théorie des groupes*, Résumé des course et travaux 81e (1980 - 1981), 75 - 87, Collège de France, Paris.

46. J. Tits, *Théorie des groupes*, Résumé des course et travaux 82e (1981 - 1982), 91 - 106, Collège de France, Paris.

47. J. Tits, *Uniqueness and presentation of Kac-Moody groups over fields*, J. Algebra **105** (1987), 542 - 573.

48. J. Tits, *Groupes associés aux algèbres de Kac-Moody*, Astérisque **177 - 178** (1989), 7 - 31.

Keywords. Kac-Moody group, Covering group, Schur multiplier, K_2

1980 *Mathematics subject classifications*: (1985 *Revision*). 17B67,19C09

CHARACTERIZATION OF LINEARIZATIONS OF A DIFFERENTIAL OPERATOR OF SECOND ORDER AND FUNCTION THEORY

KIYOHARU NÔNO

Department of Mathematics, Fukuoka University of Education, Munakata, Fukuoka, 811-41, Japan

Abstract. One of linearizations of Laplacian of two real variables (n-variables) is well known as the Cauchy-Riemann operator (the generalized Cauchy-Riemann operator). In this paper,we give a characterization of linearizations of a partial differential operator of second order with constant coefficients. Also,we give a characterization of a partial differential operator D of first order with coefficients (Clifford valued functions) such that any solution of the differential equation Df=0 satisfies Cauchy type's integral formula.

1. Introduction

The Clifford algebra was constructed by W.K.Clifford in1878 as a generalization of the algebra of quaternions. It was studied in the 1930s in connection with the theory of spinors([2-5,15-17,23-25]) and has been studied in Mathematics and Physics([1,6-14,18-22,26-30]).

The generalized Clifford algebra was constructed as a generalization of the ordinary Clifford algebra and it was well studied by many authors([16,17,23,24]). Also,T.Nôno([23,24]) has generalized the concept of the linearization of the wave equation from another point of view.

In [8], R.Fueter introduced a quaternionic differential operator (generalized Cauchy-Riemann operator) as a generalization of the Cauchy-Riemann operator in the complex function theory and developed a theory of quaternionic regular functions defined as smooth solutions of the generalized Cauchy-Riemann equation. Also, many authors([1,6,18,19,30]) have developed the Clifford regular function theory. In this function theory, the regularity of functions have been defined by a linearization of Laplacian of several real or complex variables.

In [9,10], B. Goldschmit has considered a partial differential equation $\sum_{i=0}^{n} e_i \partial w/\partial x_i + \sum_A c_A(x) J_A w = 0$ and has given Cauchy integral formula,Series expansion and Maximum principle for solutions of the differential operator.

In [22], we gave a characterization of a differential operator $D = \sum_{i=0}^{n} \alpha_i \frac{\partial}{\partial x_i}$ with coefficients α_i $(i = 0, 1, ..., n)$ of complex Clifford numbers such that any solution of the differential equation Df=0 is always a solution of the differential equation $\Box f = 0$, where $\Box = \sum_{i,j=0}^{n} g_{ij} \partial^2/\partial x_i \partial x_j$.

In this paper, we give a characterization of linearizations of the following partial differential operator:

$$\square = \sum_{i,j=0}^{n} c_{ij} \frac{\partial^2}{\partial x_i \partial x_j} + \sum_{i=0}^{n} h_i \frac{\partial}{\partial x_i} + g,$$

where $c_{ij} = c_{ji}$ $(i, j = 0, 1, ..., n)$.

Also, we consider a linearization $D_\alpha^\gamma = \sum_{i=0}^{n} \alpha_i(x)\partial/\partial x_i + \gamma(x)$ of a partial differential operator of second order and give a characterization of the differential operator D_α^γ such that any solution of the differential equation $D_\alpha^\gamma f = 0$ satisfies Cauchy type's integral formula.

2.Preliminaries

Let $A_n(R)$ be a real Clifford algebra over an n-dimensional vector space with basis $\{e_1, e_2, ..., e_n\}$. Then,it is well known that $A_n(R)$ is a real 2^n-dimensional associative,but non-commutative algebra and its basis $\{e_0, e_1, ..., e_n, ..., e_1 e_2 \ldots e_n\}$ satisfy the following:

$$e_i e_j + e_j e_i = -2\delta_{ij} e_0 \quad (i, j = 1, 2, ..., n),$$

e_0 is the identity of $A_n(R)$,where δ_{ij} is the Kronecker's delta.Then, every element x in $A_n(R)$ is of form:

$$x = e_0 x_0 + e_1 x_1 + ... + e_n x_n + ... + e_1 e_2 ... e_n x_{12...n}, \tag{2.1}$$

where $x_0, x_1, ..., x_{12...n}$ are real numbers.

By taking the tensor product of $A_n(R)$ with the field C of complex numbers:

$$A_n(R) \otimes C,$$

we obtain a complex Clifford algebra which we denote by $A_n(C)$.It is a complex 2^n-dimensional associative,but non-commutative algebra. Every element z in $A_n(C)$ has the form:

$$z = e_0 z_0 + e_1 z_1 + ... + e_n z_n + ... + e_1 e_2 ... e_n z_{12...n}, \tag{2.2}$$

where $z_0, z_1, ..., z_{12...n}$ are complex numbers. For any element z of the form (2.2), we define the following:

$$N(z) = z_o^2 + z_1^2 + ... + z_n^2 + ... + z_{12...n}^2. \tag{2.3}$$

Since the subspace of $A_n(R)$ $(A_n(C))$ spanned by the base elements $\{e_0, e_1, ..., e_n\}$ is identified with R^{n+1} (C^{n+1}),we denote the subspace by R^{n+1} (C^{n+1}).

For each element $z = e_0z_0 + e_1z_1 + ... + e_nz_n$ in C^{n+1},the conjugate number z^* of z is given by

(2.4) $$z^* = e_0z_0 - e_1z_1 - ... - e_nz_n.$$

Also,we define the following:

(2.5) $$N(z) = z_0^2 + z_1^2 + ... + z_n^2, \quad z^{-1} = \frac{z^*}{zz^*} = \frac{z^*}{N(z)} \quad (N(z) \neq 0).$$

For elements z, w in C^{n+1}, we obtain the following:

(2.6) $$N(zw) = N(z)N(w).$$

For any x in R^{n+1}, put $|x| = \sqrt{N(x)}$. Then we have the following:

(2.7) $$|xy| = |x||y|$$

for any $x, y \in R^{n+1}$.

Let $\partial/\partial x_i (i = 0, 1, ..., n)$ be the real differential operators. We consider the following differential operator with complex Clifford coefficients:

(2.8) $$D = \sum_{j=0}^{n} \alpha_j \frac{\partial}{\partial x_j} + \gamma,$$

where $\alpha_j = \sum_{i=0}^{n} e_i a_{ij} (j = 0, 1, ..., n)$ and γ are elements in C^{n+1}. We define the following conjugate differential operator D^* of D :

(2.9) $$D^* = \sum_{j=0}^{n} \alpha_j^* \frac{\partial}{\partial x_j} + \gamma^*,$$

where α_j^* and γ^* are the conjugate number of $\alpha_j (j = 0, 1, ..., n)$ and γ, respectively. We consider the following partial differential operator:

(2.10) $$\Box = \sum_{i,j=0}^{n} c_{ij} \frac{\partial^2}{\partial x_i \partial x_j} + \sum_{i=0}^{n} h_i \frac{\partial}{\partial x_i} + g,$$

where c_{ij},h_i $(i, j = 0, 1, ..., n)$ and g are real numbers.

Let f be a function of $(n + 1)$-real variables $x = (x_0, x_1, ..., x_n)$ with values in $A_n(C)$.

Definition 2.1. *A differential operator D is a linearization of $\Box$ if and only if any solution of the differential equation $Df = 0$ is always a solution of the differential equation $\Box f = 0$.*

3. Linearizations of $\Box = \sum_{i=0}^{n} c_{ii} \frac{\partial^2}{\partial x_i^2} + \sum_{i=0}^{n} h_i \frac{\partial}{\partial x_i} + g$

In this section, we consider the following differential operators:

$$D = \sum_{i=0}^{n} \alpha_i \frac{\partial}{\partial x_i} + \gamma, \tag{3.1}$$

$$\Box = \sum_{i=0}^{n} c_{ii} \frac{\partial^2}{\partial x_i^2} + \sum_{i=0}^{n} h_i \frac{\partial}{\partial x_i} + g, \tag{3.2}$$

where $\alpha_i, \gamma \in R^{n+1}, h_i$ and g are real numbers, c_{ii} $(i = 0, 1, ..., n)$ are positive numbers and there exists at least an integer j such that $\alpha_j \neq 0, \quad 0 \leq j \leq n$.

We consider the following function:

$$E(\alpha t) = \sum_{m=0}^{\infty} \frac{\alpha^m}{m!} t^m, \quad t \in R, \tag{3.3}$$

where $\alpha \in A_n(R)$. The function $E(\alpha t)$ has the following properties:

Property 3.1.

$$\frac{d}{dt} E(\alpha t) = \alpha E(\alpha t) = E(\alpha t)\alpha, \quad \alpha \in A_n(R).$$

Property 3.2. *Let α, β be elements in $A_n(R)$. If $\alpha\beta = \beta\alpha$, then we have that*

$$E(\alpha t) E(\beta t) = E(\beta t) E(\alpha t).$$

This function $E(\alpha t)$ plays an important role in this paper.

Lemma 3.1. *Let D be a linearization of $\Box$. Then, we have*

$$\alpha_i \neq 0 \quad (i = 0, 1, ..., n).$$

Proof. Assume that there exists an integer i such that $\alpha_i = 0$. From the condition of D, there exists an integer j such that $\alpha_j \neq 0$. In case $h_i \neq 0$, we consider the following function:

$$f_0(x) = E(-\alpha_j^{-1} \gamma x_j) x_i \quad (i \neq j).$$

Then,we have that

$$\begin{aligned} Df_0 &= \alpha_i \frac{\partial}{\partial x_i} f_0(x) + \alpha_j \frac{\partial}{\partial x_j}(E(-\alpha_j^{-1}\gamma x_j))x_i + \gamma E(-\alpha_j^{-1}\gamma x_j)x_i \\ &= 0. \end{aligned}$$

Since D is a linearization of $\square$, we have that

$$\begin{aligned} \square f_0 &= \{c_{jj}(\alpha_j^{-1}\gamma)^2 - h_i(\alpha_j^{-1}\gamma) + g\} f_0 + h_i E(-\alpha_j^{-1}\gamma x_j) \\ &= 0. \end{aligned} \tag{3.4}$$

Also,since $f_1(x) = E(-\alpha_j^{-1}\gamma x_j)$ is a solution of the differential equation $Df = 0$ and D is a linearization of $\square$, it follows that

$$c_{jj}(\alpha_j^{-1}\gamma)^2 - h_j\alpha_j^{-1}\gamma + g = 0. \tag{3.5}$$

From (3.4) and (3.5), we have that

$$h_i E(-\alpha_j^{-1}\gamma x_j) = 0.$$

Since $h_i \neq 0$, this is a contradiction. In case $h_i = 0$, we consider the following function

$$f_2(x) = E(-\alpha_j^{-1}\gamma x_j)x_i^2.$$

Then, since $Df_2 = 0$ and $\square f_2 = 0$, it follows that

$$E(-\alpha_j^{-1}\gamma x_j)\{c_{ii} + h_i x_i\} = 0.$$

Putting $x_i = x_j = 0$, this is a contradiction.

LEMMA 3.2. *Let D be a linearization of $\square$. If there exist two integers $i, j (i \neq j)$ such that $\alpha_k^*\gamma = \gamma^*\alpha_k \quad (k = i, j)$, we have $\gamma = 0$.*

Proof. Assume that $\gamma \neq 0$. From $\alpha_k^*\gamma = \gamma^*\alpha_k$, it follows that $\alpha_k^{-1}\gamma$ is a real number. From Lemma 3.1, We can consider the following functions:

$$f(x) = E(-\frac{1}{2}\alpha_i^{-1}\gamma x_i)E(-\frac{1}{2}\alpha_j^{-1}\gamma x_j).$$

Then,we see that $Df = 0$. Since D is a linearization of $\square$, it follows that

$$c_{ii}(\alpha_i^{-1}\gamma)^2 + c_{jj}(\alpha_j^{-1}\gamma)^2 - 2h_i\alpha_i^{-1}\gamma - 2h_j\alpha_j^{-1}\gamma + 4g = 0. \tag{3.6}$$

Also,since $c_{kk}(\alpha_k^{-1}\gamma)^2 - h_k\alpha_k^{-1}\gamma + g = 0 \quad (k = i, j)$, from (3.6) and $\gamma \neq 0$, it follows that

$$c_{ii}\alpha_i^{-1}\gamma\alpha_i^{-1}\gamma + c_{jj}\alpha_j^{-1}\gamma\alpha_j^{-1}\gamma = 0. \tag{3.7}$$

From (3.7), we have the following:

$$\frac{|\alpha_i|^2}{c_{ii}} = \frac{|\alpha_j|^2}{c_{jj}}, \qquad c_{ii}\gamma\alpha_i^{-1}\alpha_j + c_{jj}\alpha_i\alpha_j^{-1}\gamma = 0. \tag{3.8}$$

Also, from (3.8) and $\alpha_j^*\gamma = \gamma^*\alpha_j$, it follows that

$$\alpha_i^*\gamma + \gamma^*\alpha_i = 0.$$

This is a contradiction.

LEMMA 3.3. *Let D be a linearization of $\Box$. If there exists an integer i such that $\alpha_i^*\gamma = \gamma^*\alpha_i$,then we have that for any j $(j \neq i)$,*

$$\frac{|\alpha_i|^2}{c_{ii}} = \frac{|\alpha_j|^2}{c_{jj}} = \frac{|\gamma|^2}{g}, \qquad \gamma^*\alpha_j + \alpha_j^*\gamma = 0. \tag{3.9}$$

Proof. We consider the function given in Lemma 3.2. From (3.8) and $\alpha_i^*\gamma = \gamma^*\alpha_i$, it follows that

$$\frac{c_{jj}}{|\alpha_j|^2}(\gamma^*\alpha_j + \alpha_j^*\gamma)\alpha_i^* = 0. \tag{3.10}$$

From (3.6),(3.7) and (3.8), we have that

$$\alpha_j^*\gamma(\frac{c_{jj}}{|\alpha_j|^2} - \frac{g}{|\gamma|^2}) = h_j. \tag{3.11}$$

Taking the conjugation of (3.11), we have

$$(\alpha_j^*\gamma - \gamma^*\alpha_j)(\frac{c_{jj}}{|\alpha_j|^2} - \frac{g}{|\gamma|^2}) = 0. \tag{3.12}$$

From $\alpha_j^*\gamma - \gamma^*\alpha_j \neq 0$, (3.10),(3.11) and (3.12), it follows that (3.9).

LEMMA 3.4. *Let D be a linearization of $\Box$.*

(1) If $g \neq 0$, then $\gamma \neq 0$.

(2) If $g = 0$, then $\gamma = 0$.

Proof. We assume that $\gamma = 0$. Then the function $f_0(x) = 1$ is a solution of the differential equation $Df = 0$. Since D is a linearization of $\Box$, it follows that

$$0 = \Box f_0(x) = g.$$

This is a contradiction. Therefore (1) is valid. In next, $\gamma \neq 0$. From $g = 0$ and (3.5), it follows that

$$c_{ii}\alpha_i^{-1}\gamma = h_i \quad (i = 0, 1, ..., n). \tag{3.13}$$

Taking the conjugation of (3.13), we have that

$$\alpha_i^*\gamma - \gamma^*\alpha_i = 0 \quad (i = 0, 1, ..., n).$$

From Lemma 3.2,we see that this is a contradiction. Therefore (2) is valid.

LEMMA 3.5. *Let D be a linearization of $\Box$. If $g = 0$, then $h_i = 0$ $(i = 0, 1, ..., n)$.*

Proof. Assume that there exists an integer i such that $h_i \neq 0$. From Lemma 3.4, we see that $\gamma = 0$. Then, the following functions

$$f_1(x) = \alpha_i^{-1}x_i - \alpha_j^{-1}x_j,$$

$$f_2(x) = \alpha_i^{-1}\alpha_j x_i^2 - 2x_ix_j + \alpha_j^{-1}\alpha_i x_j^2$$

are solutions of the differential equation $Df = 0$. Since D is a linearization of $\Box$, we have that

$$\alpha_i^{-1}h_i - \alpha_j^{-1}h_j = 0 \quad (j = 0, 1, ..., n), \tag{3.14}$$

$$c_{ii}\alpha_i^{-1}\alpha_j + c_{jj}\alpha_j^{-1}\alpha_i = 0 \quad (j = 0, 1, ..., n). \tag{3.15}$$

From (3.15), it follows that

$$\frac{|\alpha_i|^2}{c_{ii}} = \frac{|\alpha_j|^2}{c_{jj}}, \quad \alpha_i^*\alpha_j + \alpha_j^*\alpha_i = 0 \quad (j = 0, 1, ..., n, j \neq i). \tag{3.16}$$

Also, from (3,14) and $h_i \neq 0$, it follows that

$$h_j\alpha_i = \alpha_j h_i. \tag{3.17}$$

From (3.16) and (3.17), we have that

$$\frac{2h_j}{h_i}|\alpha_i|^2 = 0.$$

Since $|\alpha_i| \neq 0$, we have $h_j = 0$. From (3.17),we have $h_i = 0$. This is a contradiction.

LEMMA 3.6. *Let D be a linearization of $\Box$. If $g \neq 0$, we have that*

$$\alpha_i + \alpha_j \neq 0$$

for all i, j.

Proof. From Lemma 3.3, we obtain that $\gamma \neq 0$. Assume that there exist two integers i and j such that $\alpha_i + \alpha_j = 0$. We consider the function $f(x)$ given in Lemma 3.2. From (3.7) and $\alpha_i + \alpha_j = 0$, it follows that

$$(c_{ii} + c_{jj})(\alpha_i^{-1}\gamma)^2 = 0.$$

This is a contradiction.

Theorem 3.1. *Let D and $\Box$ be differential operators of (3.1) and (3.2), respectively. The following conditions are all equivalent.*

(1) D is a linearization of $\Box$.

(2) (i) $\alpha_i^*\alpha_j + \alpha_j^*\alpha_i = 2pc_{ii}\delta_{ij} \quad (i,j = 0,1,...,n)$,

(ii) $\alpha_i^*\gamma + \gamma^*\alpha_i = ph_i \quad (i = 0,1,...,n)$,

(iii) $\gamma\gamma^* = pg,\ p > 0$.

(3) $D^*D = DD^* = p\Box \quad (p > 0)$.

Proof. In case $g = 0$, from Lemma 3.4 and Lemma 3.5 it follows that $\gamma = 0, h_i = 0\ (i = 0,1,...,n)$. In this case, from Theorem 1 in [22] we see that the conditions (1),(2) and (3) are equivalent. Therefore,we assume that $g \neq 0$. By the direct calculation, we have that

$$D^*D = \sum_{i=0}^{n} \alpha_i^*\alpha_i \frac{\partial^2}{\partial x_i{}^2} + \sum_{i<j} (\alpha_i^*\alpha_j + \alpha_i^*\alpha_j^*) \frac{\partial^2}{\partial x_i \partial x_j}$$

$$+ \sum_{i=0}^{n} (\alpha_i^*\gamma + \gamma^*\alpha_i) \frac{\partial}{\partial x_i} + \gamma^*\gamma.$$

Hence the implication (2)→(3) is valid. Also, the implication (3)→(1) is obtained by the definition of linearization of $\Box$. We prove the implication (1)→(2). From $g \neq 0$ and Lemma 3.4,it follows that

$$\gamma \neq 0$$

We cosider a function $f_0(x) = E(-\alpha_k\gamma x_k)$. Since f_0 is a solution of $Df = 0$, we have that

$$\frac{c_{kk}}{|\alpha_k|^2}\alpha_k^*\gamma + \frac{g}{|\gamma|^2}\gamma^*\alpha_k = h_k, \quad (k = 0,1,...,n). \tag{3.18}$$

Taking the conjugation of (3.18), we have the following:

$$(\alpha_k^*\gamma - \gamma^*\alpha_k)\left(\frac{c_{kk}}{|\alpha_k|^2} - \frac{g}{|\gamma|^2}\right) = 0, \quad (k = 0,1,...,n). \tag{3.19}$$

If there exists an integer i such that $\alpha_i^*\gamma - \gamma^*\alpha_i = 0$, from Lemma 3.2, it follows that for any integer $j\ (j \neq i, 0 \leq j \leq n)$,

$$\alpha_j^*\gamma - \gamma^*\alpha_j \neq 0.$$

Also, from Lemma 3.3 and (3.19), we have

$$\frac{|\alpha_i|^2}{c_{ii}} = \frac{|\alpha_j|^2}{c_{jj}} = \frac{|\gamma|^2}{g}, \quad (j \neq i, 0 \leq j \leq n). \tag{3.20}$$

Put $p = |\gamma|^2/g$. From(3.18), we obtain the conditions (ii) and (iii). In the next, we prove the condition (i). From Lemma 3.6, we consider the following function:

$$f_1(x) = E(-\mu x_i)E(-\mu x_j), \quad (i, j = 0, 1, ..., n),$$

where $\mu = (\alpha_i + \alpha_j)^{-1}\gamma$. Then we have that

$$\begin{aligned} Df_1 &= -\alpha_i \mu f_1 - \alpha_j \mu f_1 + \gamma f_1 \\ &= -(\alpha_i + \alpha_j)(\alpha_i + \alpha_j)^{-1}\gamma f_1 + \gamma f_1 \\ &= 0. \end{aligned}$$

Since D is a linearization of $\Box$, it follows that

$$(c_{ii} + c_{jj})\mu + g\mu^{-1} = h_i + h_j. \tag{3.21}$$

From (3.18),(3.20) and (3.21), we have that

$$(c_{ii} + c_{jj})\mu + g\mu^{-1} = \frac{(\alpha_i^*\gamma + \gamma^*\alpha_i)|\gamma|^2 c_{ii}}{|\gamma\alpha_i|^2} + \frac{(\alpha_j^*\gamma + \gamma^*\alpha_j)|\gamma|^2 c_{jj}}{|\gamma\alpha_j|^2}. \tag{3.22}$$

From (3.22), it follows that

$$(c_{ii} + c_{jj})\gamma = (\alpha_i + \alpha_j)c_{ii}\frac{\alpha_i^*\gamma}{|\alpha_i|^2} + (\alpha_i + \alpha_j)c_{jj}\frac{\alpha_j^*\gamma}{|\alpha_j|^2}. \tag{3.23}$$

Also,from (3.23), we have that

$$(\alpha_j\alpha_i^* + \alpha_i\alpha_j^*)\frac{\gamma}{p} = 0.$$

Since $\gamma/p \neq 0$, we have that $\alpha_j\alpha_i^* + \alpha_i\alpha_j^* = 0$. Then, we obtain the condition (iii).

4. Linearizations of $\Box_C = \sum_{i,j=0}^n c_{ij}\frac{\partial^2}{\partial x_i \partial x_j} + \sum_{i=0}^n h_i \frac{\partial}{\partial x_i} + g$

In this section, we consider the following differential operators;

$$D = \sum_{i=0}^n \alpha_i \frac{\partial}{\partial x_i} + \gamma, \tag{4.1}$$

$$\Box_C = \sum_{i,j=0}^n c_{ij}\frac{\partial^2}{\partial x_i \partial x_j} + \sum_{i=0}^n h_i \frac{\partial}{\partial x_i} + g, \tag{4.2}$$

where $\alpha_i, \gamma \in C^{n+1}, N(\alpha_i) \neq 0 \quad (i = 0, 1, ..., n)$.

By the argument in section 3, we can assume that $g \neq 0$. We put

$$\alpha_j = \sum_{i=0}^{n} e_i a_{ij} \quad (j = 0, 1, ..., n).$$

For the differential operator D, the matrix $A = (a_{ij})$ is said to be the matrix of D. Also, put

$$C = (c_{ij})$$

for the coefficients c_{ij} $(i, j = 0, 1, ..., n)$ of the differential operator $\Box$.

We can obtain easily the following lemma:

Lemma 4.1. *Let D be a differential operator of (4.1), A be the matrix of D and I be the identity matrix. Then,the following conditions are equivalent.*

(1) $\alpha_i^* \alpha_j + \alpha_i^* \alpha_j = 2r\delta_{ij}$, $(r > 0)$, δ_{ij} is the Kronecker's delta.

(2) ${}^tAA = rI$, $(r > 0)$.

Theorem 4.1. *Let D be a differential operator of the form (4.1), A be the matrix of D and $C = (c_{ij})$ be a real symmetric regular matrix. Then, the following conditions (1),(2) and (3) are all equivalent.*

(1) D is a linearization of $\Box_C$.

(2) (i) $\alpha_i^* \alpha_j + \alpha_j^* \alpha_i = 2pc_{ij}, p \in C$, $\quad (p \neq 0, i, j = 0, 1, ..., n)$.

(ii) $\alpha_i^* \gamma + \gamma^* \alpha_i = ph_i$, $p \in C$, $\quad (p \neq 0, i = 0, 1, ..., n)$.

(iii) $\gamma^* \gamma = pg$, $p \in C$ $(p \neq 0)$.

(3) $D^* D = DD^* = p\Box_C$, $p \in C$ $(p \neq 0)$.

Proof. From the same way as in the proof of Theorem 3.1,it follows that the conditions (2) and (3) are equivalent and the implication (3)→(1) is valid. Hence, we shall prove the implication (1)→(3). Since C is a real symmetric regular matrix, there exists a regular matrix $P = (p_{ij})$ such that

$$({}^tP)^{-1} C P^{-1} = I. \tag{4.3}$$

Put $B = AP^{-1}$, $B = (b_{ij})$, $\beta_k = \sum_{j=0}^{n} e_j b_{jk}$ and $Q = P^{-1}, Q = (q_{ij})$. From $A = BP$, we have that

$$a_{ji} = \sum_{k=0}^{n} b_{jk} p_{ki} \quad (j, i = 0, 1, ..., n).$$

Hence we have that

$$\alpha_i = \sum_{j=0}^{n} e_j a_{ji} = \sum_{k=0}^{n}(\sum_{j=0}^{n} e_j b_{jk}) p_{ki} = \sum_{k=0}^{n} \beta_k p_{ki} \quad (i = 0, 1, ..., n).$$

In the next, we put

$$x_i = \sum_{k=0}^{n} p_{ki} y_k \quad (i = 0, 1, ..., n).$$

Since $\partial x_i / \partial y_k = p_{ki}$, we have that

$$(4.4) \qquad \sum_{i=0}^{n} p_{ki} \frac{\partial}{\partial x_i} = \sum_{i=0}^{n} \frac{\partial x_i}{\partial y_k} \frac{\partial}{\partial x_i} = \frac{\partial}{\partial y_k} \quad (k = 0, 1, ..., n),$$

$$(4.5) \qquad \frac{\partial}{\partial x_i} = \sum_{j=0}^{n} q_{ij} \frac{\partial}{\partial y_j} \quad (i = 0, 1, ..., n).$$

From (4.4),it follows that

$$(4.6) \qquad D_\beta = \sum_{k=0}^{n} \beta_k \frac{\partial}{\partial y_k} + \gamma = \sum_{i=0}^{n}(\sum_{k=0}^{n} \beta_k p_{ki}) \frac{\partial}{\partial x_i} + \gamma = \sum_{i=0}^{n} \alpha_i \frac{\partial}{\partial x_i} + \gamma = D.$$

From (4.3), we have that

$$(4.7) \qquad c_{ij} = \sum_{k=0}^{n} p_{ki} p_{kj} \quad (i, j = 0, 1, ..., n).$$

From (4.4) and (4.7),it follows that

$$(4.8) \qquad \begin{aligned} \square_C &= \sum_{i,j=0}^{n} c_{ij} \frac{\partial^2}{\partial x_i \partial x_j} + \sum_{i=0}^{n} h_i \frac{\partial}{\partial x_i} + g \\ &= \sum_{i,j=0}^{n} (\sum_{k=0}^{n} p_{ki} p_{kj}) \frac{\partial^2}{\partial x_i \partial x_j} + \sum_{i=0}^{n} h_i (\sum_{k=0}^{n} q_{ik} \frac{\partial}{\partial y_k}) + g \\ &= \sum_{k=0}^{n} (\sum_{i=0}^{n} p_{ki} \frac{\partial}{\partial x_i})(\sum_{j=0}^{n} p_{kj} \frac{\partial}{\partial x_j}) + \sum_{k=0}^{n}(\sum_{i=0}^{n} h_i q_{ik} \frac{\partial}{\partial y_k}) + g \\ &= \sum_{k=0}^{n} \frac{\partial^2}{\partial y_k \partial y_k} + \sum_{k=0}^{n}(\sum_{i=0}^{n} h_i q_{ik}) \frac{\partial}{\partial y_k} + g \\ &= \sum_{k=0}^{n} \frac{\partial^2}{\partial y_k^2} + \sum_{k=0}^{n} s_k \frac{\partial}{\partial y_k} + g \\ &= \square, \end{aligned}$$

where

$$s_k = \sum_{i=0}^{n} h_i q_{ik} \; (k = 0, 1, ..., n). \tag{4.9}$$

From (4.6), it follows that

$$\sum_{k=0}^{n} (\beta_k \frac{\partial}{\partial y_k} + \gamma) f(\sum_{k=0}^{n} p_{k0} y_k, ..., \sum_{k=0}^{n} p_{kn} y_k) = 0$$

implies $Df(x_0, x_1, ..., x_n) = 0$. Since D is a linearization of $\Box_C$, we have that $\Box_C f(x_0, x_1, ..., x_n) = 0$. Hence, from (4.8),it follows that

$$\Box f(\sum_{k=0}^{n} p_{k0} y_k, ..., \sum_{k=0}^{n} p_{kn} y_k) = 0$$

Therefore, D_β is a linearization of $\Box$. From Theorem 3.1 and Lemma 4.1,it follows that

$$^tBB = pI, \; p \in C \; (p \neq 0), \tag{4.10}$$

$$\beta_j^* \gamma + \gamma^* \beta_j = p s_j, \; p \in C \; (p \neq 0, j = 0, 1, ..., n), \tag{4.11}$$

$$\gamma^* \gamma = pg, \; p \in C \; (p \neq 0). \tag{4.12}$$

From (4.3) and (4.10), it follows that

$$^tAA =^t (BP)(BP) =^t P(pI)P = pC.$$

Therefore, we obtain the conditions (i) and (iii). In the next,from (4.9),(4.11), $\alpha_i = \sum_{j=0}^{n} \beta_j p_{ji}$ and $QP = I$, it follows that

$$\alpha_i^* \gamma + \gamma^* \alpha_i = \sum_{j=0}^{n} (\beta_j^* \gamma + \gamma^* \beta_j) p_{ji} = \sum_{j=0}^{n} p s_j p_{ji} = \sum_{k=0}^{n} p h_k \{\sum_{j=0}^{n} q_{kj} p_{ji}\} = p h_i$$

for any i.

5. Regularity of Clifford valued functions

At first, we consider the following differential forms:

$$\omega_i = dx_0 \wedge dx_1 \wedge ... dx_{i-1} \wedge dx_{i+1} \wedge ... \wedge dx_n \; (i = 0, 1, ..., n), \tag{5.1}$$

$$(5.2) \qquad \omega = \sum_{i=0}^{n}(-1)^i\alpha_i(x)\omega_i, \quad \Omega = dx_0 \wedge dx_1 \wedge \ldots \wedge dx_n,$$

where $\alpha_i(x)$ $(i = 0, 1, ..., n)$ are non-zero smooth functions defined on R^{n+1} with values in $A_n(R)$. Let G be an open set in R^{n+1} and $f(x)$ be a function defined in G with values in $A_n(R)$. Then, we say that $f(x)$ is a Clifford valued function in G.

DEFINITION 5.1. *$f(x)$ is said to be regular in G iff*
(1) $f(x)$ is smooth in G (i.e. $f_i(x)$ is smooth in G , $i = 0, 1, ..., n$),
(2) $d(\omega f) = 0$ in G.

From Stokes' theorem, we can easily obtain the following proposition.

PROPOSITION 5.1. *Let G be a bounded domain in R^{n+1} with smooth boundary and $f(x)$ be a Clifford valued function defined in G. If $f(x)$ is regular in G and continuous on the closure $\overline{G}$ of G, then*

$$\int_{\partial G} \omega f = 0.$$

Next, we consider the following differential operator:

$$(5.3) \qquad D_\alpha = \sum_{i=0}^{n} \alpha_i(x)\frac{\partial}{\partial x_i} + \sum_{i=0}^{n} \frac{\partial \alpha_i(x)}{\partial x_i},$$

where $\alpha_i(x)$ $(i = 0, 1, ..., n)$ are coefficients of the differential form ω of (5.2).

From the definition of regular functions, we have the following proposition.

PROPOSITION 5.2. *Let G be an open set in R^{n+1} and $f(x)$ be a smooth function in G. Then,the following conditions are equivalent:*
(1) $d(\omega f) = 0$ in G.
(2) $D_\alpha f(x) = 0$ in G.

Proof. From

$$d(\omega f) = \sum_{i=0}^{n}\{\alpha_i(x)\frac{\partial f}{\partial x_i} + \frac{\partial \alpha_i(x)}{\partial x_i} f\}\Omega = D_\alpha f \Omega,$$

it follows that (1) and (2) are equivalent.

Let $\beta_i(x)$ $(i = 0, 1, ..., n)$ be non-zero smooth functions defined in R^{n+1} with values in $A_n(R)$ and $y = e_0y_0 + e_1y_1 + ... + e_ny_n$ be a point in R^{n+1}. We consider the following function $H_\beta(y; x)$:

$$(5.4) \qquad H_\beta(y;x) = \sum_{i=0}^{n} \frac{\beta_i(y)(y_i - x_i)}{|y - x|^{n+1}}, \quad x \in R^{n+1} - \{y\}.$$

THEOREM 5.1. *Let ω be the differential form of (5.2) and $H_\omega(x; y)$ be the function of (5.4). Let G be an open set in R^{n+1}. Then, the following conditions are equivalent.*

(1) $\beta_i(x)\alpha_j(x) + \beta_j(x)\alpha_i(x) = 2p(x)\delta_{ij}, \quad p(x) \neq 0 \ (i, j = 0, 1, ..., n),$

$\sum_{j=0}^{n} \frac{\partial \beta_i(x)}{\partial x_j}\alpha_j(x) = 0 \quad (i = 0, 1, ..., n),$

in G.

(2) For every point y in G and for every Clifford function $f(y)$ on G with values in $A_n(R)$,

$$d(H_\beta(x;y)) \wedge (\omega f(x)) = 0$$

on $G - \{y\}$.

Proof. By a direct calculation, we have

$$(5.5) \qquad d(H_\beta(x;y)) \wedge (\omega f(x)) = \sum_{i=0}^{n}\{\sum_{j=0}^{n} \frac{\partial \beta_i(x)}{\partial x_j}\alpha_j(x)\}\frac{x_i - y_i}{|x - y|^{n+1}} f\Omega$$

$$+\{\sum_{i=0}^{n} \frac{\beta_i(x)\alpha_i(x)}{|x - y|^{n+1}} - (n+1)\sum_{i=0}^{n} \frac{\beta_i(x)\alpha_i(x)(x_i - y_i)^2}{|x - y|^{n+3}}\}f\Omega$$

$$-(n+1)\{\sum_{\substack{i,j=0 \\ i<j}}^{n} \frac{(\beta_i(x)\alpha_j(x) + \beta_j(x)\alpha_i(x))}{|x - y|^{n+1}}(x_i - y_i)(x_j - y_j)\}f\Omega.$$

From (5.5), we have the implication (1)→(2). In the next,we prove the implication (2)→(1). For each index i, we put

$$x_k - y_k \neq 0 \ (k = i), \quad x_k - y_k = 0 \ (k \neq i).$$

From (5.5) and $d(H_\beta(x; y)) \wedge (\omega f(x)) = 0$, it follows that for each i,

$$\sum_{j=0}^{n} \frac{\partial \beta_i}{\partial x_j}\alpha_j \frac{y_i - x_i}{|y - x|^{n+1}} = \frac{1}{|y - x|^{n+1}}\{\sum_{j=0}^{n} \beta_j\alpha_j - (n+1)\beta_i\alpha_i\}$$

in G. Therefore, we have that

$$\sum_{j=0}^{n} \frac{\partial \beta_i}{\partial x_j} \alpha_j (y_i - x_i) = \sum_{j=0}^{n} \beta_j \alpha_j - (n+1)\beta_i \alpha_i, \ (i = 0, 1, ..., n) \tag{5.6}$$

in $G - \{y\}$. Since $\partial \beta_i / \partial x_j$ and α_j $(i, j = 0, 1, ..., n)$ are continuous in G, from (5.6), we have that for every point x in G,

$$\sum_{j=0}^{n} \beta_j(x)\alpha_j(x) = (n+1)\beta_i(x)\alpha_i(x), \quad \sum_{j=0}^{n} \frac{\partial \beta_i(x)}{\partial x_j} \alpha_j(x) = 0, \tag{5.7}$$

$(i = 0, 1, ..., n)$. In next, for each i and j, we put that

$$x_i - y_i \neq 0, \ x_j - y_j \neq 0, \ x_k - y_k = 0 \ (k \neq i, j).$$

By the similar way as in the above proof, from (5.7) we have that for every point x in G,

$$\beta_i(x)\alpha_j(x) + \beta_j(x)\alpha_i(x) = 0 \ (i \neq j). \tag{5.8}$$

From (5.7) and (5.8), we have the condition (1).

THEOREM 5.2. *Let G be a domain in R^{n+1}. Then the following conditions are equivalent.*

(1) $\beta_i(x)\alpha_j(x) + \beta_j(x)\alpha_i(x) = 2p(x)\delta_{ij}$, $\quad p(x) \neq 0 \ (i, j = 0, 1, ..., n)$,

$\sum_{j=0}^{n} \frac{\partial \beta_i(x)}{\partial x_j} \alpha_j(x) = 0 \ (i = 0, 1, ..., n)$,

in G.

(2) For every continuously differentiable solution f of $D_\alpha f = 0$ and for every ball B such that $B \subset \overline{G}$,

$$\frac{1}{S_{n+1}} \int_{\partial B} H_\beta(y; x)\omega f(y) = \begin{cases} 0, & x \in G - \overline{B} \\ p(x)f(x), & x \in B. \end{cases} \tag{5.9}$$

Proof. At first, we prove the implication (1)→(2). If $x \in G - \overline{B}$, then the function $H_\beta(x; y)$ is continuously differentiable on B. From Stokes' theorem, we have

(5.10)

$$\begin{aligned} \int_{\partial B} H_\beta(y; x)\omega f(y) &= \int_B d(H_\beta(y; x)\omega f(y)) \\ &= \int_B d(H_\beta(y; x)) \wedge (\omega f(y)) + \int_B H_\beta(y; x) d(\omega f(y)). \end{aligned}$$

From Theorem 5.1 and the condition (1), it follows that

$$d(H_\beta(y;x)) \wedge (\omega f(y)) = 0$$

on B. From Proposition 5.2, we have that $d(\omega f) = (Df)\Omega$ on B. Hence,it follows that

$$\int_{\partial B} H_\beta(y;x)\omega f(y) = 0$$

on $G - \overline{B}$. In the next, let x be a point in B and V_ϵ be an open ball of radius ε about x such that $\overline{V_\epsilon(x)} \subset B$. Put $B_\epsilon = B - \overline{V_\epsilon}$. Applying Stokes' theorem to the form $H_\beta(y;x)(\omega f(y))$ on the domain B_ϵ , from Theorem 5.1 and Proposition 5.1, we have

$$\int_{\partial B} H_\beta(y;x)(\omega f(y)) - \int_{\partial V_\epsilon} H_\beta(y;x)(\omega f(y)) = \int_{B_\epsilon} H_\beta(y;x)d(\omega f(y)) = 0.$$

Again,applying Stokes' theorem, we have that

$$\int_{\partial V_\epsilon} H_\beta(y;x)(\omega f(y)) = \frac{n+1}{\varepsilon^{n+1}} \int_{V_\epsilon} p(y)f(y)\Omega$$

$$+ \frac{n+1}{\varepsilon^{n+1}} \int_{V_\epsilon} \{\sum_{i=0}^{n} \alpha^*(y_i - x_i)\} d(\omega f(y)),$$

where $p(y)$ is the function in Theorem 5.1. Since f is continuously differentiable on B, we have that

$$\lim_{\epsilon \to 0} \int_{\partial V_\epsilon} H_\beta(y;x)(\omega f(y)) = S_{n+1}p(x)f(x).$$

Hence,from (5.10) we have (5.9). In the next, we prove the implication (2)→(1). Let x be an arbitrary point in G and ζ be an arbitrary element in $G - \{x\}$. Let $B = B(\zeta)$ be a neighbourhood of ζ such that $x \in G\text{-}\overline{B}$. Then, the condition (2) is that

$$\int_{\partial B} H_\beta(y;x)\omega f(y) = 0. \tag{5.11}$$

From Stokes' theorem and (3.11),it follows that

$$\int_B d(H_\beta(y;x)) \wedge (\omega f(y)) + \int_B H_\beta(y;x)d(\omega f(y)) = 0.$$

Also,from $d(\omega f) = 0$, we have that $\int_B d(H_\beta(y;x)) \wedge (\omega f(y)) = 0$. Since the functions in (5.5) and f are continuously differentiable on B,it follows that $d(H_\beta(y;x)) \wedge (\omega f(y)) = 0$ on B. Therefore, we have that

$$d(H_\beta(y;x)) \wedge (\omega f(y)) = 0,\ y \in G - \{x\}.$$

From Theorem 5.1, we have the condition (1).

Since $\alpha_i(x)$ and $\beta_i(x)$ $(i = 0, 1, ..., n)$ are functions with values in R^{n+1}, from Theorem 5.2, we obtain the following Cauchy type's integral formula.

COROLLARY(INTEGRAL FORMULA). *Let G be a bounded domain in R^{n+1} with smooth boundary and $f(x)$ be a continuous function on the closure of G and regular in G. If $\alpha_i(x)$, $\beta_i(x)$ $(i = 0, 1, ..., n)$ satisfy the following conditions:*

$$\beta_i(x)\alpha_j(x) + \beta_j(x)\alpha_i(x) = 2p(x)\delta_{ij},\ p(x) \neq 0\ (i, j = 0, 1, ..., n),$$

$$\sum_{j=0}^{n} \frac{\partial \beta_i(x)}{\partial x_j} \alpha_j(x) = 0\ (i = 0, 1, ..., n),$$

in R^{n+1}, then we have that

$$f(x) = \frac{p^{-1}(x)}{S_{n+1}} \int_{\partial B} H_\beta(y; x)\omega f(y)$$

in G,where S_{n+1} is the area of unit sphere.

REMARK. *(1) If $p(x)$ is a r-times differentiable function,then $f(x)$ is r-times differentiable.*

(2) If $p(x)$ is an analytic function, then $f(x)$ is analytic.

REFERENCES

1. F.Brackx,R.Delanghe and F.Sommen, "Clifford Analysis," Research Notes in Mathematics, 76, Pitman Books Ltd., 1982.
2. R.Brauder and H Weyl, *Spinors in n-dimensions*, Amer. J. Math., **57** (1935), 425–449.
3. C.Chevalley, "The algebraic theory of spinors," Columbia Univ. Press, New York, 1954.
4. E.Cartan, "Theory of spinors," Dover, New York, 1966.
5. J.S.R.Chisholm and R.S.Farwell, *Spin gauge theory of electric magnetic spinors*, Proc. R. Soc. London, **A337** (1981), 1–23.
6. C.A.Deavours, *The quaternion calculas*, Amer. Math. Mon., **80** (1973), 138–162.
7. R.Delanghe, *On regular analytic functions with values in a Clifford algebra*, Math. Ann., **185** (1970), 91–111.
8. R.Fueter, *Die funktionentheorie der differentialgeleichungen $\Delta u = 0$ und $\Delta\Delta u = 0$ mit vier reellen variablen*, Comment. Math. Helv., **7** (1935), 303–330.
9. B.Goldschmit, *Regularity properties of generalized analytic vectors in R^{m+1}*, Math. Nachr., **103** (1980), 245–254.
10. B.Goldschmit, *A Cauchy integral formula for a class of elliptic systems of partial differential equations of first order in the space*, Math. Nachr., **108** (1982), 167–178.

11. F.Gürsey and H.C.Tze, *Complex and quaternionic analyticity in chiral and gauge theories I*, Ann. Phys., **128** (1980), 29–130.
12. K.Imaeda, *A new formulation of electromagnetism*, Nuovo Cimento, **32B** (1976), 138–162.
13. P.Lounesto, *Spinor valued regular functions in hypercomplex analysis*, Doctoral thesis, Helsinki Univ. of Tech. (1979).
14. P.Lounesto, *Sur les idéaux á gauche des algebres de Clifford et les produits scalaires des spineurs*, Ann. Inst. H. Poincaré, Sect. A **33** (1980), 53–61.
15. A.Micalli and O.E.Villamator, *Sur les algebres de Clifford*, Ecole Normales Sup.,4é serie **1** (1968), 271-304.
16. A.O.Morris, *On generalized Clifford algebra*, Q. J. Math., **18** (1967), 7–12.
17. K.Morinaga and T.Nôno, *On the linearization of a form of higher degree and its representation*, J. Hiroshima Univ. (A) **16** (1952), 13–41.
18. M.Naser, *Hyperholomorphic functions*, Siberian Math. J., **12** (1971), 959–968.
19. K.Nôno, *On the quaternion linearization of Laplacian Δ*, Bull. Fukuoka Univ. of Ed. Part III **35** (1985), 5–10.
20. K.Nôno, *Characterization of domains of holomorphy by the existence of hyper-conjugate harmonic functions*, Rev. Roumaine Math. Pures et Appl. **31** (1986), 159–161.
21. K.Nôno, *On the octonionic linearization of Laplacian and octonionic function theory*, Bull. Fukuoka Univ. of Ed. part III **37** (1988), 1–15.
22. K.Nôno, *On the linearization of a partial differential operator*, Bull. Fukuoka Univ. of Ed. part III **40** (1991), 1–11.
23. T.Nôno, *On the linearization of partial differential equation $\nabla^m\psi = c^m\psi$*, Bull. Fukuoka Univ. of Ed. part III **21** (1971), 35–41.
24. T.Nôno, *Generalized Clifford algebra and linearization of a partial differential equation: $g^{i_1 i_2 \ldots i_m}\nabla_{i_1}\nabla_{i_2}\ldots\nabla_{i_m}\psi = c^m\psi$*, Proc. Conf. on Clifford algebra, it generalization and applications (1971), 1–24, MATSIENCE, Madras.
25. M.Riesz, "Clifford numbers and spinors," Lecture Notes 38, Institute for Fluid Dynamics and Applied Mathematics , Univ. of Maryland, 1958.
26. J.Ryan, *Complexified Clifford analysis*, Complex Variables **1** (1982), 119–149.
27. J.Ryan, *Properties of isolated singularities of some functions taking values in real Clifford algebra*, Math. Proc. Camb. Phil. Soc., **95** (1984), 227–298.
28. F.Sommen, *Some connections between Clifford. analysis and complex analysis*, Complex Variables, **1** (1982), 97–118.
29. E.M.Stein and G.Weiss, *Generalizations of the Cauchy-Riemann equations and representations of the rotation group*, Amer. J. Math., **90** (1968), 163–196.
30. A.Sudbery, *Quaternionic analysis*, Math. Proc. Camb. Phil. Soc., **85** (1979), 199–225.

THE CLASS OF COMMUTATIVE ALGEBRAS

Robert H. Oehmke

Department of Mathematics, The University of Iowa, Iowa City, IA 52242-1466

Abstract

Considerable attention has been given to special subclasses of the class C of commutative algebras. In particular, the class of commutative power associative algebras and the smaller class of Jordan algebras have been studied extensively. In its entirety the class C seems to be rather intractable with our present set of tools. Therefore it seems reasonable that other special subclasses be identified for further study. However, it is appropriate that such a selection be motivated by some external impetus as well as the desire to have the class amenable to a study of its structure. Thus, here we will present a relationship between our class C and certain geometric considerations on differential manifolds. We will then present some results on n-dimensional algebras in C with special attention being given to $n = 8$. The interest in $n = 8$ arises from an interest in $sl(3)$. The methods we present here are particularly wedded to the finite dimensional case.

Most of the results presented here have appeared elsewhere and thus we will not include their proofs. Any other results are quite straightforward and should be easily seen by the reader.

1. Geometric Considerations

We can begin with a differentiable manifold M and a fiber bundle $E \equiv E(M, G, \pi)$. We have π as a differential map $\pi : E \to M$. If $x \in M$ then the fiber at x is $\pi^{-1}(x)$.

The group G stands for a group that acts transitively on each fiber of M. If $T_x(M)$ and $T_u(E)$ represent, respectively, the tangent spaces to M at x and to E at u then π induces a surjective, linear map $\pi_* : T_u(E) \to T_{\pi(u)}(M)$. We denote the kernel of π_* at u by G_u and call it the subspace of vertical vectors to E at u.

For each u of E we select a subspace Q_u of $T_u(E)$ subject to the restrictions:

1. $T_u(E) = G_u \oplus Q_u$ (a direct sum as vector spaces).
2. If $R_g : u \to ug$ for $g \in G$ then

$$(R_g)_* : Q_u \to Q_{ug} \quad .$$

3. Q_u depends differentiably on u.

Such a selection of subspaces Q_u is called a connection. For a given connection, Γ, the subspace, Q_u, is called the set of horizontal vectors to E at u. For each u, π_* is injective and surjective when restricted to Q_u. Thus for each vector field X on M we can define a unique "horizontal lift" to a vector field on E. In turn, if α is a path on M from x_0 to x_1 and u is an element of $\pi^{-1}(x_0)$ then α can be uniquely lifted to a path on E that starts at u. Such paths will define a mapping, as we let u vary, from the fiber at x_0 to the fiber at x_1. This gives rise to holonomic considerations of closed curves on M, i.e., a group of mappings of the fiber $\pi^{-1}(x_0)$ into itself.

We will restrict our attention to the fiber bundle $T(M) = UT_x(M)$ taken over all $x \in M$. In a standard manner each connection relates to a covariant differentiation on M. We let $\mathcal{X}$ be the space of vector fields on M and $\mathcal{E}(\mathcal{X})$ the set of endomorphisms on $\mathcal{X}$. Covariant differentiation is a mapping $v : \mathcal{X} \to \mathcal{E}(\mathcal{X})$ such that if $\nabla(X) = \nabla_X$ for $X \in \mathcal{X}$ we have:

1. $\nabla_{Y+Z}(X) = \nabla_Y(X) + \nabla_Z(X)$
2. $\nabla_Y(X + Z) = \nabla_Y(X) + \nabla_Y(Z)$
3. $\nabla_{fY}(X) = f \nabla_Y (X)$
4. $\nabla_Y(fX) = f \nabla_Y (X) + (Yf)X$

where X, Y, Z are in $\mathcal{X}$ and f is a differentiable function on M.

More complete expositions of this material are available in a number of places, some of which are listed in the bibliography.

2. Multiplications and Algebras

Let V be a vector space over a base field F. A binary map α on V, denoted by $\alpha(x, y) = x \cdot y$ is a multiplication on V if

1. $(x + y) \cdot z = x \cdot z + y \cdot z$
2. $z \cdot (x + y) = z \cdot x + z \cdot y$
3. $(x \cdot y) = (ax) \cdot y = x \cdot (ay)$

for $x, y, z, \in, V$ and $a \in F$. The system $A \equiv (V, \cdot)$ is called an algebra over V. Right and left multiplications by x are, respectively, the maps

$$R_x : y \to y \cdot x$$

$$L_x : y \to x \cdot y \quad .$$

The properties of the multiplication guarantee the linearity of these maps. Given the algebra $A \equiv (V, \cdot)$ we are interested also in two other related algebras, $A^+ \equiv (V, \circ)$ and $A^- \equiv (V, *)$ where the multiplications are defined by

$$x \circ y = \frac{1}{2}(x \cdot y + y \cdot x)$$

$$x * y = x \cdot y - y \cdot x \quad .$$

From the relations

$$R_x^- = R_x - L_x = -L_x^-$$

$$R_x^+ = \frac{1}{2}(R_x + L_x) = L_x^+$$

we see that A is completely determined by A^- and A^+.

If V is finite dimensional over F with a basis $u_1 \ldots , u_n$ then A is also completely determined by the $n - tuple$ $(R_{u_1} \ldots , R_{u_n})$ of right multiplications. This follows from noting that if $y \in V$ and $y = \Sigma \alpha_i u_i$ then

$$R_y = \Sigma \alpha_i \, R_{u_i} \quad .$$

3. Some Restrictions

It is easily seen that a covariant differentiation on the set $\mathcal{X}$ of vector fields defines an algebra on $\mathcal{X}$ in keeping with the above definitions. However, $\mathcal{X}$ is infinite dimensional over $\mathbb{R}$. Thus, the set of connections relates to a class of infinite dimensional algebras. In order to keep our study tractable we will place some restrictions on M and on the connections we wish to examine. We first assume there is a group that acts transitively on M. Thus, we can reduce our considerations to vector fields that are invariant under the group action. This reduces our study to a class of finite dimensional algebras. Next, we will only consider connections that have 0 torsion. thus, we can assume that for each algebra A in our restricted class, the multiplication of A^- is lie differentiation. With this restriction, A is completely determined by A^+, i.e., by the n-tuple

$$(R^+_{u_1}, \ldots, R^+_{u_n})$$

where

$$u_i R^+_{u_j} = u_j R^+ u_i \quad .$$

We let $\mathbb{C}(M)$ be the set of all such n-tuples. We can easily obtain the following algebraic results:

1. $\mathbb{C}(M)$ is an $\frac{n^2(n+1)}{2}$-dimensional vector space over F.
2. $\mathbb{C}(M)$ is an $M_n(F)$-module.
3. There is a multiplication $*$ such that $(\mathbb{C}(M), *)$ is a special Jordan algebra.
4. $(\mathbb{C}(M), *)$ has a multiplicative identity element.

5. $(\mathbb{C}(M), *)$ has n^2 primitive pairwise orthogonal idempotents.

6. The 1-spaces of these idempotents are isomorphic to F.

7. The $\frac{1}{2}$-spaces are either 0- or 1-dimensional (completely determinable).

8. $(\mathbb{C}(M), *)$ is unique and only dependent on n.

9. If S is a nonsingular $n \times n$ matrix then the map

$$(A_1, \dots, A_n) \to (SA, S^{-1}, \dots, SA_nS^{-1})$$

is an automorphism of $(G(M), *)$.

10. If p is a permutation of $(1, \dots, n)$ and P the corresponding $n \times n$ permutation matrix then the map

$$(A_1, \dots, A_n) \to (PA_{p(1)}, \dots, PA_{p(n)})$$

is an automorphism of $(G(M), *)$.

11. Let $J = (D_1, \dots, D_n)$ where all entries d_{ij}^k are 0 except for d_{kj}^k which are equal to 1. J is the identity element of $(\mathbb{C}(M), *)$.

If $(A_1, \dots, A_n)$ is an n-tuple in $\mathbb{C}(M)$ we let a_{ij}^k be the ij-entry of A_k. The product $(A_1, \dots, A_n) * (B_1, \dots, B_n) = (C_1, \dots, C_n)$ of two elements of $\mathbb{C}(M)$ is defined by

$$c_{ij}^k = \frac{1}{2} \sum_{s=1}^{n} (a_{ij}^s b_{kj}^s + b_{ij}^s, a_{kj}^s) \quad .$$

These results are only about the algebraic relationship between the objects and not on the nature of the objects. Thus, these results will translate to results on the "geometric" relationships between our connections and not on the geometric nature of the individual connections.

4. The Manifold SL(3)

In this section we give special attention to the manifold $SL(3)$ and its attached lie algebra $s\ell(3)$. For $s\ell(3)$ we will use the basis $u_1 = E_{11} - E_{22}, u_2 = E_{11} - E_{33}, u_3 = E_{13}, u_4 = E_{12}, u_5 = E_{21}, u_6 = E_{13}, u_7 = E_{31}$ and $u_8 = E_{32}$ where E_{ij} is the 3×3 matrix whose only nonzero entry is a 1 as the ij-entry. Again, we are interested in 8-tuples of 8×8 matrices. $(R_1, \dots, R_8)$ where R_i is to represent a right multiplication by the basal element u_i and for which the multiplication is commutative. We add still another restriction on our 8-tuples, namely, that the corresponding algebra admits automorphisms that are induced by conjugation by elements of the Heisenberg group, i.e., the mappings

$$x \to \alpha x \alpha^{-1}$$

where

$$\alpha = \begin{bmatrix} 1 & a & b \\ 0 & 1 & c \\ 0 & 0 & 1 \end{bmatrix}$$

and $x \in so(3)$ is an automorphism. These algebras have been studied in an earlier paper [3]. The set of all such algebras forms a 13- dimensional subspace of $\mathbb{C}(M)$. In the cited paper a basis is given for this subspace along with a delineation of the automorphism group of each algebra in this subspace. It has not yet been determined if this subspace is a subalgebra of the Jordan algebra $(\mathbb{C}(M), *)$.

For each of our algebras A we designate a subalgebra, $H(A)$, of the algebra of all endomorphisms of A that is generated by certain polynomials in the right multiplications of A. The algebra $H^-(A)$ is a lie algebra and its corresponding lie group is the holonomy group of $SL(3)$. We shall call $H(A)$ the holonomy algebra of A. Since the above polynomials (related to the curvature tensor) are quadratic, it is difficult to use any linearity property on them. Thus, even though we have an extensive computer run on the description of $H(A)$ for A equal to any one of our basal algebras we have not been able to extend these results to more general algebras.

As an attempt to minimize the computations we have contracted $so(3)$ to a four-dimensional algebra generated by the elements $u_1 + u_2, u_3 + u_4, u_5 + u_7$ and $u_6 + u_8$. Not all of our choices of connections are compatible with this contraction. In fact, this reduces our study to a 3-dimensional subspace of $\mathbb{C}(M)$. This subspace is currently under study.

Bibliography

1. S. Helgason, *Differential Geometry and Symmetric Spaces*, Academic Press, New York, 1962.
2. S. Kobayashi and K. Nomizu, *Foundations of Differential Geometry*, John Wiley & Sons, New York, Vol. 1, 1963.
3. J. F. Oehmke and R. H. Oehmke, Lie-admissible Algebras with Specified Automorphism Groups, *Hadronic J.* **4** (1981), 550–579.
4. R. H. Oehmke, On the Geometry of Lie-admissible Algebras, *Hadronic J.* **5** (1982), 518–546.
5. R. H. Oehmke, On the Geometry of Lie-admissible Algebras II, *Hadronic J.* **7** (1984), 62–72.

6. R. H. Oehmke, *Holonomy Algebras of Certain Connections, Proc. 14th Int. Colloquium on Group Theoretical Methods in Physics* (1985), 205–208.

7. R. H. Oehmke, *On Connections and Holonomy Algebras, Proc. XV Int. Conference on Differential Geometry Methods in Theoretical Physics* (1987), 535–543.

8. W. Sagle and R. Walde, *Introduction to Lie Groups and Lie Algebras*, Academic Press, New York, 1973.

Stability theorems of K_1-functor for Chevalley groups

E. B. Plotkin

Department of Mathematics

Riga Technical University, Riga, Latvia

Abstract

The aim of the paper is to consider not studied cases of inclusions of root systems and to display the complete list of conditions providing surjective stabilization of K_1-functor. We proceed from fundamental M. Stein's results and use methods of weight diagrams developed in his work.

Introduction

The aim of the paper is to consider cases not studied in [21] of inclusions of root systems and to display the complete list of conditions providing surjective stabilization for all maximal inclusions of root systems. We proceed from fundamental M. Stein's results [20],[21] and use methods developed there. Besides that, we suppose that the consequent application of basic representations and weight diagrams technique [16],[21] may be useful in various questions of structure of Chevalley groups [31],[29],[37],[30].

The author is grateful to the organizers of the conference "Nonassociative Algebras and Related Topics" for stimulating talks, which led to progress in this field.

1. The main definitions

Suppose that Φ is a reduced irreducible root system and $G(\Phi,\ )$ is the Chevalley-Demazure group scheme. If R is a commutative ring, then $G(\Phi, R)$ is called a

Chevalley group of type Φ over R. Let $\Pi = \{\alpha_1, \cdots, \alpha_n\}, \Phi^+, \Phi^-$ will denote subsystems of simple, positive and negative roots respectively. To each root $\alpha \in \Phi$ there correspond elementary root unipotent elements $x_\alpha(t), \alpha \in \Phi, t \in R$.

For a set X a symbol $\langle X \rangle$ will denote the operation of generating in the following sense: if $X \subset G(\Phi, R)$, then $\langle X \rangle$ denotes the subgroup generated by X; if $X \subset R$, then $\langle X \rangle$ denotes the ideal in R generated by X; and if $X \subset \Phi$, then $\langle X \rangle$ will be the minimal closed subsystem of roots in Φ which contains X.

Let, as usual,

$$E(\Phi, R) = \langle x_\alpha(t), \alpha \in \Phi, t \in R \rangle,$$
$$V(\Phi, R) = \langle x_\alpha(t), \alpha \in \Phi^-, t \in R \rangle,$$
$$U(\Phi, R) = \langle x_\alpha(t), \alpha \in \Phi^+, t \in R \rangle.$$

A group $E(\Phi, R)$ is called an elementary subgroup of the Chevalley group $G(\Phi, R)$.

It is well known that $G(\Phi, R) = E(\Phi, R)$ if $G(\Phi, \ )$ is simply connected and R is a field or, more generally, a semilocal ring.

Let $T(\Phi, R)$ be a splittable maximal torus in $G(\Phi, R)$. Then, for semilocal rings holds ([1])

$$G(\Phi, R) = E(\Phi, R) T(\Phi, R).$$

Denote in the standard way,

$$w_\alpha(t) = x_\alpha(t) x_\alpha(-t^{-1}) x_\alpha(t), \ t \in R^*,$$
$$h_\alpha(t) = w_\alpha(t) w_\alpha(1)^{-1},$$

where R^* is the multiplicative group of R. Let

$$N(\Phi, R) = \langle w_\alpha(t), \alpha \in \Phi, t \in R^* \rangle,$$
$$H(\Phi, R) = \langle h_\alpha(t), \alpha \in \Phi, t \in R^* \rangle.$$

Then the Weyl group $W(\Phi)$ of the system Φ is isomorphic to $N(\Phi, R)/H(\Phi, R)$. Starting from this moment we shall assume that $G(\Phi, R)$ is simply connected.

The question whether the subgroup $E(\Phi, R)$ is the normal subgroup in $G(\Phi, R)$ had been opened for a long period. The negative result in the case of $\mathrm{rk}\Phi = 1$ has been made by Cohn [9], Swan [26], Suslin [24]. Moreover, for the case of associative rings there exists a ring A, such that the subgroup $E_n(A)$ of all elementary matrices in a general linear group is not a normal subgroup there [11], [12]. However, for any Chevalley group $G = G(\Phi, R), \mathrm{rk}\Phi \geq 2$ holds

$$E(\Phi, R) \triangleleft G(\Phi, R)$$

[27], [36], [2], [25], [15]. Therefore, one can consider

$$K_1(\Phi, R) = G(\Phi, R)/E(\Phi, R).$$

This is the definition of K_1-functor for a Chevalley group. If $\Phi = A_n$, then in this way appears usual SK_1-functor. The inclusion of root systems $\Delta \subset \Phi$ induces homomorphisms $\nu : G(\Delta, R) \longrightarrow G(\Phi, R)$; $\nu : E(\Delta, R) \longrightarrow E(\Phi, R)$ and $\overline{\nu} : K_1(\Delta, R) \longrightarrow K_1(\Phi, R)$.

The surjective (injective) stability question is to find conditions on the ring R, depending on $\Delta \subset \Phi$, which provides the surjectivity or injectivity of the homomorphism $\overline{\nu}$.

The surjective stabilization is equivalent to the existence of the decomposition

$$G(\Phi, R) = E(\Phi, R)G(\Delta, R)$$

and injective one is equivalent to the formula

$$E(\Delta, R) = G(\Delta, R) \cap E(\Phi, R).$$

Therefore, speaking about stability of K_1-functor for Chevalley groups, we may simply take in mind the existence of above formulas.

The problem of stabilization for K_1-functor for Chevalley groups over rings is the natural extension of the similar problem for classical groups. It was mentioned in the book by Bass [5]. After the works by Bass, Vaserstein and others for classical groups [4], [32], [33] etc. (where a ring is not necessary commutative), the essential progress from the general positions of Chevalley groups had been made in [16] and in the Stein's papers [21], [22]. The definition of K_1-functor and definitions below follow [21].

A commutative ring R satisfies the absolute stable range condition ASR_n if, given $(r_1, \cdots, r_n) \in R^n$, there exist $t_1, \cdots, t_{n-1} \in R$ such that every maximal ideal of R containing the ideal $\langle r_1 + t_1 r_n, \cdots, r_{n-1} + t_{n-1} r_n\rangle$ also contains the ideal $\langle r_1, \cdots, r_n\rangle$.

Evidently, if $(r_1, \cdots, r_n) \in R^n$ is unimodular, then the restricted row $(r'_1, \cdots, r'_{n-1}) \in R^{n-1}$, $r'_i = r_i + t_i r_n$ is also unimodular and ASR_n condition is transformed to the stable range condition SR_n. It is known that if the dimension of the maximal spectrum $\dim \mathrm{Max}(R) = n - 2$ then both these conditions are fulfilled [10], [4], [32].

At last, the condition ASR_n satisfies the usual properties, i.e. for every ideal $I \subset R$ it may be lifted to R/I and if $n \geq m$ then ASR_m implies ASR_n.

2. Basic representations

We remind that from the structure of the scheme G it follows that to each complex representation π of a Lie algebra g of type Φ there corresponds a representation π of the Chevalley group $G(\Phi, R)$ over a free R-module V [23], [16]. It is extremely important to choose the representation π in some optimal way.

Let $\Lambda(\pi), \Lambda^*(\pi), \Delta(\pi)$ be the sets of all weights of π, of all non-zero weights of π and of all simple roots, which are simultaneously the weights of π, respectively.

The natural way to transfer approaches used for classical groups in the general case is to choose such a representation π that all weight submodules $V^\lambda, \lambda \in \Lambda^*(\pi)$ have multiplicity one.

An irreducible representation π is called a basic representation if the Weyl group $W(\Phi)$ acts transitively on the set $\Lambda^*(\pi)$, [16].

This definition is equivalent to the following one: if $\lambda_1 - \lambda_2 = \alpha$, $\lambda_1, \lambda_2 \in \Lambda^*(\pi), \alpha \in \Pi$, then there exists a fundamental reflection $w_\alpha \in W(\Phi)$ such that $w_\alpha \lambda_1 = \lambda_2$.

Each basic representation satisfies the condition on multiplicity of $\Lambda^*(\pi)$. Besides that, $\dim V^0 = \mathrm{card}\Delta(\pi)$. It will be useful further to speak about different zero weights $\widehat{\alpha_i}, \alpha_i \in \Delta(\pi)$.

The module $V = \sum V^\lambda \oplus V^0, \lambda \in \Lambda^*(\pi)$ has a special basis $v^\lambda \in V^\lambda, \lambda \in \Lambda^*(\pi), v^0_\alpha \in V^0, \alpha \in \Delta(\pi)$ in which elementary root unipotent elements act in the simplest way. This action is described by Lemma 2.3 of [16], which is our basic tool.

Instead of $\pi(x_\alpha(t))v^\lambda$, we shall write simply $x_\alpha(t)v^\lambda$. Remind now, that an element v of an R-module V is called unimodular, if there exists $f \in V^* = \mathrm{Hom}(V, R)$ such that $f(v) \in R^*$. For a free R-module V this means that f belongs to some basis of the module V.

LEMMA ([16]). *Elements $x_\alpha(t)$ act in the chosen basis of V as follows:*
1. *If $\lambda \in \Lambda^*(\pi), \lambda + \alpha \notin \Lambda(\pi)$, then $x_\alpha(t)v^\lambda = v^\lambda$,*
2. *If $\lambda, \lambda + \alpha \in \Lambda^*(\pi)$, then $x_\alpha(t)v^\lambda = v^\lambda \mp tv^{\lambda+\alpha}$,*
3. *If $\alpha \notin \Lambda(\pi)$, then $x_\alpha(t)v^0 = v^0$ for each $v^0 \in V^0$,*
4. *If $\alpha \in \Lambda(\pi)$, then $x_\alpha(t)v^{-\alpha} = v^{-\alpha} \mp tv^0(\alpha) \mp t^2 v^\alpha$, $x_\alpha(t)v^0 = v^0 \mp t\alpha_*(v^0)v^\alpha$,*

where α_* is the unimodular element of $(V^0)^* = \mathrm{Hom}(V^0, R)$, and $v^0(\alpha)$ is the unimodular element of V^0. More precisely, $x_\alpha(t)v^0_\beta = v^0_\beta \pm \delta_{\alpha,\beta}v^\beta$; $\alpha, \beta \in \Lambda(\pi)$. Besides that, the elements $\alpha_*, \alpha \in \Delta(\pi)$ form the basis in $(V^0)^*$.

Let μ be the highest weight of the representation π. If the group $G(\Phi, R)$ is considered in this specific representation, then we denote it by $(G(\Phi, R), \mu)$. In the chosen basis of π every $g \in (G(\Phi, R), \mu)$ can be considered as a matrix of the form $g = (g_{\lambda\nu}); \lambda, \nu \in \Lambda^*(\pi) \cup \Delta(\pi)$, i.e. rows and columns are enumerated by the weights of representation π. Denote $g_{\lambda\nu}$ by $\lambda(g)$.

We remind the list of basic representations [30]. It is known that every $G(\Phi, R)$ has the unique basic representation with the zero weight. This is the representation on short roots of Φ and the highest weight of it is the dominant short root, i.e. simply maximal root for the root systems in which all roots are of the same length. So, using the notations of [7] we have the following table:

A_l	$\mu = \varepsilon_1 - \varepsilon_{l+1}$	π is the adjoint representation
B_l	$\mu = \omega_2$	π is the representation of minimal dimension
C_l	$\mu = \omega_2$	
D_l	$\mu = \omega_2$	π is the adjoint representation
E_7	$\mu = \omega_1$	π is the adjoint representation
F_4	$\mu = \omega_4$	
G_2	$\mu = \omega_4$	

Let $\Lambda(\pi) = \Lambda^*(\pi)$, i.e. $\Lambda(\pi) = W(\Phi)\mu$. Then for any basic representation, μ is a microweight.

A_l	$\mu = \omega_k, k = 1, \cdots, l$	π is the k-external power of the minimal representation
B_l	$\mu = \omega_l$	π is the spinorial representation
C_l	$\mu = \omega_l$	π is the minimal representation
D_l	$\mu = \omega_l$	π is the minimal representation
	$\mu = \omega_{l-1}, \omega_l$	π is the half-spin representation
E_7	$\mu = \omega_7$	

For E_6, F_4, G_2 root systems there are no such representations. So, all basic representations are fundamental, besides the adjoint representation for $\Phi = A_l$.

The key role in the theory plays the fact, that the weighted Stein's diagram can be associated with each basic representation [21], [29], [30], etc.

A diagram consists of nodes and edges.

1. The node labeled by λ corresponds to each weight $\lambda \in \Lambda^*(\pi)$. The diagram is read from left to right, therefore the leftmost node corresponds to the highest weight.

2. If $\lambda_1, \lambda_2 \in \Lambda(\pi)$ and $\lambda_1 = \lambda_2 - \alpha_i, \alpha_i \in \Pi$ then the corresponding nodes are connected by the edge labeled by α_i or simply i.

3. The set of nodes $\{1, 2, \cdots, k\}, k = \dim V^0$ corresponds to a zero weight. To each "zero weight" $\widehat{\alpha_i}, \alpha_i \in \Delta(\pi)$ corresponds the chain of length three:

$$\overset{\alpha_i}{\circ} \!\!-\!\!-\!\!-\!\!-\!\! \overset{\widehat{\alpha}_i}{\circ} \!\!-\!\!-\!\!-\!\!-\!\! \overset{-\alpha_i}{\circ}$$

and all $\widehat{\alpha_i}$ are not connected with any other weights.

If the highest weight and labels on edges are known, then one can uniquely reconstruct the labels on nodes. Therefore, the last are often omitted. Note finally, that in terms of diagrams the irreducibility of the representation means that the diagram is simply connected.

Let us move to the Chevalley-Matsumoto theorem. We fix $\alpha_k \in \Pi$ and let $\Delta = \langle \Pi \setminus \{\alpha_k\} \rangle$. Let, further

$$\Sigma = \Phi \setminus \Delta; \; \Sigma^+ = \Phi^+ \cap \Sigma; \; \Sigma^- = \Phi^- \cap \Sigma,$$

and

$$U(\Sigma, R) = \langle x_\alpha(t), \alpha \in \Sigma^+, t \in R \rangle,$$
$$V(\Sigma, R) = \langle x_\alpha(t), \alpha \in \Sigma^-, t \in R \rangle.$$

Remind that for a basic representation π there exists a unique root $\alpha_k \in \Lambda(\pi)$ such that $\mu - \alpha_k \in \Sigma$. We take this α_k for the definition of Δ. If $\Phi = A_n$, then $\alpha_k = \alpha_1$ or $\alpha_k = \alpha_n$.

The Chevalley-Matsumoto theorem [8], [16] says that if $g \in (G(\Phi, R), \mu)$ satisfies $g_{\mu\mu} \in R^*$ then $g = vhg_1u, v \in V(\Sigma, R), u \in U(\Sigma, R), hg_1 \in T(\Phi, R)G(\Delta, R)$ and all multiples v, u, hg_1 are uniquely defined. Moreover, if $g_{\mu\mu} = 1$, then $g = vg_1u = eg_1, e \in (E(\Phi, R), \mu)$. If finally, $g_{\mu\mu} = 1, g_{\lambda\mu} = 0, \lambda \neq \mu$, then $g = g_1u$.

It is easy to understand that the Chevalley-Matsumoto decomposition is the first step to the important Gauss decomposition

$$G(\Phi, R) = U(\Phi, R)V(\Phi, R)T(\Phi, R)U(\Phi, R).$$

Note that if α_k is an arbitrary root, the theorem is still true. We must only replace the condition $g_{\mu\mu} \in R^*$ by the condition on certain principal minors of the matrix $g \in (G(\Phi, R), \mu)$. Namely, the principal minor $(g_{\lambda\nu})$ such that $\lambda - \nu = \sum \alpha_s, \; \lambda, \nu \in \Lambda(\pi), \alpha_s \in \Pi, \alpha_s \neq \alpha_k$ has to be invertible. It is clear, that if the root α_k is such that there exists the basic representation with the highest weight μ, such that $\mu - \alpha_k \in \Lambda(\pi)$, then if $g_{\mu\mu} \in R^*$, the defined principal ideal for any representation has to be invertible. Indeed, using the Chevalley-Matsumoto theorem for such $(G(\Phi, R), \mu)$, we see that components u, v, hg_1 are independent from the representation.

The Chevalley-Matsumoto theorem admits diagram illustration. Cutting off the root α_k leads to sharing of the diagram into several connected components, and the leftmost of them corresponds to the invertible minor.

One more illustration of the theorem can be observed for the classical groups. Almost all fundamental representations of these groups besides the spinorial one for $\Phi = B_k$ and half-spins for $\Phi = D_n$ can be obtained as external powers of natural representations of universal groups or as their subrepresentations ($\Phi = C_n$). Therefore, the block form $g \in (G(\Phi, R), \omega_1), \Phi = A_n, B_n, C_n, D_n$ implies the form of the n-th external power, which belongs to $(G(\Phi, R), \omega_k)$ ($k \neq n$ for $\Phi = B_n, k \neq n, n-1$ for $\Phi = D_n$).

3. Stability theorems

Let consider maximal standard inclusions of root systems, i.e. inclusions corresponding to maximal connected subdiagrams of a Dynkin diagram. As a result, we can get a stabilization theorem for maximal inclusions of root systems.

THEOREM. *The maps of root systems $\Delta \subset \Phi$ imply surjections of $\overline{\nu} : K_1(\Delta, R) \longrightarrow K_1(\Phi, R)$ under the following assumptions:*

1. $B_3, C_3 \longrightarrow F_4$ *under* ASR_3,
2. $A_5 \longrightarrow E_6$; $A_6 \longrightarrow E_7$; $A_7', A_7'' \longrightarrow E_8$; $A_{n-1} \longrightarrow D_n$ *under* ASR_4,
3. $D_5 \longrightarrow E_6$; $E_6 \longrightarrow E_7$; $E_7 \longrightarrow E_8$ *under* ASR_5,

where A_7' and A_7'' denote two different inclusions of the system A_7 to E_8.

Summarizing all results for surjective stability, i.e. results by Bass, Vaserstein, Stein, Bass-Milnor-Serre, Matsumoto, etc., we can get the complete list of all maximal root inclusions and of corresponding ring conditions.

We consider here, as an example, the case $D_5 \longrightarrow E_6$ of the theorem.

It follows from the Chevalley-Matsumoto theorem that the proof will be completed, if for any $g \in (G(\Phi, R), \mu)$ we can take $e \in (E(\Phi, R), \mu)$, such that

$(eg)_{\mu\mu} \in R^*$, or in other notations $\mu(eg) = 1$. Of course, only stable computations, i.e. computations which make use of one row or column of the matrix $g \in (G(\Phi, R), \mu)$ are considered. These computations use Matsumoto's lemma, which describes the action of $x_\alpha(t)$, and the basic representation diagram which allows to make visible transformations of a fixed column of the matrix g.

From the Chevalley-Matsumoto theorem also follows that if $\mu(g) = 1$ then there exists $e \in (E(\Phi, R), \mu)$ such that $\lambda_i(eg) = 0, \lambda_i \neq \mu$. Besides that, if $\lambda_i \neq \mu$ and $\lambda_i(g) = 1$, then there exists $e \in (E(\Phi, R), \mu)$ such that $\mu(eg) = 1$.

Since $D_5 = \langle E_6 \setminus \{\alpha_6\}\rangle$, let us consider the matrix $g \in (G(E_6, R), \omega_6)$. The diagram of the representation (E_6, ω_6) has the form:

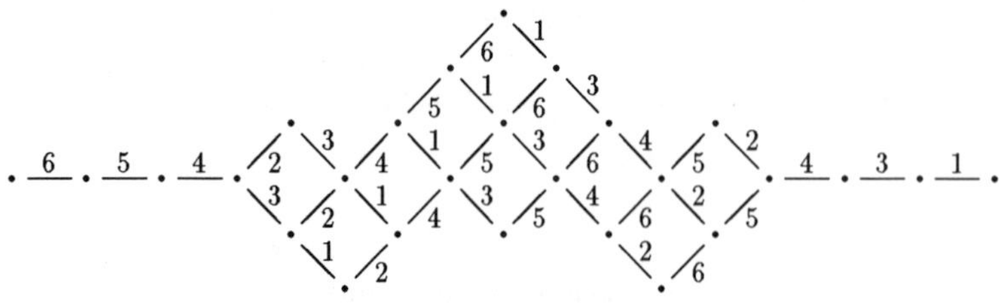

In correspondence with the inclusion $D_5 \longrightarrow E_6$ we enumerate the weights (here $\lambda_i = i$):

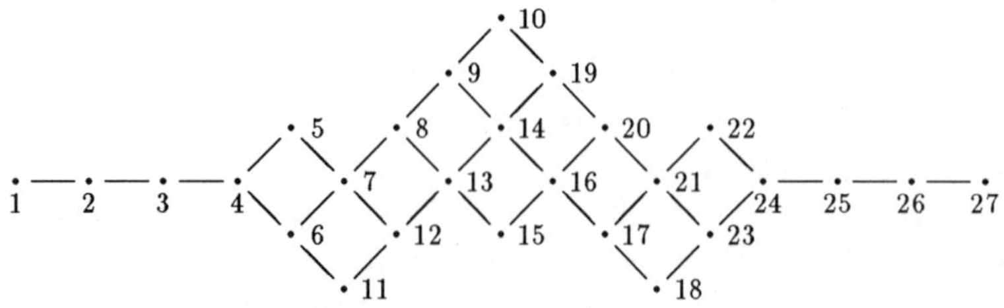

The row $(\lambda_1(g), \cdots, \lambda_{27}(g))$ is unimodular. Let $\mathfrak{a}$ be an ideal generated by $\lambda_5(g)$, $\lambda_7(g), \cdots, \lambda_{27}(g)$. Since the group $(G(A_5 \to E_6, R), \omega_1)$, (where ω_1 denotes the fundamental weight of the representation of type A_5) acts on the weights $\lambda_1, \lambda_2, \lambda_3, \lambda_4$, λ_6, λ_{11}, then using the condition SR_6 we can obtain that there is $e_1 \in (E(A_5 \to E_6, R), \omega_1)$ such that $(\lambda_2(e_1g), \lambda_3(e_1g), \lambda_4(e_1g), \lambda_6(e_1g), \lambda_{11}(e_1g))$ is unimodular mod $\mathfrak{a}$. The ideal $\langle\lambda_5(e_1g), \lambda_7(e_1g), \cdots, \lambda_{27}(e_1g)\rangle$ is still $\mathfrak{a}$. Denote $g_1 = e_1g$. Then $(\lambda_2(g_1), \cdots, \lambda_{27}(g_1))$ is unimodular. Let $\mathfrak{a}_1$ be the ideal $\langle\lambda_{10}(g_1), \lambda_{19}(g_1), \cdots,$

$\lambda_{27}(g_1)\rangle$. Then $(\lambda_2(g_1), \cdots, \lambda_9(g_1), \lambda_{11}(g_1), \cdots, \lambda_{18}(g_1))$ is unimodular mod $\mathfrak{a}_1$. As $\mu - \lambda_k \in E_6$, $k = 2, 3, \cdots, 9, 11, 12, \cdots, 18$, we can use the corresponding elementary unipotent elements. Thus there exist $t_2, t_3, \cdots, t_9, t_{11}, \cdots t_{18} \in R$ such that $\sum_{i=2,(\neq 10)}^{18} t_i\lambda_i(g_1) \equiv 1 - \lambda_1(g_1) \bmod \mathfrak{a}_1$, and an element $e_2 \in (E(E_6, R), \omega_6)$ such that $\lambda_1(e_2 g_1) \equiv 1 \bmod \mathfrak{a}_1$. Moreover, e_2 belongs to $(U(E_6, R), \omega_6)$ and therefore, the ideal $\langle\lambda_{10}(e_2g_1), \lambda_{19}(e_2g_1), \cdots, \lambda_{27}(e_2g_1)\rangle$ is still $\mathfrak{a}_1$. Thus the row $(\lambda_1(e_2g_1), \lambda_{10}(e_2g_1), \lambda_{19}(e_2g_1), \cdots, \lambda_{27}(e_2g_1))$ is unimodular. Let us consider now the weights $\lambda_{10}, \lambda_{19}, \cdots, \lambda_{27}$. The group of type $D_5 = \langle E_6 \setminus \{\alpha_6\}\rangle$ is acting on these weights. Let $g_2 = e_2g_1$ and $\mathfrak{a}_3 = \langle\lambda_1(g_2)\rangle$. Using Stein's theorem for the orthogonal case D_5 we can find, under the condition ASR_5, that there exists $e_3 \in (E(D_5 \to E_6, R), \omega_1)$ such that $\lambda_{10}(e_3g_2) \equiv 1 \bmod \mathfrak{a}_3$. Since $\alpha \notin D_5 \longrightarrow E_6$ then the ideal $\lambda_1(e_3g_2)$ is still $\mathfrak{a}_3$. Finally, we have the unimodular row $(\lambda_1(e_3g_2), \lambda_2(e_3g_2), \cdots, \lambda_{10}(e_3g_2))$. The group of type $D_5 = \langle E_6 \setminus \{\alpha_6\}\rangle$ is acting on $\lambda_i, i = 1, 2, \cdots, 10$. Similar arguments leads to the existence of $e_4 \in (E(D_5 \to E_6, R), \omega_1)$ such that $\lambda_1(e_4e_3g_2) = 1$.

The corollary of this theorem and the theorem by Stein on injective stabilization of K_1-functor give the following list of isomorphisms.

$$
\begin{array}{lll}
ASR_3 & K_1(D_n, R) \simeq K_1(B_n, R) & D_n \longrightarrow B_n \\
ASR_4 & K_1(D_5, R) \simeq K_1(E_6, R) & D_5 \longrightarrow E_6 \\
ASR_4 & K_1(E_6, R) \simeq K_1(E_7, R) & E_6 \longrightarrow E_7 \\
ASR_4 & K_1(E_7, R) \simeq K_1(E_8, R) & E_7 \longrightarrow E_8 \\
ASR_3 & K_1(D_4, R) \simeq K_1(F_4, R) & D_4 \longrightarrow F_4
\end{array}
$$

In particular, under the condition ASR_3 all these groups are isomorphic to $SK_1(R)$. Moreover, if R satisfies ASR_2, then for every root system Φ holds $K_1(\Phi, R) = 1$. Indeed, under SR_2 the functor $SK_1(R)$ is trivial and $K_1(\Phi, R) \simeq SK_1(A, R)$. Actually for the ASR_2 condition the stronger statement takes place, namely for any $G(\Phi, R)$ holds the Gauss decomposition

$$G(\Phi, R) = U(\Phi, R)T(\Phi, R)V(\Phi, R)U(\Phi, R).$$

Therefore, if R satisfies ASR_2 then for every R Theorem 1 of [28] takes place, i.e. if R satisfies ASR_2 and some small restrictions, then the standard description of parabolic subgroups holds in the group $G(\Phi, R)$.

The problem of the stabilization can be also considered for nonstandard inclusions of root systems.

PROPOSITION ([21],[17]). *The maps of root systems $\Phi_\rho \longrightarrow \Phi$ imply surjection on K_1-functors, where Φ_ρ is a twisted root system:*

$$\begin{array}{lll} G_2 \longrightarrow D_4 & \textit{under} & ASR_3 \\ B_n \longrightarrow D_{n+1} & \textit{under} & ASR_n \\ F_4 \longrightarrow E_6 & \textit{under} & ASR_4 \\ C_n \longrightarrow A_{2n-1} & \textit{under} & SR_3 \end{array}$$

4. Stabilization for twisted Chevalley groups

In this section we briefly describe the surjective stability problem for twisted Chevalley groups. We study the cases of root systems ${}^2A_{2n-1}, {}^2D_n, {}^2E_6$.

Let ρ be an involution of a ring R. Then, one can define the involution of the Chevalley group $G(\Phi, R)$. The twisted Chevalley group $G_\rho(\Phi, R)$ (or ${}^2G(\Phi, R)$) is defined as the subgroup of fixed points in $G(\Phi, R)$ (see in detail [3]). Subgroups $E_\rho(\Phi, R), U_\rho(\Phi, R), V_\rho(\Phi, R)$ are defined in a standard way. Let R_0 be the set of fixed points in R. Denote the action of automorphism ρ by $\rho x = \overline{x}, x \in R$ and $\rho\alpha = \overline{\alpha}, \alpha \in \Phi$.

The group $E_\rho(\Phi, R)$ is generated by elementary root unipotent elements $x_A(t)$:

$$x_A(t) = \begin{cases} x_\alpha(t) & \text{if } \overline{\alpha} = \alpha, t \in R_0, \\ x_\alpha(t)x_{\overline{\alpha}}(\overline{t}) & \text{if } \overline{\alpha} \neq \alpha, t \in R. \end{cases}$$

Generally speaking it is not yet proved for twisted Chevalley groups that $E_\rho(\Phi, R)$ $\triangleleft G_\rho(\Phi, R)$, although it has to be true. Therefore, we consider

$${}^2K_1(\Phi, R) = {}^2G(\Phi, R)/{}^2E(\Phi, R)$$

and the problem of stabilization is defined similarly to those for Chevalley groups of normal type.

Define a basic representation of a twisted Chevalley group as the restriction of the basic representation of $G(\Phi, R)$ to the subgroup $G_\rho(\Phi, R)$. The action of $x_A(t) \in E_\rho(\Phi, R)$ is described easily by the weight diagrams of $G(\Phi, R)$ and by the following analogue of Matsumoto's lemma [8]:

LEMMA. *Let $x_A(t) = x_\alpha(t)$.*
a) *If $\lambda \in \Lambda(\pi), \lambda + \alpha \notin \Lambda(\pi)$, then $x_\alpha(t)v^\lambda = v^\lambda$.*
b) *If $\lambda, \lambda + \alpha \in \Lambda(\pi)$, then $x_\alpha(t)v^\lambda = v^\lambda \mp tv^{\lambda+\alpha}$.*

Let $x_A(t) = x_\alpha(t)x_{\overline{\alpha}}(\overline{t})$.

c) *If* $\lambda \in \Lambda(\pi), \lambda + \alpha \notin \Lambda(\pi), \lambda + \overline{\alpha} \notin \Lambda(\pi), \lambda + \alpha + \overline{\alpha} \notin \Lambda(\pi)$, *then* $x_\alpha(t)v^\lambda = v^\lambda$.

d) *If* $\lambda, \lambda + \alpha \in \Lambda(\pi), \lambda + \overline{\alpha} \notin \Lambda(\pi), \lambda + \alpha + \overline{\alpha} \notin \Lambda(\pi)$, *then* $x_\alpha(t)v^\lambda = v^\lambda \mp tv^{\lambda+\alpha}$.

e) *If* $\lambda, \lambda + \overline{\alpha} \in \Lambda(\pi), \lambda + \alpha \notin \Lambda(\pi), \lambda + \alpha + \overline{\alpha} \notin \Lambda(\pi)$, *then* $x_\alpha(t)v^\lambda = v^\lambda \mp \overline{t}v^{\lambda+\overline{\alpha}}$.

f) *If* $\lambda, \lambda+\alpha, \lambda+\overline{\alpha} \in \Lambda(\pi), \lambda+\alpha+\overline{\alpha} \in \Lambda(\pi)$, *then* $x_\alpha(t)v^\lambda = v^\lambda \mp tv^{\lambda+\alpha} \mp \overline{t}v^{\lambda+\overline{\alpha}} \mp t\overline{t}v^{\lambda+\alpha+\overline{\alpha}}$.

Let $\alpha_k \in \Pi, \Delta = \langle \Pi \setminus \{\alpha_k, \overline{\alpha_k}\}\rangle$ and $\Sigma^\pm = \Phi^\pm \cap (\Phi \setminus \Delta)$. Denote

$$U_\rho(\Sigma, R) = \langle x_A(t), \alpha \in \Sigma^+, t \in R\rangle,$$
$$V_\rho(\Sigma, R) = \langle x_A(t), \alpha \in \Sigma^-, t \in R\rangle.$$

Then, there is an analogue of the Chevalley-Matsumoto decomposition for twisted Chevalley groups of types A_{2n-1}, D_n, E_6. Namely, if $g_{\mu\mu} \in R^*$, then there exist $v \in V_\rho(\Sigma, R), u \in U_\rho(\Sigma, R), hg_1 \in T_\rho(\Phi, R)G_\rho(\Delta, R)$ such that

$$g = vhg_1u$$

and v, u, hg_1 are uniquely defined.

Let us introduce now one more condition of ASR_n type. A commutative ring with an involution satisfies the $\overline{SR_n}$ condition if, for a given unimodular row $(r_1, \cdots, r_n, \overline{r_n}) \in R^{n+1}$, there exist $t_1, \cdots, t_{n-1} \in R$ such that $\langle r_1 + t_1r_n + \overline{t_1r_n}, \cdots, r_{n-1} + t_{n-1}r_{n-1} + \overline{t_{n-1}r_{n-1}}\rangle$ is unimodular.

THEOREM. *The maps of root systems* $\Delta \subset \Phi$ *induce surjections on* ${}^2K_1$*-functors under the following conditions*:

$$\begin{array}{ll} A_{2n-3} \longrightarrow A_{2n-1} & ASR_n \\ D_{n-1} \longrightarrow D_n & (ASR)(R_0)_{n-1}, \overline{SR_n} \\ D_4 \longrightarrow E_6 & (SR)(R_0)_3, ASR_2 \\ A_5 \longrightarrow E_6 & (SR)(R_0)_3, ASR_2 \end{array}$$

where $(ASR)(R_0)_n$ *denotes the absolute stable range condition for the ring* R_0.

Moreover, if R *satisfies* ASR_2 *and* $ASR(R_0)_2$ *then the Gauss decomposition takes place*:

$$G_\rho(\Phi, R) = U_\rho(\Phi, R)V_\rho(\Phi, R)T_\rho(\Phi, R)U_\rho(\Phi, R).$$

Finally, we remark some notes. Historically the most popular stable range condition from the point of view of Chevalley groups turns to be useful for the case of special linear groups, or more precisely to the case of classical groups for which rows and columns of their matrices are arbitrary, i.e. $SL(n,R)$ and $Sp(2n,R)$. Vaserstein showed [34], [35] that even in the cases of orthogonal and unitary groups one more complicated conditions is necessary (see also [6], [14], [19], etc.). We formulate it's specific for the orthogonal group.

A commutative ring R satisfies the condition $V_n(R)$ if, for a given unimodular row $(r_1, r_2, \cdots, r_n, r_{-n}, \cdots, r_{-2}, r_{-1}) \in R^{2n}$, there exist $t_1, t_2, \cdots, t_n, t_{-n}, \cdots, t_{-2}, t_{-1} \in R$, such that

$$(1)\quad \sum_{-1}^{1} t_i r_i = 1,$$
$$(2)\quad t_1 t_{-1} + t_2 t_{-2} + \cdots + t_n t_{-n} = 0.$$

The second equality means precisely that the row $(t_1, \cdots, t_n, t_{-n}, \cdots, t_{-1})$ is the row of a certain orthogonal matrix. It looks very surprising that the ASR_n condition automatically takes into account the equation (2) in orthogonal and unitary cases. But in the cases of exceptional Chevalley groups $G(\Phi, R)$ even in minimal representations there are many independent equations to which each row of $g \in G(\Phi, R)$ has to satisfy. Therefore, in these cases the absolute stable range condition can work "roughly". Probably, the most natural way to get surjective stability theorems on K_1-functor with precise conditions is to generalize $V_n(R)$ to all inclusions of root systems.

References

1. E. Abe, *Chevalley groups over local rings*, Tohoku Math. J. **21**, 1969, 474-494.

2. E. Abe, *Normal subgroups of Chevalley groups over commutative rings*, Contemp. Math. **83**, 1989, 1-17.

3. E. Abe, *Coverings of twisted Chevalley groups over commutative rings*, Sci. Rep. Tokyo Kyoiku Daigaku Sect. A**13**, no.366-382, 1977, 194-218.

4. H. Bass, *K-theory and stable algebra*, Publ. Math. IHES **22**, 1964, 5-60.

5. H. Bass, *Algebraic K-theory*, Benjamin, New York, 1968.

6. H. Bass, *Unitary algebraic K-theory*, Lecture Notes in Math., Springer-Verlag, 1973, 57-265.

7. N. Bourbaki, *Groupes et algèbres de Lie*, Chaptre VI, Hermann, Paris, 1968.

8. C. Chevalley, *Certain schemes des groupes semi-simples*, Sem. Bourbaki Exp. 219 (1960-1961).

9. P. Cohn, *On the structure of the GL_2 of a ring*, Publ. Math. IHES **30**, 1966, 365-413.

10. D. Eastes and J. Ohm, *Stable range in commutative rings*, J. Algebra **7**, 1967, 343-362.

11. V. Gerasimov, *On free linear groups*, XVII All-Union Algebraic Conf., Abstracts, 1983, 52-53.

12. S. Khlebutin, *The sufficiency conditions on subgroup of elementary matrices to be normal subgroup*, Uspekhi Mat. Nauk **39**, no.3, 1984, 245-246.

13. I. Klein and A. Mihalev, *Unitary Steinberg group over the ring with involution*, Algebra and Logika **9**, no.5, 1970, 510-519.

14. M. Kolster, *Surjective stability for unitary K_2-groups*, Preprint, 1975, 23pp.

15. V. Kopeiko, *The stabilization of symplectic groups over the ring of polynomials*, Matem. Sb. **106**, no.1, 1978, 94-107.

16. H. Matsumoto, *Sur les sous-groupes arithmétiques des groupes semi-simples déployés*, Ann. Sci. Ecole Norm. Sup. 4eme, no.2, 1969, 1-62.

17. E. Plotkin, *Surjective stability for K_1-functor for Chevalley groups of normal and twisted types*, Uspekhi Mat. Nauk **44**, no.2, 1989, 239-240.

18. E. Plotkin, *Decomposition of Chevalley-Matsumoto type for twisted Chevalley groups*, Topological spaces and their maps, Riga, 1985, 64-73.

19. M. Saliani, *On the stability of the unitary group*, Preprint, Zurich, 12pp.

20. M. Stein, *Matsumoto's solution of the congruence subgroup problem and stability theorems in algebraic K-theory*, Univ. of Tokyo, 1973.

21. M. Stein, *Stability theorems for K_1, K_2 and related functors modeled on Chevalley groups*, Japan. J. Math. **4**, no.1, 1978, 77-108.

22. M. Stein, *Surjective stability in dimension 0 for K_2 and related functors*, Trans. Amer. Math. Soc. **178**, 1973, 165-191.

23. R. Steinberg, *Lectures on Chevalley groups*, New Haven, 1968.

24. A. Suslin, *On one theorem by Cohn*, Proc. of LOMI **64**, 1976, 127-130.

25. A. Suslin, *On structure of special linear group over ring of polynomials*, Izv. AN. USSR **41**, no.2, 1977, 235-253.

26. R. Swan, *Generators and relations for certain special linear groups*, Advances in Math. **6**, 1971, 1-77.

27. G. Taddei, *Normalite des groupes elementaire dans les groupes de Chevalley sur un anneau*, Contemp. Math. **35**, Part 2, 1986, 639-710.

28. N. Vavilov, *Parabolic subgroups of Chevalley groups over commutative ring*, Proc. of LOMI **116**, 1982, 20-43.

29. N. Vavilov, *Bruhat decomposition of long root semisimple elements in Chevalley groups*, In Rings and Modules II, Leningrad Univ., 1989, 18-39.

30. N. Vavilov and E. Plotkin, *Net subgroups of Chevalley groups* II, Proc. of LOMI **114**, 1982, 62-76.

31. N. Vavilov, E. Plotkin and A. Stepanov, *Computations in Chevalley groups over commutative rings*, DAN USSR **307**, no.4, 1989, 788-791.

32. L. Vaserstein, *On stabilization of general linear groups over a ring*, Matem. Sb. **75**, no.3, 1969, 405-424.

33. L. Vaserstein, *K-theory and a congruence problem*, Matem. Zametki. **5**, no.2, 1969, 233-244.

34. L. Vaserstein, *Stabilization of unitary and orthogonal groups over a ring with an involution*, Matem. Sb. **81**, no.3, 1970, 328-351.

35. L. Vaserstein, *Stabilization for the classical groups over rings*, Matem. Sb. **93**, no.2, 1974, 268-295.

36. L. Vaserstein, *On normal subgroups of Chevalley groups over commutative rings*, Tohoku Math. J. **38**, 1986, 219-230.

37. A. Zalessky, *Semisimple root elements of algebraic groups*, Minsk, Preprint, 1980, 24pp.

STRUCTURE OF CHEVALLEY GROUPS OVER COMMUTATIVE RINGS

NIKOLAI A.VAVILOV
Department of Mathematics and Mechanics
University of Sanct Petersburg
Petrodvorets, 198904, Russian Republic
and
Mathematics Institute
University of Warwick
Coventry CV4 7AL, England

ABSTRACT

This paper is a survey of the theory of Chevalley groups over commutative rings, centred around two topics: explicit calculations in the groups using their minimal modules and the main structure theorems (description of the normal subgroups, normality of the elementary subgroups etc.). We give the necessary background and compare several approaches to the proofs of these results, including some new ones.

Introduction

The purpose of this talk is to describe two new approaches in the study of Chevalley groups over commutative rings. Our chief goal is to show **why** and **how** rather than **what**. So we concentrate on the methods by which the results are proven rather than on the results themselves.

Our exposition here is centered around the following two problems: normality of the elementary subgroups and classification of normal subgroups. The solutions of the problems are called the *main structure theorems*. Of course for the classical cases the solutions (by J.S.Wilson, I.Z.Golubchik, A.A.Suslin, V.I.Kopeiko and many others) are very well known while for the exceptional ones have been obtained by E.Abe, G.Taddei and L.N.Vaserstein

[227,247,6,7] fairly recently. We propose to show that with the right approach the exceptional cases do not differ substantially from the classical ones. We discuss four essentially different proofs of these results, two of which have K-theoretical and further two representation-theoretical flavour. The K-theoretical proofs are based on stability conditions and "localization and patching" respectively. The first method which gave complete solutions for the classical groups used direct matrix calculations taking into account the whole matrix and all the equations among matrix entries - what we call *general calculations* - and was not easy to imitate for the exceptional groups. As a result the first complete proofs for the exceptional groups used "localization and patching".

Here for the first time we exhibit in some details the fourth method based on decomposition of unipotents. The starting points for this approach were the Stepanov's proof [213] of Suslin normality theorem [218] and the proof by Borewicz and the author [40] of the Wilson-Golubchik normal subgroup theorem [271,102]. If one studies these proofs thoroughly one realizes that they *never* refer to the multiplication of matrices - only to the multiplication of elementary matrices and the action of elementary matrices on columns or rows, i.e. what we call *elementary calculations* and *stable calculations* respectively. Of course these things are very easy to perform for all the groups including the exceptional ones. Here we survey some necessary background facts and discuss theproofs obtained along these lines for the classical cases and the case E_6 which is already fairly difficult but still not as technically obstructive as E_7, E_8 and F_4. Even for the classical cases these proofs are often remarkably easier than the extant ones.For the exceptional groups they involve case by case analysis and so are somewhat longer than the existing ones but in any case far more elementary.

We discuss also the equations satisfied by the entries of matrices from exceptional groups in minimal representations. As a pattern we give here a new construction of the cubic form on a 27-dimensional space invariant with respect to the action of the simply connected group of type E_6 and a new proof of the theorem (due to E.Cartan - L.Dickson - C.Chevalley - H.Freudenthal - N.Jacobson - T.Springer - ... - M.Aschbacher) which says that the Chevalley group coincides with the isometry group of the form. We discuss also some of the traditional constructions of the form which come from the theory of Jordan algebras as well as analogous constructions of other exceptional groups. Such realizations give a solid ground for general calculations in the exceptional groups and indeed recently the author produced analogues of the original matrix proofs for some of the exceptional cases as well.

The exposition here is rather sketchy: we skip many details - virtually all of them for the types E_7, E_8, F_4 - and do not formulate results in the strongest possible form. Our excuse is that we attach an extensive bibliography and the whole contents of the talk with the minutest technical details will be covered in a series of six forthcoming papers under the same title four of which are joint with E.B.Plotkin. The present paper serves as an informal introduction to the field and as an invitation to these much more specialized and technical works. The basic facts and notations related to the Chevalley groups over rings are collected here as a common background for our subsequent publications on the subgroup structure of Chevalley groups, related Steinberg groups, unstable K-theory, non-split groups etc., which are not related directly to the main structure theorems presented here but are based on the same techniques.

The understanding of the Chevalley groups which we present here is a further development of the viewpoint introduced by H.Matsumoto [157] and M.R.Stein [206]. We believe that it gives a richer and deeper picture than the traditional approaches. It is also more elementary in many aspects and my teaching experience suggests that large parts of the proofs are accessible to a good second year undergraduate student.

Part of the contents of the present paper was announced in [256,265].

Contents

CH.1. GENERALITIES

In this chapter we describe the common background of the Chevalley groups over rings for the rest of the paper. Most of the facts we mention are quite familiar but to the best of my knowledge have never been collected in one place. Our general references for the root systems and semi-simple Lie algebras are [41,42,124]; for the classical groups [11,82,114,159,160,167,168,273]; for the Chevalley groups [39,50-53,124,209,190,226]; for the algebraic groups [38,58,113,198,199]; for the affine group schemes [3,74,75,76,132,270]; for the algebraic K-theory [20,23-25,163,219].

§1. Chevalley groups

Let Φ be a reduced root system of rank ℓ, P – a lattice lying between the root lattice $Q(\Phi)$ and the weight lattice $P(\Phi)$. From this data one can construct an affine group scheme $G_P(\Phi,\ )$ over $\mathbb{Z}$, (i.e. a representable covariant functor from the category of commutative rings with 1 to the category of groups), such that for any algebraically closed field K the value $G_P(\Phi,K)$ of this functor on K is the semisimple algebraic group over K corresponding to Φ,P. The existence of this group scheme was first proven by C.Chevalley [59], and its uniqueness – by M.Demazure [73]. We will call it the *Chevalley-Demazure group scheme of type* (Φ,P), and its value $G_P(\Phi,R)$ on a commutative ring R with 1 ("the group of rational points $G_P(\Phi,\ )$ with the coefficients in R") – *the Chevalley group of type* (Φ,P) over R. Usually we'll omit P in the notation and speak about "*a Chevalley group* $G=G(\Phi,R)$ *of type* Φ *over* R". When we want to stress that we are talking about the *simply connected* group (i.e. $P=P(\Phi)$) we write $G=G_{sc}(\Phi,R)$, and if the group G is *adjoint* (i.e. $P=Q(\Phi)$) we write $G=G_{ad}(\Phi,R)$. Usually we may (and will) assume that the group G is simply connected.

Recall the construction of Chevalley groups.

1°. Chevalley algebras. Let $L=L_{\mathbb{C}}$ be a complex semisimple Lie algebra of type Φ, $[\ ,\ ]$ - Lie bracket in L, H - a Cartan subalgebra of L. Then L admits the root decomposition $L=H\oplus\sum L_\alpha$, $\alpha\in\Phi$, where the L_α's are the root subspaces i.e. one-dimensional H-invariant subspaces. For a root α we denote by the same letter a linear functional $\alpha\in H^*$ on H, such that $[h,e_\alpha]=\alpha(h)e_\alpha$, for any $h\in H$ and $e_\alpha\in L_\alpha$. Restriction of the Killing form of L to H is nondegenerate and will be denoted by $(\ ,\ )$. This inner product allows us to identify H with H^*. For a root $\alpha\in H$ we denote by $h_\alpha=2\alpha/(\alpha,\alpha)$ the corresponding coroot. Let us fix an order on Φ. We denote by Φ^+,Φ^-, and $\Pi=\{\alpha_1,\dots,\alpha_\ell\}$ the corresponding sets of positive, negative and fundamental roots respectively. If for $\alpha\in\Phi^+$ we fix elements $e_\alpha\in L_\alpha$, $e_\alpha\neq0$, then there is a unique choice of $e_{-\alpha}\in L_{-\alpha}$, $\alpha\in\Phi^+$, such that $[e_\alpha,e_{-\alpha}]=h_\alpha$. Then the set $\{e_\alpha,\ \alpha\in\Phi;\ h_\alpha,\ \alpha\in\Pi\}$ is called a *Weyl base* of the Lie algebra L. All the structure constants in this base, apart from probably the $N_{\alpha\beta}$'s, where $[e_\alpha,e_\beta]=N_{\alpha\beta}e_{\alpha+\beta}$, are integers. Let p be such that $-p\alpha+\beta,\dots,\beta,\dots,q\alpha+\beta$ is the α-series of roots passing through β. C.Chevalley [57] has shown that one can choose e_α in such a manner that $N_{\alpha\beta}=\pm(p+1)$, i.e. *all the structure constants are integers* - this fact is called a *Chevalley theorem.* (see [42,208,52,124] for the proof). The set e_α, $\alpha\in\Phi$, satisfying this condition is called a *Chevalley system* and a Weyl base with integral structure constants - a *Chevalley base.* An explicit choice of the signs of the structure constants is rather tricky (see [233,181,182,52,46,192,163,164]). For what follows we fix the same choice of signs as in [101]. Of course the only cases which present real difficulties are the algebras of types E_6,E_7 and E_8 (the signs for F_4 may be deduced from those for E_6). The most elegant way to explicitly control the signs for these cases is via the Frenkel-Kac cocycle (see [97,197,98]).

Let now $L_{\mathbb{Z}}$ be the integral span of a Chevalley base. Then $L_{\mathbb{Z}}$ is a Lie algebra over $\mathbb{Z}$, which is a $\mathbb{Z}$-form of L, i.e. $L=L_{\mathbb{Z}}\otimes_{\mathbb{Z}}\mathbb{C}$. This $\mathbb{Z}$-form is called an *admissible $\mathbb{Z}$-form* or a *Chevalley order* in L. Let now R be an arbitrary commutative ring Set $L_R=L_{\mathbb{Z}}\otimes_{\mathbb{Z}}R$. In other words L_R– is a Lie algebra over R, which is a free R-module with base $e_\alpha=e_\alpha\otimes 1$, $h_\beta=h_\beta\otimes 1$, with the multiplication given by (1). The algebra L_R is called a *split semisimple Lie algebra of type* Φ over R or a *Chevalley algebra of type* Φ *over* R. At this stage one can construct the *adjoint* Chevalley groups (see [57,50,52,124,190]) to construct *simply-connected* groups one also needs to choose integral bases in the finite dimensional representations of L.

2°. **Weyl modules**. Let again $L=L_{\mathbb{C}}$ be a complex semisimple Lie algebra, $\pi:L\to gL(V)$ its representation in a finite dimensional vector space V over $\mathbb{C}$. For an element $\lambda\in H^*$ we denote by V^λ the corresponding *weight subspace* of the space V viewed as an H-module, i.e. $V^\lambda=\{v\in V\mid \pi(h)v=\lambda(h)v, h\in H\}$. Then λ is called a *weight* of the representation π, if $V^\lambda\neq 0$. The dimension $m_\lambda=\mathrm{mult}(\lambda)$ of V^λ is called the *multiplicity* of the weight λ. Let's denote by $\bar{\Lambda}(\pi)$ the *set of weights of the representation* π, (all the weights in $\bar{\Lambda}(\pi)$ are distinct) and by $\Lambda(\pi)$ – the *set of weights with multiplicities*. This means that we assign to each weight $\lambda\in\bar{\Lambda}(\pi)$ a set of m distinct "weights" $\lambda_1,\dots,\lambda_m\in\Lambda(\pi)$, where $m=\mathrm{mult}(\lambda)$. By $\bar{\Lambda}^*(\pi)$ and $\bar{\Lambda}^*(\pi)$ we denote the corresponding sets of non-zero weights. Let $P=P(\pi)$ be the lattice of weights of the representation π, i.e. the subgroup of $P(\Phi)$ spanned by $\bar{\Lambda}(\pi)$. Then $V=\oplus V^\lambda$, $\lambda\in\Lambda(\pi)$. In the case of the adjoint representation $\pi=\mathrm{ad}$ one has $V=L$, $\Lambda^*(\pi)=\Phi$, $\Lambda(\pi)=\Phi\cup\{o_1,\dots,o_\ell\}$; $P=Q(\Phi)$, $V^\alpha=L_\alpha$ for $\alpha\in\Phi$ and $V^o=H$.

Let $\omega\in\Lambda(\pi)$ and $v^+\in V$. Then the weight $\omega=\omega(\pi)$ is called the *highest weight*, and the vector v^+ a *highest weight vector* (or a *primitive element*), if $\pi(e_\alpha)v^+=0$ for all $\alpha\in\Phi^+$. Obviously this notion depends on the choice of the order on the root system Φ. The representation π is irreducible if and only if V is generated as an L-module by a primitive element. The multiplicity of the highest weight of an irreducible representation is 1, so that a primitive vector v^+ is unique up to a nonzero scalar multiple. It is well known that the correspondence $\pi\mapsto\omega(\pi)$ establishes a bijection between the set of isomorphism classes of finite-dimensional irreducible L-modules and the set $P(\Phi)_{++}$ of *integral dominant* (with respect to a given order) *weights*. Recall that $P(\Phi)_{++}=\{\omega\in P(\Phi)\mid(\omega,\alpha)>0,\ \forall\alpha\in\Pi\}$.

The *Chevalley-Ree theorem* asserts that every finite-dimensional L-module V contains a $\mathbb{Z}$-lattice M invariant with respect to all $\pi(e_\alpha)^m/m!$, $\alpha\in\Phi$, $m\in\mathbb{Z}^+$, and that such a lattice is a direct sum of its its weight components $M^\lambda=M\cap V^\lambda$, see [59,183,208,39,124]. Such a lattice $L_{\mathbb{Z}}$ is called an *admissible $\mathbb{Z}$-form* of the module V, and a base v^λ, $\lambda\in\Lambda(\pi)$ of the lattice $V_{\mathbb{Z}}$, consisting of weight vectors such that for any $\alpha\in\Phi$, $m\in\mathbb{Z}^+$, $\mu\in\Lambda(\pi)$ the vector $\pi(e_\alpha^{(m)})v^\mu$ is an integral linear combination of the base vectors is called an *admissible base*. This theorem can be proven by a direct check [183], but the current approach is to use *Kostant's theorem*, [144] (see also [39,204,124,132,226]), which says that the divided powers $e_\alpha^{(m)}=e_\alpha^m/m!$, $\alpha\in\Phi$, $m\in\mathbb{Z}^+$ generate a $\mathbb{Z}$-form $U(L)_{\mathbb{Z}}$ of the universal enveloping algebra U(L) of L (the so called *Kostant form*). Now it is straightforward to construct an admissible $\mathbb{Z}$-form of V. One has just to take a primitive element $v^+\in V$ and set $V_{\mathbb{Z}}=U(L)_{\mathbb{Z}}v^+$.

Now again let R be a commutative ring with 1. Set $V_R=V_{\mathbb{Z}}\otimes_{\mathbb{Z}}R$. In other words, V_R is a free R-module with base $v^\lambda=v^\lambda\otimes 1$, $\lambda\in\Lambda(\pi)$. Obviously V_R is a L_R-module: e_α's and h_i's act on the first component of $v\otimes\xi$, $v\in V_{\mathbb{Z}}$, $\xi\in R$, while the scalars from R act on the second. If V – is an irreducible L-module with highest weight ω, then V_R is called the *Weyl module* of the Chevalley algebra R corresponding to the highest weight ω.

3°. **Elementary Chevalley groups**. Let v^λ, $\lambda\in\Lambda(\pi)$ be an admissible $\mathbb{Z}$-base of an L-module V, $\alpha\in\Phi$ and $\xi\in\mathbb{C}$. Then the linear operator $\pi(\xi e_\alpha)\in GL(V)$ is nilpotent and we can define its exponential by the usual formula

$$\exp(\xi e_\alpha)=e+\xi\pi(e_\alpha)+\xi^2\pi(e_\alpha^{(2)})+\dots$$

The image of a base vector under this operator is a linear combination of the base vectors whose coefficients are linear combinations of powers of ξ with integral coefficients. This means that we can define an automorphism

$$x_\alpha(\xi)=x_\alpha^\pi(\xi)=\exp(\xi\pi(e_\alpha))\in GL(V_R)$$

of the R-module V_R by the same formula as above for any commutative ring.

The subgroup of the automorphism group of the R-module V_R, generated by all the automorphisms of the form $x_\alpha(\xi)=x_\alpha^V(\xi)$, is called the *elementary Chevalley group of type Φ over* R and is denoted by $E_\pi(\Phi,R)$. Thus we have

$$E_\pi(\Phi,R)=\left\langle x_\alpha(\xi),\ \alpha\in\Phi,\ \xi\in R\right\rangle\le GL(V_R).$$

So the elementary Chevalley group from the very start arises as a *linear group*. As an *abstract group* $E=E_\pi(\Phi,R)$ depends up to isomorphism not on the π

itself but just on its lattice of weights $P=P(\pi)$. If we want to stress this we write $E=E_P(\Phi,R)$. But according to our definition the groups $E_P(\Phi,R)$ *always arise in some particular representations* $E_\pi(\Phi,R)$. The corresponding modules are also called the *Weyl modules.*

4°. **Chevalley groups.** Let now $G=G_{\mathbb{C}}$ be the connected complex semisimple group with the Lie algebra $L=L_{\mathbb{C}}$ and the weight lattice P. Denote by $\mathbb{C}[G]$ the *affine algebra* of G, i.e. the algebra of regular complex-valued functions on G, viewed as a *Hopf algebra* in the usual fashion [38,113,3,197]. Denote by the same letter π the representation of G on $V=V_{\mathbb{C}}$ whose differential equals $\pi:L_{\mathbb{C}}\to GL(V_{\mathbb{C}})$. The choice of an admissible base v^λ, $\lambda\in\lambda(\pi)$, allows us to identify $V_{\mathbb{C}}$ with $\mathbb{C}^n$, $n=\dim V$, and thus introduces coordinate functions $x_{\lambda,\mu}$, $\lambda,\mu\in\in\Lambda(\pi)$ on $GL(V_{\mathbb{C}})$. If we identify $GL(V_{\mathbb{C}})$ with $GL(n,\mathbb{C})$ using these coordinates then their restrictions on $\pi(G_{\mathbb{C}})$ generate a subring $\mathbb{Z}[G]$ of the affine algebra $\mathbb{C}[G]$. Chevalley has shown that this subring is in fact a *Hopf subalgebra* of $\mathbb{C}[G]$ (see [59,73,76]) and thus provides an *affine group scheme over* $\mathbb{Z}$ by

$$R\to G_P(\Phi,R)=\mathrm{Hom}_{\mathbb{Z}}(\mathbb{Z}[G],R).$$

The image of a ring R under this functor is denoted by $G_P(\Phi,R)$ and is called a *Chevalley group of type* Φ *over* R. Up to isomorphism of algebraic groups it depends on Φ and P, but not on π. At the same time again by the very definition we may consider corresponding linear groups $G_\pi(\Phi,R)$ as subgroups of $GL(n,R)$, $n=\dim\pi$.

Let $T=T_P(\Phi,\ )$ be a split maximal torus of a Chevalley-Demazure group scheme $G_P(\Phi,\ )$. If R is a commutative ring $T=T_P(\Phi,R)$ is called a *split maximal torus* of the Chevalley group $G=G_P(\Phi,R)$. It is well known that

$$T=T_P(\Phi,R)=\mathrm{Hom}(\mathbb{Z}[T],R)\cong\mathrm{Hom}(P,R^*)$$

where $\mathbb{Z}[T]=\mathbb{Z}[\lambda_1,\lambda_1^{-1},\dots,\lambda_\ell,\lambda_\ell^{-1}]$ is the algebra of Laurent polynomials for some $\mathbb{Z}$-base $\lambda_1,\dots,\lambda_\ell$ of the lattice P. This means that the elements of the torus correspond bijectively to the R-characters of the weight lattice P. For such a character χ let us denote the corresponding element of the torus T by $h(\chi)$. We choose our maximal torus T to act diagonally in the chosen admissible base v^λ, $\lambda\in\Lambda(\pi)$. We refer to this torus T as *the* split maximal torus.

§2. Elementary calculations

Here we start to discuss the interrelation of Chevalley groups and elementary Chevalley groups and recall very briefly some fundamental facts about calculations in an elementary Chevalley group. We refer to the calculations based on the elementary generators $x_\alpha(\xi)$ and the Steinberg relations as the *elementary calculations.*

1°. **Chevalley groups versus elementary Chevalley groups.** Let now $\alpha\in\Phi$, and u be a variable. The homomorphism of $\mathbb{Z}[G]$ on $\mathbb{Z}[u]$ which sends a coordinate function $x_{\lambda,\mu}$ to its value on $x_\alpha(u)$ induces a homomorphism

$$G_a(R)=\mathrm{Hom}(\mathbb{Z}[u],R)\to G(\Phi,R)=\mathrm{Hom}(\mathbb{Z}[G],R)$$

from the additive group $R^+=G_a(R)$ of the ring R to the Chevalley group $G(\Phi,R)$. The image of this homomorphism is the root subgroup $X_\alpha=\{x_\alpha(\xi),\ \xi\in R\}$. Thus the elementary Chevalley group $E_\pi(\Phi,R)$ is contained in the Chevalley group $G_\pi(\Phi,R)$. The interrelations between these two groups constitute one of the major problems in the theory of Chevalley groups over rings. Whereas for an elementary Chevalley group there is a very nice system of generators $x_\alpha(\xi)$, $\alpha\in\Phi$, $\xi\in R$, and the relations among these generators are fairly well understood (see below), nothing like that is available for the Chevalley group itself.

This distinction is not that essential for fields. Let K be an *algebraically closed field.* Then always $G_\pi(\Phi,K)=E_\pi(\Phi,K)$. It is very easy to verify that this equality is not true in general even for the case of a field but if the group G is *simply-connected*, then for an *arbitrary field* one has $G_{sc}(\Phi,K)=E_{sc}(\Phi,K)$. This was proven by the method of the "*grosse cellule*", which is a particular case of the Chevalley-Matsumoto decomposition [59,76,39,157,206,8]. In fact the equality $G_{sc}(\Phi,R)=E_{sc}(\Phi,R)$ holds even in the case when R is a *semilocal ring*, see [157,1,204,10]. Recall that a commutative ring is *semilocal* if it has only finitely many maximal ideals. There are also some further cases when the groups G_{sc} and E_{sc} are known to coincide, say, the euclidean rings [208], the Dedekind rings of arithmetic type which are not totally imaginary [27,157] and the polynomial rings with coefficients in a field or a principal ideal ring [218,220,141,4,268,69,111].

If we work with fields or semilocal rings then the distinction between a Chevalley group and the corresponding elementary subgroup is easily bridged even for non simply-connected cases by adding certain semisimple generators (see 4^{0} for details). But for general rings the situation becomes much more complicated as the K_1-functor comes into play and there is no hope to give an explicit presentation for the Chevalley group itself. So one of the first questions which arise here is *whether the elementary subgroup is normal in a Chevalley group?*

It is very well-known that the elementary group $E(2,R)=E_{sc}(A,R)$ is *not necessarily normal* in the special linear group $SL(2,R)=G_{sc}(A,R)$ (see [64,224,217]), but it turns out that *if* Φ *is an irreducible root system of rank* $\ell\geq 2$, *then* $E(\Phi,R)$ *is always normal in* $G(\Phi,R)$ [218,220,141,225-227]. In fact to sketch a new direct proof of this statement is one of the main objectives of the present survey.

Thus for these cases one can define a quotient group

$$K_1(\Phi,R)=G_{sc}(\Phi,R)/E_{sc}(\Phi,R),$$

which is the famous K_1*-functor of type* Φ *over* R (see [203-206,4]). Algebraic K-theory shows that this functor is generally speaking non-trivial so that for a ring the group $E_{sc}(\Phi,R)$ *may be strictly smaller then* $G_{sc}(\Phi,R)$.

This normality statement is important also because a lot of natural questions can be comfortably answered for $E(\Phi,R)$ and if we know that it is normal in $G(\Phi,R)$ the answers can be extended to the latter group as well.

2^{0}. **Steinberg group**. The most common way to calculate in a Chevalley group is to use the elementary generators $x_\alpha(\xi)$ and the relations between those - the so called *Steinberg relations*, see [208,51].

Now we recall very briefly some properties of the elementary root unipotents $x_\alpha(\xi)$ which do not depend on a representation. It is obvious that for any $\xi,\eta\in R$ one has

$$x_\alpha(\xi+\eta)=x_\alpha(\xi)x_\alpha(\eta),$$

and thus for a fixed $\alpha\in\Phi$ the map $x_\alpha:\xi\longmapsto x_\alpha(\xi)$ is a homomorphism of the additive group R^+ of R to a one-parameter subgroup $X_\alpha=\{x_\alpha(\xi)\mid\xi\in R\}$, which is called the *elementary unipotent root subgroup* corresponding to α. In fact x_α is an isomorphism of R^+ on X_α. When it does not lead to a confusion we omit the epithets "elementary" and "unipotent" and speak about *root elements* and *root subgroups* (these expressions are given a wider sense in §5). For elements x,y of a group G we denote by $[x,y]$ their commutator $xyx^{-1}y^{-1}$. Let now $\alpha,\beta\in\Phi$,

$\alpha+\beta\neq 0$, and $\xi,\eta\in R$. Then the *Chevalley commutator formula* asserts that

$$[x_\alpha(\xi),x_\beta(\eta)]=\prod x_{i\alpha+j\beta}(N_{\alpha\beta ij}\xi^i\eta^j),$$

where the product in the right hand side is taken over all the roots of the form $i\alpha+j\beta\in\Phi$; $i,j\in\mathbb{N}$, in a fixed order and the constants $N_{\alpha\beta ij}$ do not depend on ξ,η (though they may in general depend on the order). The integral numbers $N_{\alpha\beta ij}$ are called the *structure constants* of the Chevalley group and it's easy to check that they may take just the values $\pm1,\pm2,\pm3$ (eventually $N_{\alpha\beta ij}=N_{\alpha\beta}$ are just the structure constants of L in the Chevalley base). It is a much more delicate task to determine the signs of the constants, and we refer to [57,209,76,181-183,125,52,15,16,197,193] for the details. Tables for a particular choice of the structure constants and some further related information are reproduced in a forthcoming paper by the author and E.B.Plotkin "Structure of Chevalley groups over commutative rings. I. Structure constants".

Now if $\mathrm{rk}\Phi\geq 2$, the *Steinberg group* $\mathrm{St}(\Phi,R)$ has generators $y_\alpha(\xi)$, $\alpha\in\Phi$, $\xi\in R$, subject to the relations above: additivity in ξ and the Chevalley commutator formula (for $\Phi=A_1$ the second relation is vacuous and one has to replace it by another one, involving $w_\alpha(\varepsilon)$). Since $x_\alpha(\xi)$ satisfy these relations and generate $E(\Phi,R)$, there is a natural epimorphism $\pi(\Phi,R)$: $\mathrm{St}(\Phi,R)\twoheadrightarrow E(\Phi,R)$ sending $y_\alpha(\xi)$ to $x_\alpha(\xi)$ and in fact most of the elementary calculations could be done equally well in the Steinberg group (see, for example, [203-206]). In general the kernel of $\pi(\Phi,R)$ is *very far* from being trivial *even* for the case of a field - this is the K_2*-functor of type* Φ *over* R:

$$1\longrightarrow K_2(\Phi,R)\longrightarrow \mathrm{St}(\Phi,R)\longrightarrow E(\Phi,R)\longrightarrow 1,$$

see [203,206] and its calculation in each particular case is a highly nontrivial task (compare, for example, [157,163,28] and of course a huge number - maybe something like 200 - of papers devoted to the calculation of K_2 for the rings of arithmetic and algebro-geometric nature has appeared since then which we cannot quote here). For the case of a field (or a semi-local ring) though it is rather easy to supplement the relations among root unipotents to get an explicit presentation for $E(\Phi,R)$, see below.

Another major problem in the theory of Chevalley groups over rings asks *whether* $K_2(\Phi,R)$ *is central in* $\mathrm{St}(\Phi,R)$ *when* $\mathrm{rk}\Phi\geq 4$? This is true at the stable level (see [203,206,136] and the references therein) and for $\Phi=A_\ell$ (see [135,239]), but even for the other classical cases definitive proofs are missing. It is our belief that the methods exposed in the present paper should lead also to a complete solution of this problem.

3°. **Some important elements and subgroups.** Let now $\alpha\in\Phi$ and $\varepsilon\in R^{*}$, where R^{*} is the multiplicative subgroup of R. As usual we set $w_{\alpha}(\varepsilon)=x_{\alpha}(\varepsilon)x_{-\alpha}(-\varepsilon^{-1})\cdot x_{\alpha}(\varepsilon)$ and $h_{\alpha}(\varepsilon)=w_{\alpha}(\varepsilon)w_{\alpha}(1)^{-1}$. The elements $h_{\alpha}(\varepsilon)$ – and their conjugates – are called *semisimple root elements.*

Then by a famous *Steinberg theorem* (see [208]) to get the actual presentation of the simply-connected Chevalley group over a field one has only to add to the Steinberg relations one additional type of relations,viz. the multiplicativity of $h_{\alpha}(\varepsilon)$ in ε:

$$h_{\alpha}(\varepsilon\eta)=h_{\alpha}(\varepsilon)h_{\alpha}(\eta),\ \varepsilon,\eta\in R^{*}.$$

Thus for a field the (elementary) Chevalley groups admit a very nice presentation. As we've mentioned before for an arbitrary ring there is no way even to control the generators - not to say relations - of a Chevalley group. The case of a field is particularly pleasant because there is a canonical form - the *reduced Bruhat decomposition* - which guarantees that any Chevalley group G has *finite width* in terms of the generators $x_{\alpha}(\xi)$. To state this we have to recall definitions of some very important subgroups of G.

The group $\widetilde{W}=\widetilde{W}(\Phi,R)$ generated by $w_{\alpha}(1)$, $\alpha\in\Phi$, is called the *extended Weyl group* (or else the *Tits-Demazure group) of type* Φ. For the case of a field R=K, char$K\neq2$, one has $|\widetilde{W}|=2^{\ell}|W|$ and the structure of the group has been studied in some detail in [234,76]. This group plays the crucial role in our construction of the cubic invariant forms for the exceptional groups.

Let $T=T(\Phi,R)$ be the split maximal torus and $N=N(\Phi,R)$ be the group, generated by $T(\Phi,R)$ and $\widetilde{W}(\Phi,R)$. For the case of a field K, $|K|\geq4$, it coincides with the normalizer of the torus T in G. Often N is referred to as the *torus normalizer* - in the algebraic sense. Of course for the rings the actual normalizer of T in G - in the abstract sense - can be much larger than N. The quotient group N/T is canonically isomorphic to the Weyl group W and for any $w\in W$ we fix a preimage n_w of w in N.

Recall that we've fixed an order of Φ, with Φ^{+} and Φ^{-} being the corresponding sets of positive and negative roots respectively. Set

$$U=U(\Phi,R)=\langle x_{\alpha}(\xi),\ \alpha\in\Phi^{+},\ \xi\in R\rangle,$$

$$U^{-}=U^{-}(\Phi,R)=\langle x_{\alpha}(\xi),\ \alpha\in\Phi^{-},\ \xi\in R\rangle.$$

Then $U=U(\Phi,R)$ is the product of elementary root subgroups X_α, $\alpha\in\Phi^+$, in any fixed order. In other words any element $u\in U$ may be written in the form $u=\prod x_\alpha(u_\alpha)$, where α runs over the set Φ^+ of the positive roots and the coordinates $u_\alpha\in R$ are uniquely determined by u itself and by the order on Φ^+. The product $B=B(\Phi,R)$ of the groups T and U is called the *standard Borel subgroup of* G (corresponding to the given choice of T and Φ^+), U is called the *unipotent radical of* B. The product $B^-=B^-(\Phi,R)$ of T and V is called a *Borel subgroup opposite to* B.

Now if R=K is a field the *Bruhat lemma* asserts that n_w, $w\in W$, form a system of double coset representatives for B in G, or, in other words, that any element x of $G=G(\Phi,K)$ may be written in the form $x=b_1n_wb_2$ where $b_1,b_2\in B$ and w is uniquely determined by x. This decomposition of x - referred to as its *Bruhat decomposition* - shows that actually any $x\in G$ is - to give a very rough bound - a product of not more than $2m+7\ell$ elementary generators $x_\alpha(\xi)$, where m is the number of positive roots and ℓ the rank of Φ.

Bruhat decomposition is limited to the case of a field. But if R is semilocal G still admits a very useful decomposition, namely the so called *Gauss decomposition* G=BVU (see [204,10,252]), which shows that any $x\in G$ is a product of not more than $3m+4\ell$ elementary generators. Now for a general ring R even if the groups G and E happen to coincide no upper bound for the minimal length of an elementary expression of elements from G exists (see [137,48,49,228] for some examples and positive results - one may note that the original proofs of the estimates for the elementary expressions for Hasse domains were based on the generalized Riemann's hypothesis). This is one of the reasons why the elementary calculations tend to be less and less efficient for rings which are not so close to fields. Anyhow the elementary calculations are entirely in terms of the group E and they are not suitable to study the interrelation of E and G. That's why to approach this problem we have to develop different techniques.

4°. Diagonal extensions. Here we recall construction of some semi-simple elements which together with the elementary subgroup generate the whole Chevalley group - or even some larger groups - in the case of a field (see [190,1,31,253,255,259]).

Recall that the split maximal torus is isomorphic to the group of R-characters of the weight lattice P:

$$T=T_P(\Phi,R)=\mathrm{Hom}(P,R^*)$$

At the same time the intersection of T with the elementary subgroup $E=E_P(\Phi,R)$ which is usually denoted by $H=H_P(\Phi,R)$ is generally speaking somewhat smaller:

$$H=H_P(\Phi,R)=\mathrm{Hom}(P(\Phi),R^*).$$

This means that the elements of H correspond to those characters of P which can be extended to the whole weight lattice $P(\Phi)$. Later we think of P as being fixed and suppress it in the notations. Actually one has

$$H=H(\Phi,R)=\langle h_\alpha(\varepsilon),\ \alpha\in\Phi,\ \varepsilon\in R^*\rangle.$$

Following [1] we may introduce the subgroup

$$G_o=G_o(\Phi,R)=T(\Phi,R)E(\Phi,R).$$

Then one has $G_o(\Phi,R)=G(\Phi,R)$ for every group - not just the simply-connected one - if R is a field or, more generally, a semi-local ring [1,157,10].

The elements $h(\chi)$ are related to the elementary generators $x_\alpha(\xi)$ by the following formula

$$h(\chi)x_\alpha(\xi)h(\chi)^{-1}=x_\alpha(\chi(\alpha)\xi).$$

One may think of χ here as being an arbitrary R-character of the lattice $Q(\Phi)$, not just a character of P. Such a map realizes a *diagonal automorphism* of the group G (see [204,51]) which is not necessarily internal (often one understands under diagonal automorphisms the *cosets* of diagonal automorphisms by internal diagonal automorphisms). Sometimes it is convenient to look at extensions of the Chevalley group where all the diagonal automorphisms become internal. It is very easy to construct such an extension for the adjoint groups. One has just to consider linear operators on the Chevalley algebra which act on a Chevalley base as follows: $h(\chi)h_i=h_i$ and $h(\chi)x_\alpha=\chi(\alpha)x_\alpha$ (see [190,51]). Then the group $\overline{T}_{ad}=\overline{T}_{ad}(\Phi,R)$ consisting of all such $h(\chi)$ for $\chi\in\mathrm{Hom}(Q(\Phi),R^*)$ normalizes E_{ad} and the product $\overline{G}_{ad}=\overline{T}_{ad}E_{ad}$ is the *extended adjoint Chevalley group of type* Φ *over* R. In [190] one may find identification of these groups for the classical series. It is much more difficult to construct such an extension for the simply-connected case since here to keep the maximal torus connected one has to increase its dimension. Such an extension was constructed only in [31]. As examples of the extended Chevalley group one may think of $GL(n,R)$ for the series A_ℓ and $Sp(2\ell,R)$ for the series C_ℓ. Below we occasionally refer to the weight elements $h_\omega(\varepsilon)$, $\omega\in P(\Phi^\vee)$, $\varepsilon\in R^*$. These are the elements acting as $h(\chi)$ for the character $\chi=\chi_{\omega,\varepsilon}$ defined by $\chi_{\omega,\varepsilon}(\alpha)=\varepsilon^{(\alpha,\omega)}$. Look [253,255,259] for details.

5°. Reduction to smaller ranks. Usually the elementary calculations lead to a complete success if the problem may be reduced to the groups of smaller rank.

Let first S be any *closed* set of roots in Φ i.e. such a subset that if $\alpha,\beta\in S$ and $\alpha+\beta\in\Phi$, then $\alpha+\beta\in S$. One can associate with this set a subgroup $G(S,R)$ of the Chevalley group $G=G(\Phi,R)$ which is the group of points of a certain group scheme (see [157] for details). This group is very close to the group $G_0(S,R)=T(\Phi,R)E(S,R)$, where as usual $T=T(\Phi,R)$ is a split maximal torus of $G(\Phi,R)$ and $E(S,R)$ is the subgroup generated by all the elementary root subgroups X_α, $\alpha\in S$, with respect to T:

$$E(S,R)=\langle x_\alpha(\xi),\ \alpha\in S,\ \xi\in R\rangle.$$

For example $G_0(S,R)=G(\Phi,R)$ when R is semilocal (look [264] where it is proven for a more general class of subgroups, the so called "net subgroups" of G which are defined in terms of certain congruences). The groups $E(S,R)$ are particularly important when the set S is *special* (alias *unipotent*), i.e. $S\cap(-S)=\varnothing$.

Two sets of roots $S_1,S_2\subseteq\Phi$ are called *conjugated* if there exists an element w of the Weyl group $W=W(\Phi)$ such that $wS_1=S_2$. If the sets S_1 and S_2 are conjugated then there exists an $n\in\widetilde{W}(\Phi)$ such that $nG(S_1,R)n^{-1}=G(S_2,R)$. Recall that we have fixed an order on the root system which determines Π,Φ^+ and Φ^-. A *standard parabolic* subset P is a closed set of roots containing Φ^+. A *parabolic* subset Q is a subset conjugated to a standard parabolic one.

It is well known that the parabolic subsets fall into 2^ℓ conjugacy classes, where $\ell=rk(\Phi)$ is the rank of Φ. The standard parabolic subsets are pairwise not conjugated and correspond bijectively to all of the subsets $J\subseteq\Pi$ of the fundamental system. Namely if $J\subseteq\Pi$ is such a subset then we may define P_J to be the smallest closed set of roots containing Φ^+ and $-\Pi$. The most important parabolic subsets are the maximal ones. A maximal parabolic subset corresponds to a set $J=J_r$, $1\le r\le\ell$, which contains all the fundamental roots apart from α_r. The corresponding parabolic set P_{J_r} is maximal among the closed subsets and will be denoted P_r. Thus there are precisely ℓ conjugacy classes of the maximal parabolic subsets.

In the sequel we will use only the parabolic subgroups of the form $G(Q,R)$, where Q is a parabolic subset. We commonly practice the following abusive expression: when we say that an element or a subgroup is contained in a proper parabolic subgroup we actually mean that it is contained in one of the $G(Q,R)$'s for a $Q\neq\Phi$. Since any such subgroup $G(Q,R)$ is conjugated to a standard parabolic subgroup $G(P,R)$ by an element from $\tilde{W}(\Phi)$, we may consider only the standard parabolic subgroups.

Now recall the structure of $G(P,R)$. Let again $S\subseteq\Phi$ be any closed set of roots. Then S is the disjoint union of its *reductive* (alias *symmetric*) part S^r which consists of $\alpha\in S$ such that $-\alpha\in S$ and its *unipotent* part S^u which consists of $\alpha\in S$ such that $-\alpha\notin S$. The set S^r is a closed subsystem of roots while the set S^u is special. Moreover S^u is an *ideal* in S, i.e. if $\alpha\in S$, $\beta\in S^u$ and $\alpha+\beta\in\Phi$, then $\alpha+\beta\in S^u$. The group $G(S,R)$ is the semidirect product of the reductive subgroup $G(S^r,R)$ (a *Levi subgroup* of $G(S,R)$) and the unipotent subgroup $E(S^u,R)$ (the *unipotent radical* of $G(S,R)$). In the case when $S=P$ is parabolic the Levi subgroup and the unipotent radical of the parabolic subgroup $G(P,R)$ are usually denoted by $L(P,R)$ and $U(P,R)$ respectively.

By $U^-(P,R)$ we denote the unipotent radical of the *opposite parabolic subgroup* (recall that all of our parabolic subgroups contain the same T and thus there is just one opposite subgroup corresponding to the opposite parabolic subset $P^-=P^r\cup(-P^u))$. In other words $U^-(P,R)$ is generated by the root subgroups X_α where $\alpha\in(-S^u)$. An obvious but very important consequence of the Chevalley commutator formula is that both $U(P,R)$ and $U^-(P,R)$ are normalized by $L(P,R)$. Usually we adhere to the following notation. For a parabolic subset $P\subseteq\Phi$ we denote P^r by Δ and P^u by Σ. Thus $U(P,R)=E(\Sigma,R)$ and $U^-(P,R)=E(-\Sigma,R)$. By the definition of a parabolic subset $\Phi=\Sigma\cup\Delta\cup(-\Sigma)$ is a partition of Φ.

Now of course for a proper parabolic subset P the system Δ has smaller rank than Φ. One is in a very good position when one can reduce to a smaller rank since in that case one can either argue by induction on the rank or sometimes even finish the proof immediately. An inclusion $\Delta\subseteq\Phi$ of root systems induces inclusions of the corresponding groups $G(\Delta,R)\leq G(\Phi,R)$, $E(\Delta,R)\leq E(\Phi,R)$

and so on. In turn these inclusions induce a map of the corresponding K_1-functors $K_1(\Delta,R)\longrightarrow K_1(\Phi,R)$ and as we'll see later a lot can be said about the structure of the group $G(\Phi,R)$ if there exists a proper subsystem $\Delta\subset\Phi$ such that this map is bijective.

§3. Minimal modules

In fact the equations which determine the Chevalley groups may be written down explicitly so that these groups may be identified with stabilizers of certain systems of tensors. For example the *Ree-Dieudonné theorem* establishes isomorphisms between Chevalley groups of classical series and split classical groups in the usual sense, i.e. the stabilizers of certain bilinear forms. We believe that it is quite natural also to think of the exceptional Chevalley groups of types G_2,F_4,E_6,E_7,E_8 as just certain groups of $7\times7, 26\times26, 27\times27, 56\times56$, 248×248 matrices respectively. Another purpose of this survey is to show how to easily control the equations on the entries of these matrices and why it might be useful in the study of the exceptional groups. In this paragraph we introduce the main tool which we use to study Chevalley groups - the so called "minimal", or "basic" representations.

1°. **Basic representations**. Return to the notations of §1. Let us recall that an irreducible representation π of the complex semi-simple Lie algebra L is called *basic* if the Weyl group $W=W(\Phi)$ acts transitively on the set $\overline{\Lambda}^*(\pi)$ of non-zero weights of the representation π, This is equivalent to saying that if for any two non-zero weights λ,μ their difference is a fundamental root $\alpha=\lambda-\mu$, then $w_\alpha\lambda=\mu$ for the corresponding fundamental reflection $w_\alpha\in W$. Such representations were first considered in [58] and first used to study Chevalley groups over rings by H.Matsumoto [156,157].

It is straightforward to enumerate all the basic representations. It is clear that all the non-zero weights of such a representation have multiplicity 1 (they are in the Weyl orbit of the highest weight). Thus $\Lambda^*(\pi)=\overline{\Lambda}^*(\pi)$. It is easy to show that the multiplicity of the zero weight is $m=|\Delta(\pi)|$, where $\Delta(\pi)=\Pi\cap\Lambda^*(\pi)$ is the set of fundamental roots which are the weights of the representation π. Thus we may speak about m "zero-weights" $\hat{\alpha}_1,\dots,\hat{\alpha}_m$, where $\Delta(\pi)=\{\alpha_1,\dots,\alpha_m\}$.

Now if π actually has zero weight then all the remaining weights of π

must be the short roots of the root system Φ. Thus every complex semisimple Lie algebra has a unique such representation, called the "short-root representation". Its highest weight ω coincides with the short dominant root of Φ. If there is just one root length then ω is the maximal root and this representation is just the adjoint representation of L. If there is no zero weight then $\Lambda(\pi)=\Lambda^{*}(\pi)$ and all the weights of π form one Weyl orbit. Such a representation is called a *microweight representation* and of course a list of such these representations is very well known (see [42]).

Let's give the list of possible highest weights ω for the basic representations. With the sole exception of the adjoint representation for A_ℓ all these weights are fundamental. Our numbering of the simple roots follows Bourbaki [41]

A_ℓ: $\omega=\bar\omega_k, k=1,\dots,\ell$; the k-th external power of the usual representation;
$\omega=\varepsilon_1-\varepsilon_{\ell-1}$, the adjoint representation;
B_ℓ: $\omega=\bar\omega_1$, the usual representation;
$\omega=\bar\omega_\ell$, the spinorial representation;
C_ℓ: $\omega=\bar\omega_1$, the usual representation;
$\omega=\bar\omega_2$, the short root representation;
D_ℓ: $\omega=\bar\omega_1$, the usual representation;
$\omega=\bar\omega_2$, the adjoint representation;
$\omega=\bar\omega_{\ell-1},\bar\omega_\ell$, the two half-spinorial representations;
E_6: $\omega=\bar\omega_1,\bar\omega_6$, the two minimal dimensional representations;
$\omega=\bar\omega_2$, the adjoint representation;
E_7: $\omega=\bar\omega_1$, the adjoint representation;
$\omega=\bar\omega_7$, the minimal dimensional representation;
E_8: $\omega=\bar\omega_8$, the adjoint representation;
F_4: $\omega=\bar\omega_4$, the short root representation;
G_2: $\omega=\bar\omega_1$, the short root representation.

Thus the total number of basic representations of the Lie algebra L of type Φ equals $|P(\Phi):Q(\Phi)|$.

2°. **Weight diagrams**. Now we'll describe a very useful device to visualize the action of elements of a Chevalley groups on vectors of a given representation - the corresponding *weight diagrams*. It is very difficult to trace their origin. The Moscow State University folklore tells that they were systematically drawn by the Dynkin's school in early fifties (though never appeared in the published works) and that Dynkin has even coined a special word referring to their form, something like "shuttleness" (this is what is now called unimodality in the theory of posets). The earliest appearance of these or similar pictures in print which I was able to trace had been [72]. They were systematically used by M.Stein in his stability paper [205,206] and has appeared many times since then in different places (see some references at the end of this subsection).

Let's associate with a representation a graph which is *almost* the Hasse diagram of the set $\overline{\Lambda}(\pi)$ of its weights with respect to the usual partial order defined by the choice of a fundamental system Π, viz. $\lambda\geq\mu$ if and only if $\lambda-\mu$ is a linear combination of the fundamental roots with non-negative coefficients. Actually, for the basic representations with $\mathrm{mult}(0)\leq 1$ it will be *precisely* this Hasse diagram.

Namely let's construct a marked graph in the following way. Its vertices correspond to the weights $\lambda\in\Lambda(\pi)$ *with multiplicities* of the representation π, and the vertex corresponding to λ is actually marked by λ (often the marks are omitted). Usually we read the diagram from right to left and from bottom to top, which means that a larger weight tends to stand to the left of and higher then a smaller one, with the landscape orientation being primary. The vertices corresponding to $\lambda,\mu\in\Lambda(\pi)$ are linked by a bond marked α_i (or just i) if and only if $\lambda-\mu=\alpha_i\in\Pi$. When λ and μ are non-zero weights this definition is unambiguous. We have to explain how to understand the equality when λ or μ is a zero weight. If $\lambda=\hat{\alpha}$, $a\in\Delta(\pi)$, then we stipulate $\mu=-\alpha$ and $\alpha_i=\alpha$, so that $\hat{\alpha}=(-\alpha)+\alpha$. If $\mu=\hat{\alpha}$, $\alpha\in\Delta(\pi)$, then $\lambda=\alpha_i=\alpha$ and $\alpha=\hat{\alpha}+\alpha$. This means that to any root $\alpha\in\Delta(\pi)$ there corresponds the following weight chain of length three:

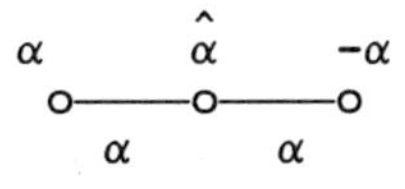

and $\hat{\alpha}$ is not adjacent to any other vertex. In fact to really calculate with the zero weights we have to introduce also another sort of bonds, which we denote by dotted lines and which join $\hat{\alpha}$ to $\pm\beta$ if $\alpha,\beta\in\Delta(\pi)$, $\alpha\neq\beta$, are not orthogonal. But these bonds have to be read in one direction, from a zero weight to a non-zero one and we omit the details here. We try to draw the diagrams in such a way that the marks on the opposite sides of a parallelogram are equal and in that case at least one of them is omitted.

For the case of a microweight representation there is another natural way to look at these diagrams. Let $\omega=\bar{\omega}_k$ be the highest weight of a microweight representation. Then all the other weights lie in the Weyl orbit of ω and thus correspond bijectively to the cosets W/W_k, where W_k is the Weyl subgroup of the Weyl group $W=W(\Phi)$ generated by reflections in all the fundamental roots except α_k. Now of course there is a usual way to introduce a partial order on the set of such cosets, viz. the (*induced*) *Bruhat order*. Namely in each coset there is a unique element of the smallest length (the *distinguished coset representative*) and one takes the ordinary Bruhat order of W on these representatives. What we' ve defined before corresponds rather to the *weak Bruhat order*, but a well-known combinatorial result (see [177]) guarantees that for a microweight these two definitions coincide. When there is a zero-weight the dotted lines occur precisely because the corresponding Bruhat order on the non-zero weights is actually stronger then the weak order (a pair of an ordinary and a dotted line with common vertex corresponds to a bond in the Hasse diagram of the Bruhat order which does not come from a fundamental reflection).

In this form the diagrams appeared in [206]. To show the relevance of the microweights and the corresponding posets we attach some references picked up almost at random out of the huge literature of the subject [14,29,34,35,36,37,61,72,78-80,116,120,121,127,128,133,156,157,164,165,170-175,176-178,184,191,197,201,205,206,254,255,257,263,264,265,267,272,275].

3°. **Action on a minimal module.** Fix a basic representation π of a Chevalley group $G=G(\Phi,R)$ on the free R-module $V=V_R=V_{\mathbb{Z}}\otimes_{\mathbb{Z}}R$. We tend to identify G with its image $\pi(G)=G_{\pi}(\Phi,R)$ under this representation and often omit the symbol π in the action of G on V. Thus for an $x\in G$ and $v\in V$ we write xv for $\pi(x)v$. Decompose the module R into the direct sum of its weight submodules

$$V=\sum V^{\lambda}\oplus V^{o},\quad \lambda\in\Lambda^{*}(\pi).$$

H.Matsumoto [157], lemma 2.3, has shown that one may choose a base of weight vectors $v^{\lambda}\in V^{\lambda}$, $\lambda\in\Lambda^{*}(\pi)$, $v^{o}_{\alpha}\in V^{o}$, $\alpha\in\Delta(\pi)$, in which the action of the root unipotents $x_{\alpha}(\xi)$, $\alpha\in\Phi$, $\xi\in R$, is described by the following very nice formulas:

i. If $\lambda\in\Lambda^{*}(\pi),\lambda+\alpha\notin\Lambda(\pi)$, then $x_{\alpha}(\xi)v^{\lambda}=v^{\lambda}$;

ii. If $\lambda,\lambda+\alpha\in\Lambda^{*}(\pi)$, then $x_{\alpha}(\xi)v^{\lambda}=v^{\lambda}\pm\xi v^{\lambda+\alpha}$;

iii. If $\alpha\notin\Lambda^{*}(\pi)$, then $x_{\alpha}(\xi)v^{o}=v^{o}$, for any $v^{o}\in V^{o}$;

iv. If $\alpha\in\Lambda^{*}(\pi)$, then $x_{\alpha}(\xi)v^{-\alpha}=v^{-\alpha}\pm\xi v^{o}(\alpha)\pm\xi^{2}v^{\alpha}$,

$$x_\alpha(\xi)v^o=v^o\pm\xi\alpha_*(v^o)v^\alpha;$$

where α_* is a certain unimodular element of the dual space $(V^o)^*=\mathrm{Hom}_R(V^o,R)$ and $v^o(\alpha)$ is a unimodular element of V^o (recall that an element v of a free R-module V is *unimodular* if the submodule generated by v is a direct summand of V, or, equivalently, if there exists a $\phi\in V^*=\mathrm{Hom}_R(V,R)$, such that $\phi(v)\in R^*$). We refer to this fact as the *Matsumoto lemma.* In fact since the only basic representations which actually have zero weights come from the adjoint ones, it would be easy to give explicit formulas for $v^o(\alpha)$, $\alpha_*(v)$ as well, but we will not use them here. For the sake of brevity we write $v^{\hat\alpha}$ instead of v^o_α. Then our base $\{v^\lambda\}$ of V is indexed by all the weights $\lambda\in\Lambda(\pi)$ *with multiplicities.*

Now we may expand any $v\in V$ in the chosen base, $v=\sum c_\lambda v^\lambda+ +\sum c^o_\alpha v^o_\alpha$, $\lambda\in\Lambda^*(\pi)$, $\alpha\in\Delta(\pi)$. If we prefer to suppress the distinction between zero and non-zero weights we write simply $v=\sum c_\lambda v^\lambda$, $\lambda\in\Lambda(\pi)$ and refer to c_λ as the *λ-th coordinate* of v. Then of course the Matsumoto lemma provides explicit formulas for the action of $x_\alpha(\xi)$ on v and on its coordinates. This action is most suggestively described in the following way. Conceive a vector $v\in V$ as the marked graph which is obtained by putting marks c_λ and c^o_α to the corresponding vertices of the weight diagram of type (Φ,π). Expand a root $\alpha\in\Phi$ in the fixed base of the root system: $\alpha=\sum m_i\alpha_i$, $\alpha_i\in\Pi$. Then the action of $x_\alpha(\xi)$ on v looks as follows: it adds the λ-th coordinate of v multiplied by $\pm\xi$ to the coordinate standing in the vertex μ such that there is a directed path (we go in the positive/negative direction if m_i are positive/negative) from λ to μ having precisely $|m_i|$ bonds with the mark i for any $i=1,\dots,\ell$. There are slightly more complicated rules if the path starts/ stops at zero and the path which has $2|m_i|$ bonds with mark i has to be taken onto account too.

§4. Stable and general calculations

In this section we describe calculations produced using the minimal modules of a Chevalley group.

1°. **Stable calculations**. Fix the base v^λ, $\lambda\in\Lambda(\pi)$, described in the preceding section. Then any element g of the Chevalley group $G_\pi(\Phi,R)$ is presented by its matrix $g=(g_{\lambda\mu})$, $\lambda,\mu\in\Lambda(\pi)$, in this base. This means that $g_{\lambda\mu}$ is the

μ-th coefficient in the expansion of gv^{λ}. Thus the columns and rows of the matrix are indexed by the weights of π - as we see in the next paragraph it is not expedient to index them by the natural numbers.

By *stable calculations* we understand calculations involving just one column or one row of the matrix $g=(g_{\lambda\mu})$. This expression refers to the fact that it is precisely the sort of calculations one needs to solve the problems *at the stable level*, i.e. when the rank of the group is large with respect to the "dimension" of the ground ring, whatever it should mean. These calculations were used in proving stability theorems for the functors $K_1(\Phi,R)$ and $K_2(\Phi,R)$, see [157,206].

Actually the notions of "dimension" which were most commonly used here were Krull dimension, dimMax(R), "stable rank" and "absolute stable rank". Recall that the last two notions are defined as follows. One says that the *stable rank* of the ring R does not exceed d and writes $sr(R)\le d$ if for any *unimodular* row $(a_1,\dots,a_d,a_{d+1})$ of length d+1 with coordinates in R there exist $c_1,\dots,c_d\in R$ such that the row $(a_1+a_{d+1}c_1,\dots,a_d+a_{d+1}c_d)$ is again unimodular. One says that the *absolute stable rank* of the ring R does not exceed d and writes $asr(R)\le d$ if for *any* row $(a_1,\dots,a_d,a_{d+1})$ of length d+1 with coordinates in R there exist $c_1,\dots,c_d\in R$ such that every maximal ideal of R containing the ideal $(a_1+a_{d+1}c_1,\dots,a_d+a_{d+1}c_d)$ contains also the ideal $(a_1,\dots,a_d,a_{d+1})$. The notion of the stable rank was introduces by H.Bass and then studied by D.Estes and J.Ohm [86], L.N.Vaserstein [243], A.A.Suslin and many others in some detail. The absolute stable rank appears naturally when one passes from the linear and symplectic groups to other classical and exceptional cases and was implicitly studied by D.Estes and J.Ohm [86] and explicitly introduced by M.R.Stein [206].

A column of a matrix $g=(g_{\lambda\mu})$ is obtained by freezing the second index. Thus the columns may be represented by some vectors from V. Analogously the rows are obtained by freezing the first index and correspond to the vectors from the dual module V^*. Now if we multiply g on the right by an $x_\alpha(\xi)$ the resulting transformation of the columns of g may be easily recovered with the help of the Matsumoto lemma. The same applies of course to the rows of $gx_\alpha(\xi)$. These calculations are very similar in spirit to the "matrix problems" as developed by P.Gabriel, L.A.Nazarova, A.V.Roiter and others in the non-classical representation theory, say in the study of representations of posets, graphs and analogous combinatorial objects (see [99] for some details and further references). Calculations of analogous nature appeared also in the enumeration of Borel-orbits by A.G.Elashvili, W.H.Hesselink and H.Bürgstein (see [119,47]). W.Hesselink compared the whole business to a sort of chessboard game.

2°. Equations on columns. It should be noticed that not every vector from V corresponds to a column of a matrix from G. There are of course two obvious restrictions. First, any column of an invertible matrix is unimodular. Second, its coordinates should satisfy some quadratic equations. Namely for a vector $v \in V$ the condition to be the λ-th column of a matrix from G means precisely that v lies in the G-orbit of v^{λ}. In our case there are essentially two possibilities for λ: a zero weight and a non-zero weight. Look at the case of a field R=K first. Then it is well known that the orbit of a vector of highest weight is an intersection of quadric in V, see [147]. These equations can easily be written also using the Bruhat decomposition. Since these equations are "characteristic free" the matrix entries of a matrix $g \in G$ over any ring satisfy the same polynomial equations as for the case of fields.

For the groups of types A_{ℓ} and C_{ℓ} in the usual representations there are no equations: any unimodular vector can be a column of a unimodular matrix over a field and any unimodular vector of even length can be a column of a symplectic matrix. In all the other cases it is not true. It is obvious that to be a column of an orthogonal matrix a vector has to satisfy a quadratic equation, see §6 below. Let us give another classical example of such equations. Look at the fundamental representation of the group SL(n,K) with the highest weight $\bar{\omega}_k$. These representations are furnished by the k-th external power $\bigwedge^k V$ of the natural n-dimensional module V. Then the orbit of the highest weight vector in this representation corresponds to *decomposable k-vectors*. Thus the image of the orbit in the corresponding projective space $P(\bigwedge^k V)$ is the Grassman variety $Gr_{n,k}$ of k-dimensional subspaces in an n-dimensional space in the *Plücker embedding* [110,117]. It is classically known that this image is defined by a system of quadratic equations - the *Plücker equations* - and using this fact and the standard realizations of the fundamental representations of the classical groups [42] it is immediate to check that the orbit of the highest weight vector is cut out by a system of quadrics. Analogous equations for the 27-dimensional representation of E_6 and the

56-dimensional representation of E_7 (see §§6,7 below) were written by J.Tits and H.Freudenthal [231,232,91]. In fact as we'll see the orbit of the highest weight vector for the first of these cases corresponds to the elements of rank 1 in the exceptional Jordan algebra. The rank 1 condition amounts to the fact that all the minors of degree 2 vanish - which is of course a system of quadratic equations again. An extremely beautiful system of quadratic equations in the general case using the Casimir operator has been written by W.Lichtenstein [147].

In general for the case of a field any unimodular vector satisfying the desired equations actually can be a column of a matrix from G. This is not any more true for rings since there is a much less obvious K-theoretical obstacle. In general not every unimodular column over a ring may be included in an invertible matrix (whether it was so for the polynomial rings constituted the famous Serre problem). In our approach we ignore this K-theoretical obstruction, but the quadratic equations play a very important role.

2°. **General calculations.** By the *general calculations* we understand the usual matrix calculations taking into account the matrix $g=(g_{\lambda\mu})$ as a whole and all the equations among its matrix entries. Such calculations are very easy to produce for the classical groups since the dimensions of the minimal modules are fairly small and there are very few dependencies among matrix entries, and we recall some basic facts in the next paragraph. For the classical groups most of the results were first proven by direct matrix calculations.

One is much less happy to calculate with matrices in the exceptional cases and it is easy to attach obvious reasons for that. Apart from the G_2 case the dimensions of the minimal modules are rather large for the direct calculations. The groups themselves are sort of thinly spread in the corresponding GL(V) and so there is a lot of equations among the matrix entries to look after. The simplest elements of the groups have a pretty large *residue* resx= =rk(x-e) - whereas for the classical cases there are elements of residue 1 or 2, for the exceptional cases you cannot find anything smaller than 6 or 10.

Nevertheless we think that with the right approach the general calculations may be very useful even in the exceptional cases and one of the goals of the present paper is to illustrate an easy way to control all the equations among the matrix entries. But what is even more remarkable, is that one does not need general calculations to solve the above-mentioned problems and a lot of further ones! In fact we outline here some procedures which allow us to reorganize the calculations necessary to prove normality of E in G, to describe normal subgroups of G, etc. in such a way as to *completely* avoid the general calculations! Everything will be performed as a combination of steps involving just the elementary and stable calculations! What is really striking about this is the fact that even in the case of a field and for the classical groups our proofs are sometimes *very much simpler* than the previously known ones (this is precisely what was stated in the Russian original of [265] and mistranslated in the following bizarre way "We note that for the classical types and for the field case our proofs have already been in the literature for a considerable time" - traduttóre-traditóre!). It is precisely on the classical examples that we illustrate our methods first.

4°. Chevalley-Matsumoto decomposition. Since it has been known for some time that a simply-connected Chevalley group and the corresponding elementary Chevalley group over a field coincide, there should have been some way to relate the Chevalley group with the elementary subgroup. Such a techniques is provided by what M.R.Stein has christened the Chevalley-Matsumoto decomposition theorem (compare [157], theorem 4.3 and further) which in turn is a further development of the method of "grosse cellule" [59].

Recall that if π is a basic representation of a Chevalley group $G(\Phi,R)$ with the highest weight ω then with the sole exception of the adjoint representation for the group of type A_ℓ there exists a unique fundamental root $\alpha_r\in\Pi$ such that $\omega-\alpha_r$ is a weight of the representation π. For the adjoint representation of A_ℓ there are two such roots α_1 and α_ℓ. Now let Δ be the subsystem of Φ generated by $\Pi\backslash\{\alpha_r\}$ i.e. the smallest subsystem of Φ containing all the fundamental roots but α_r. Then Δ is precisely the reductive part of the standard parabolic subset P corresponding to $J=\Pi\backslash\{\alpha_r\}$. Set further $\Sigma=\Phi^+\backslash\Delta$ and $-\Sigma=\Sigma^-=\Phi^-\backslash\Delta$. Then Σ and Σ^- are the unipotent parts of P and the opposite parabolic set P^- respectively and Φ is a disjoint union of Σ,Δ and Σ^-. As we know from the §2 the groups

$$U(\Sigma,R)=\langle x_\alpha(\xi),\ \alpha\in\Sigma,\ \xi\in R\rangle.$$

$$U^-(\Sigma,R)=\langle x_\alpha(\xi),\ \alpha\in\Sigma^-,\ \xi\in R\rangle.$$

are normalized by the Chevalley group $G(\Delta,R)$.

Now the Chevalley-Matsumoto decomposition theorem says that if an element $g\in G(\Phi,R)$ of the group $G(\Phi,R)$ in the representation π satisfies $g_{\omega\omega}\in R^*$, then $g=vzu$, where $v\in U^-(\Sigma,R)$, $u\in U(\Sigma,R)$, and $z\in T(\Phi,R)G(\Delta,R)$ where v,u,z are uniquely determined by g. Since $G(\Delta,R)$ normalizes $v\in U^-(\Sigma,R)$ and $u\in U(\Sigma,R)$ the element g can be written also in the form $g=z(z^{-1}vz)u=v(zuz^{-1})z$, where $z^{-1}vz\in U^-(\Sigma,R)$ and $zuz^{-1}\in U(\Sigma,R)$. In other words if one denotes by the Ω_π the set of all $g\in G(\Phi,R)$ such that $g_{\omega\omega}\in R^*$, then

$$\Omega_\pi=U^-(\Sigma,R)T(\Phi,R)G(\Delta,R)U(\Sigma,R)=$$

$$=U^-(\Sigma,R)U(\Sigma,R)T(\Phi,R)G(\Delta,R)=$$

$$=T(\Phi,R)G(\Delta,R)U^-(\Sigma,R)U(\Sigma,R).$$

If moreover $g_{\omega\omega}=1$, then $z\in G(\Delta,R)$ so that the factor $T(\Phi,R)$ in the above decomposition could be omitted.

Finally if g stabilizes a primitive vector or, what is the same, if $g_{\omega\omega}=1$ and $g_{\lambda\omega}=0$, for all the weights $\lambda\neq\omega$, then $g=zu$, where as above $z\in G(\Delta,R)$ and $u\in U(\Sigma,R)$.

The thing which really matters here is that the factors in the decomposition are defined internally, without any reference to the representation. Thus this decomposition gives a good start with the reduction of questions concerning $G(\Phi,R)$ to the groups of smaller ranks. Let, say, $R=K$ is a field. Then only occasionally the element $g_{\omega\omega}$ is not invertible and if it is not then some of the elements $g_{\lambda\omega}$, where λ is such a weight that $\alpha=\omega-\lambda$ is a root, should be nonzero (since the column $(g_{\lambda\omega})$, $\lambda\in\Lambda(\pi)$, is unimodular). Then $(x_\alpha(1)g)_{\omega\omega}\neq 0$ and thus the element $x_\alpha(1)g$ factors as above. Continuing this process we get the equality $G(\Phi,K)=E(\Phi,K)T(\Phi,K)$.

Now if α_r is any fundamental root (not necessarily the one for which $\omega-\alpha_r$ is a weight) then the same decomposition holds mutatis mutandis under condition that an appropriate minor of the matrix g is invertible (for the preceding case this minor happens to have order 1). In fact invertibility of a minor of order m on a module is equivalent to invertibility of an one matrix entry in the m-th external power of the module. With this idea in mind Chevalley has taken care of all these conditions simultaneously looking at one entry in a very large representation. The set of elements $g\in G(\Phi,R)$ for which this entry is invertible is called the *grosse cellule* and denoted by $\Omega=\Omega(\Phi,R)$. Actually $\Omega=\{g\in G(\Phi,R),\ f(g)\in R^*\}$ for some function f from the affine algebra $\mathbb{Z}[G]$. Chevalley then proves that

$$\Omega(\Phi,R)=U^{-}(\Phi,R)T(\Phi,R)U(\Phi,R)$$

and this is one of his starting points in the construction of the group scheme $G(\Phi,\ )$, see [59,39].

§5. Relative groups and unipotents

Here we introduce the congruence subgroups of the Chevalley groups and the relative elementary groups which are necessary to formulate the structure theorems. Also we describe informally the distinction between unipotent root elements and unipotent elements of root type.

1°. Unipotent root elements. In what follows we consider several generalizations of the elementary root unipotents $x_\alpha(\xi)$. Some of these generalizations are defined entirely in terms of the Chevalley group G itself and some depend on a particular choice of representation.

The first of these generalizations is quite familiar. Let us recall that x is called a *(unipotent) root element* if it is conjugate to $x_\alpha(\xi)$ for some $\alpha\in\Phi$ and some $\xi\in R$. It is called a *long root element* if α is long and a *short root element* if α is short. Let I be an ideal in R. If $\xi\in I$ we say that x *has level* I (classically one would say that the level of x *is contained in* I).

The normal subgroup of the elementary group $E=E(\Phi,R)$ generated by all the *elementary* root unipotents of level I (i.e. the smallest normal subgroup of E which contains all of these elements) is called the *elementary subgroup of level* I and is denoted by $E_I=E(\Phi,R,I)$:

$$E(\Phi,R,I)=\langle x_\alpha(\xi),\alpha\in\Phi,\xi\in I\rangle^{E(\Phi,R)}.$$

When $I\neq R$ we refer to E_I as a *relative elementary subgroup* as contrasted to the *absolute* one. For a given I the group E_I actually does depend on a particular choice of R.

It is not true in general that E_I is generated by its *elementary* root unipotents. However the Chevalley commutator formula readily implies that E_I is generated by the elements of the form $x_\alpha(\xi)x_{-\alpha}(\zeta)x_\alpha(-\xi)$, $\alpha\in\Phi$, $\xi\in R$, $\zeta\in I$ (see for example [251,237,4,247,259]). In particular this shows that E_I is generated by all of its root unipotents.

Actually with a bit of extra work one can even show that E_I is generated by its *long* root unipotents, see [259]. It suffices to show that any *elementary* short root unipotent $x_\beta(\xi)$ is a product of long root unipotents $x_1,\dots,x_m$ from $E(\Phi,R,I)$ (not necessarily elementary any more). In fact then for any $z\in E(\Phi,R)$ one concludes that $zx_\beta(\xi)z^{-1}=(zx_1z^{-1})\dots(zx_mz^{-1})$ is also a product of long root unipotents. Now if $\Phi\neq G_2$ then there exists a long root α and a short root bγ such that $\beta=\alpha+\gamma$.Then the root $\alpha+2\gamma=\beta+\gamma$ is long and the Chevalley formula implies that

$$x_\beta(\pm\xi)=x_\alpha(\xi)(x_\gamma(1)x_\alpha(-\xi)x_\gamma(-1))x_{\beta+\gamma}(\pm\xi^2),$$

where the right hand side is a product of long root unipotents from $E(\Phi,R,I)$. For the case $\Phi=G_2$ the proof is somewhat more tricky since the ingenuous use of the Chevalley commutator formula provides the result only under the additional assumption $2I=I$.

Now normality of E_I in the Chevalley group G amounts to the fact that *any* (or actually *any long*) root unipotent of level I is contained in E_I.

2°. **Congruence subgroups.** Let again I be an ideal in R. Then I defines the corresponding *reduction homomorphism*

$$\phi_I\colon\ G(\Phi,R)\longrightarrow G(\Phi,R/I),$$

see [1,202,203]. If we look at the group $G=G(\Phi,R)$ in a particular representation π then ϕ_I is very easy to describe: one has just to reduce every entry of a matrix representing an element $g\in G$ modulo I, i.e. take $\bar{g}_{\lambda\mu}=\phi_I(g_{\lambda\mu})=g_{\lambda\mu}+I$. The kernel of ϕ_I is denoted by $G_I=G(\Phi,R,I)$ and is called the *principal congruence subgroup of level* I. By the definition G_I is a normal subgroup of G and one has the following exact sequence:

$$1\longrightarrow G(\Phi,R,I)\longrightarrow G(\Phi,R)\longrightarrow G(\Phi,R/I)\longrightarrow 1.$$

Similarly the inverse image under ϕ_I of the centre of $G(\Phi,R/I)$ is denoted by $C_I=C(\Phi,R,I)$ (sometimes by $G'_I=G'(\Phi,R,I)$) and is called the *full congruence subgroup of level* I. To stress that $G(\Phi,R,I)$ and $C(\Phi,R,I)$ are considered in a particular representation we write $G_\pi(\Phi,R,I)$ and $C_\pi(\Phi,R,I)$ respectively. Actually the group $G(\Phi,R,I)$ does not depend on R but the group $C(\Phi,R,I)$ does.

Clearly one has $E_I\le G_I\le C_I$. In the next section we discuss the *standard description of the normal subgroups* in G which says that (with some minor restrictions on Φ and R) all of them lie between E_I and C_I for some (uniquely determined) ideal I of R. In fact for most of the excluded cases E.Abe could recover a *parastandard description* of normal subgroups in terms of generalized congruence subgroups [1,10,6,7] but we cannot reproduce their definition here.

When the standard description holds the classification of normal subgroups in G is reduced to the computation of factor-groups C_I/E_I or, since the structure of C_I/G_I is clear, to the computation of the *relative* K_1*-functors*

$$K_1(\Phi,R,I)=G(\Phi,R,I)/E(\Phi,R,I).$$

Of course this last problem is highly non-trivial and was solved only for some very special classes of rings.

3°. **Relativization.** There is a very simple general trick which usually allows to reduce problems about relative groups to those for the absolute ones. It goes back to the works of R.Swan and M.R.Stein [202] and was first applied to the normality problem by A.A.Suslin and V.I.Kopeiko [220] and used later by L.N.Vaserstein [244] in the context of Chevalley groups. Although our new approach does not distinguish the relative and absolute cases we find it very instructive to show this idea here.

Define the *double* $R\times_I R$ of a ring R with respect to an ideal $I\trianglelefteq R$ by the cartesian square

$$\begin{array}{ccc} R\times_I R & \xrightarrow{\pi_1} & R \\ {\scriptstyle\pi_2}\downarrow & & \downarrow{\scriptstyle\pi} \\ R & \xrightarrow[\pi]{} & R/I \end{array}$$

In other words $R\times_I R$ consists of all the pairs $(a,b)\in R\times R$ such that $a\equiv b(\mathrm{mod} I)$ with the component-wise operations and $\pi_1(a,b)=a$, $\pi_2(a,b)=b$. Clearly $\mathrm{Ker}\pi_1=(0,I)$ and $\mathrm{Ker}\pi_2=(I,0)$. The diagonal embedding δ: R $R\times_I R$ given by $\delta(a)=(a,a)$ splits both π_1 and π_2. In fact $R\times_I R$ is the semidirect product of $\delta(R)\cong R$ and $\mathrm{Ker}\pi_1\cong I$. This means that $R\times_I R$ is isomorphic to the set of pairs (a,c), $a\in R$, $c\in I$, with the component-wise addition and the multiplication given by the following formula: $(a,c)(b,d)=(ab,ad+cb+cd)$. In the sequel we identify I with $\mathrm{Ker}\pi_1$.

Now Stein has proven that the relative groups defined in terms of the following exact sequences coincide with the usual ones:

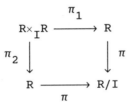

$$1\longrightarrow E(\Phi,R,I)\longrightarrow E(\Phi,R\times_I R)\underset{\delta}{\overset{\pi_1}{\rightleftarrows}}E(\Phi,R)\longrightarrow 1,$$

$$1\longrightarrow G(\Phi,R,I)\longrightarrow G(\Phi,R\times_I R)\underset{\delta}{\overset{\pi_1}{\rightleftarrows}}G(\Phi,R)\longrightarrow 1.$$

Or, in other words, $E(\Phi,R\times_I R,I)=E(\Phi,R,I)$. For the elementary groups this means that

$$\begin{array}{ccccccccc} 1 & \longrightarrow & E(\Phi,R\times_I R,I) & \longrightarrow & E(\Phi,R\times_I R) & \xrightarrow{\pi_1} & E(\Phi,R) & \longrightarrow & 1 \\ & & \downarrow\approx & & \downarrow\pi_2 & & \downarrow\pi & & \\ 1 & \longrightarrow & E(\Phi,R,I) & \longrightarrow & E(\Phi,R) & \xrightarrow[\pi]{} & E(\Phi,R/I) & \longrightarrow & 1, \end{array}$$

The natural embeddings $E(\Phi,R,I)\longrightarrow E(\Phi,R)$ and $G(\Phi,R,I)\longrightarrow G(\Phi,R)$ are defined by π_2. But now we are in a much better position because unlike the original exact sequences coming from the projection $R\longrightarrow R/I$ the new exact sequences *split.*

Using this it is easy to show that

$$G(\Phi,R\times_I R,I)\cap E(\Phi,R\times_I R)=E(\Phi,R\times_I R,I)$$

(see [221,247] for details). Now if we already know that the absolute elementary groups are normal in the corresponding Chevalley groups for all commutative rings we may deduce that the relative subgroups are normal too (being intersections of two normal subgroups).

4°. **Unipotent elements of root type**. Unfortunately the root elements introduced above are not sufficient for our purposes. Actually to prove that a root unipotent lies in E_I we will write it as a product of factors which are not root unipotents themselves - though have formal properties very similar to those of root unipotents.

Let us look at the matrix representing a root unipotent x in a minimal representation π. Its residue $m=\mathrm{res}(x)$ is small with respect to the dimension $n=\dim(\pi)$ and we may write x in the form $x=e+u_1v_1+\ldots+u_mv_m$, where $u_1,\ldots,u_m$ are columns of height n (elements of the module V) while $v_1,\ldots,v_m$ are rows of length n (elements of the dual module V^*). Now for *any* root unipotent the resulting columns and rows satisfy some equations. A typical example of such an equation is, of course, $v_iu_j=0$. Another typical example is when the module $V\cong V^*$ bears an orthogonal/symplectic structure and $v_i=\tilde{u}_{m-i}$ for a given isomorphism $u\longmapsto\tilde{u}$ of V and V^*. In turn these equations on $u_1,\ldots, u_m$ and $v_1,\ldots,v_m$ guarantee that the matrix $x=e+u_1v_1+\ldots+u_mv_m$ does actually belong to the Chevalley group $G_\pi(\Phi,R)$. We refer to the resulting class of matrices as the *(unipotent) elements of root type*. The resulting class of elements depends a priori on the choice of π and to stress this we sometimes say "*elements of root type in the representation* π".

If all the coordinates of $u_1,\ldots,u_m$ or $v_1,\ldots,v_m$ belong to an ideal I of R then we refer to x as an element *of level* I. Clearly in this case x belongs to the principal congruence subgroup $G_\pi(\Phi,R,I)$. In most cases one has to look only on those elements for which the coordinates have a common factor belonging to I. In other words one may think that $u_1,\ldots,u_m$ and $v_1,\ldots,v_m$ are arbitrary columns and rows subject to the equations but consider $x(\xi)=e+u_1\xi v_1+ +\ldots+u_m\xi v_m$, $\xi\in I$, rather than x.

5°. Geometry of root subgroups. If R=K is a field the classes of unipotent root elements and unipotent elements of root type coincide. This is not the case for an arbitrary ring R. For most of our considerations we use only some very special elements of root type. Actually almost all of our needs are covered by what we call *fake root unipotents*, i.e. those elements of root type which are conjugated to an element of $U=U(\Phi,R)$. Obviously a root unipotent is also a fake root unipotent but for $rk(\Phi)\geq 2$ the converse is true only if R is a *Bezout ring*, (i.e. every finitely generated ideal in R is principal). However the class of fake root unipotents does not depend on the choice of π and may be characterized in internal terms. Later we describe in some details the situation for the classical examples and the case $\Phi=E_6$.

The information necessary to check that an element z from U is an element of root type is available at the level of fields. Let K be a field. Write z as a product of $x_\beta(z_\beta)$, $z_\beta\in K$, over all roots $\beta\in\Phi$ in a fixed order. If we want the element $x\in U$ to be conjugated to an elementary unipotent $x_\alpha(\xi)$ its coefficients z_β have to satisfy some polynomial equations. They are particularly easy when z has very few non-zero coefficients and are readily expressed in terms of geometry of root subgroups (see [65,66,258] for further references).

A typical result here is that for a field any pair of *long* root subgroups X,Y in $G=G(\Phi,K)$ is simultaneously conjugated to a pair of elementary root subgroups X_α, X_β where α and β are long roots, i.e there exists such a $g\in G$ that $gXg^{-1}=X_\alpha$ and $gYg^{-1}=X_\beta$. This means that we may speak about the *angle* between the root subgroups X and Y which takes one of the following values: $0,\pi/3,\pi/2,2\pi/3,\pi$. The geometry of short root subgroups is much more cumbersome since they may correspond not only to *short* roots but to certain *pairs* of *long* roots (see [260]).

Let us give two patterns which cover most of our needs. If $X_1,\dots,X_m$ are long root subgroups every pair of which forms the angle $\pi/3$ (in the Cooperstein's terminology they generate a "*singular subspace*" of dimension $\le m-1$) then *any* element $x\in X_1\dots X_m$ from their product is a long root unipotent. This means that even over a ring *absolutely any* element of the corresponding product is an element of long root type.

Another archetypical configuration is when each of the root subgroups $X_1,\dots,X_m$ forms the angle $\pi/3$ with all the others but one and is orthogonal to that one. In that case the long root elements form a quadric in $X_1\dots X_m$. A necessary condition for an element $x\in X_1\dots X_m$ to be a long root element is that its components in each of the groups $X_1,\dots,X_m$ satisfy a certain quadratic equation (which is as we shall see closely related to the equations defining the orbit of the highest weight vector). But then an element of the corresponding product over a ring satisfying the equation will be a long root type element.

These two configurations correspond to the Dynkin diagrams (or rather Carter graphs [51]) of types A_m and $D_{m/2}$ respectively. We encounter some more sophisticated patterns when there are two different root lengths.

CH.2. EXAMPLES

In this chapter we discuss the Chevalley groups and some of their minimal modules. We give some details on the action of elementary unipotents on these modules and the equations satisfied by the entries of matrices representing elements of the Chevalley groups. The proofs are given only for the case $\Phi=E_6$.

§5 The classical cases

Here we recall the familiar identifications for the Chevalley groups of classical types and the corresponding minimal modules and illustrate some of the notions introduced before on these examples.

1°. **The case** $\Phi=A_\ell$. Here the group $G_{sc}(A_\ell,R)$ is isomorphic to the *special linear group* $SL(\ell+1,R)$ of degree $\ell+1$ over R, and the group $G_{ad}(\Phi,R)$ to the corresponding *projective special linear group* $PSL(\ell+1,R)$. In turn $E_{sc}(\Phi,R)=E(\ell+1,R)$ is the *elementary group*, to wit the group generated by all the *elementary transvections* $t_{ij}(\xi)=e+\xi e_{ij}$, $\xi\in R$, $i\neq j$. Here as usual e is the identity matrix and e_{ij} is a standard matrix unit, i.e e_{ij} has 1 in the position (i,j), all other entries being 0. A transvection $t_{ij}(\xi)$ is a root unipotent corresponding to the root e_i-e_j, where e_i is the standard base of the $(\ell+1)$-dimensional euclidean space $\mathbb{R}^\ell$.

The quotient group

$$K_1(\Phi,R)=SK_1(\ell+1,R)=SL(\ell+1,R)/E(\ell+1,R)$$

is the usual (*linear*) *unstable* K_1-*functor* [23,27,24,163,114].

The weight diagram of type $(A_\ell,\bar{\omega}_1)$ is a chain:

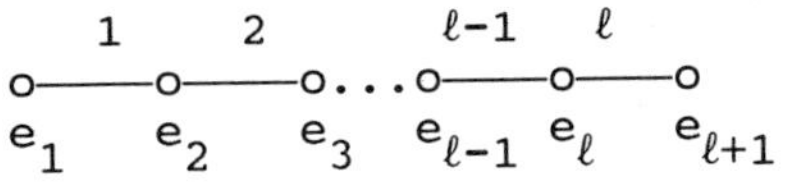

where the bond marked by i corresponds to the simple root $\alpha_i= =e_i-e_{i+1}$. We put the i-th component v_i of a vector v from the $(\ell+1)$-dimensional representation module V in the i-th node of the diagram. Then action of $t_{ij}(\xi)=x_\alpha(\xi)$, $\alpha=e_i-e_j$, on v has the following effect: it adds $\pm\xi v_j$ to v_i and leaves all the other coordinates unchanged.

2°. **The case** $\Phi=B_\ell$. Let's number the indices from 1 to $2\ell+1$ as follows: $1,\dots,\ell,0,-\ell,\dots,-1$. Introduce on the free R-module $V=R^n$ of rank $n=2\ell+1$ a quadratic form by

$$Q(x_1,\dots,x_{-1})=x_0^2+x_1x_{-1}+\dots+x_\ell x_{-\ell}.$$

In other words Q is a form of the maximal Witt index, which makes V to a *split orthogonal space of dimension* $2\ell+1$ (*"Artin space"*).

Then $G_{sc}(\Phi,R)$ is precisely the *spinorial group* Spin(n,R), defined by the form Q, and $G_{ad}(\Phi,K)$ coincides with the *special orthogonal group* SO(n,R)= =PSO(n,R). Recall the equations defining SO(n,R). As usual for a matrix $a\in GL(n,R)$ we denote by a_{ij} its entry in the position (i,j), i.e. $a=(a_{ij})$, $1\le i,j\le n$. Further $a^{-1}=(a'_{ij})$ is the inverse of a and a^t – the transpose. For the sake of simplicity we assume that $2\in R^*$ although strictly speaking this is not necessary. Denote by $\mathrm{sdiag}(\varepsilon_1,\dots,\varepsilon_n)$ the matrix which has $\varepsilon_1,\dots,\varepsilon_n$ on the skew diagonal (in the NE→SW direction) and zeros everywhere else. Define for an odd $n=2\ell+1$ the matrix $F=F_n$ by

$$F=\mathrm{sdiag}(1,\dots,1,2,1,\dots,1),$$

where the series of 1-s consist of ℓ members each. Now $SO(2\ell+1,R)$ consists of those matrices $a\in GL(n,R)$, for which $\det(a)=1$ and $aF_na^t=F_n$. In terms of the matrix entries the orthogonality condition for a matrix a is expressed by the following equations

$$a'_{ij} = \begin{cases} a_{-j,-i}, & \text{if } i,j\neq 0 \text{ or } i=j=0,\\ 2a_{-j,0}, & \text{if } i=0,\ j\neq 0,\\ a_{0,-i}/2, & \text{if } j=0,\ i\neq 0. \end{cases}$$

To be a column of an orthogonal matrix a vector $x=(x_1,\dots,x_{-1})$ has to satisfy one of the following equations:

$$x_0^2+x_1x_{-1}+\dots+x_\ell x_{-\ell}=0 \text{ or } x_0^2+x_1x_{-1}+\dots+x_\ell x_{-\ell}=1$$

depending on whether x has to be a column with a non-zero or the zero index.

Recall the usual geometric interpretation of the definition of the orthogonal groups which we will use in chapter 3. Start with the orthogonal groups. Let $V\cong R^n$ be a free R-module of rank n with the standard base $e_1,\dots,e_n$, which we again denote by $e_1,\dots,e_{-1}$. We may identify GL(V) with G=GL(n,R) via this base. Thus a matrix $a\in G$ acts on V by left multiplication. Fix an R-bilinear form on V whose matrix in the base $e_1,\dots,e_{-1}$ (which is called a *Witt base*) coincides with $f_{2\ell}$ or $F_{2\ell+1}$, depending on whether n is even or odd. In other words $(u,v)=B(u,v)=u^tgv$, where $g=f_{2\ell}$, if $n=2\ell$ is even and $g=F_{2\ell+1}$, if $n=2\ell+1$ is odd. This means that products of the base vectors $e_1,\dots,e_{-1}$ are given by the formula $(e_i,e_j)=\delta_{i,-j}+\delta_{i,0}\delta_{j,0}$. It is clear that the form B is *symmetric* i.e. (u,v)=(v,u) and *nondegenerate* i.e. the map $u\mapsto\phi_u$

from V to the dual module V^*=Hom(V,R) defined by the formula $\phi_u(v)=(u,v)$, is an isomorphism of V on V^*. Then for a ring R such that $2\in R^*$ the group O(n,R) is precisely the isometry group of B, i.e. the set of all transformations $g\in G$ such that $(gu,gv)= (u,v)$, while SO(n,R) is the intersection of O(n,R) with SL(n,R).

Recall that the root system of type B_ℓ consists of the roots $\pm e_i\pm e_j$ and $\pm e_i$ for $1\le i,j\le \ell$, where e_i is the canonical base of the ℓ-dimensional euclidean space $\mathbb{R}^\ell$. Further we set $e_{-i}=-e_i$. The *elementary root unipotents* have the form

$$T_{ij}(\xi)=T_{-j,-i}(-\xi)=e+\xi e_{ij}-\xi e_{-j,-i},$$

if $i,j\neq 0$ and $i\neq\pm j$ (these are the *long root unipotents* corresponding to the long roots e_i-e_j – in our notation $e_i+e_j=e_i-e_{-j}$), and the form

$$T_{io}(\xi)=T_{o,-i}(-\xi)=e+\xi e_{io}-2\xi e_{o,-i}-\xi^2 e_{i,-i}$$

if one of the indices i,j equals 0 (these are the *short root unipotents*, corresponding to the short roots $e_i=-e_{-i}$).

Let $i\neq\pm j$. All the elements $T_{ij}(\xi)$ form the subgroup $X_{ij}=\{T_{ij}(\xi),\ \xi\in R\}$, which is called an *elementary unipotent root subgroup* (sometimes just a *root subgroup*). If $i,j\neq 0$ then X_{ij} is called a *long root subgroup, and if* i=o, *or* j=0 – *a short root subgroup.*

The subgroup of SO(n,R) generated by all the root subgroups X_{ij} is called the *elementary subgroup* and denoted by EO(n,R). In an analogous fashion one could define the elementary spinorial group Epin(n,R). The elementary Chevalley groups $E_{sc}(\Phi,R)$ and $E_{ad}(\Phi,R)$ coincide with Epin(n,R) and EO(n,R) respectively. Let us note that although Spin(n,K)=Epin(n,K) for *any* field K, the group EO(n,K) is not necessarily identical with SO(n,K). In fact it is precisely the *kernel of the spinorial norm* which is, generally speaking, a *proper* subgroup of SO(n,K) and (apart from the very small dimensions) equals the commutator subgroup $\Omega(n,K)$ of this latter group. Thus the functor

$$K_1(B_\ell,R)=\mathrm{Spin}(2\ell+1,R)/\mathrm{Epin}(2\ell+1,R),$$

defined above is distinct from the Bass orthogonal K_1-functor

$$KO_1(n,R)=SO(n,R)/EO(n,R),$$

see [26,20,206,114]

The weight diagram of type $(B_\ell,\bar{\omega}_1)$ is a chain:

$$\begin{array}{ccccccccccccccccccc} & 1 & & 2 & & & & \ell & & \ell & & & & 2 & & 1 & \\ \circ & \text{---} & \circ & \text{---} & \circ & \dots & \circ & \text{---} & \circ & \text{---} & \circ & \dots & \circ & \text{---} & \circ & \text{---} & \circ \\ e_1 & & e_2 & & e_3 & & e_\ell & & 0 & & e_{-\ell} & & e_{-3} & & e_{-2} & & e_{-1} \end{array}$$

where bonds marked by i correspond to the simple root $\alpha_i=e_i-e_{i+1}$ if $1\le i\le \ell-1$ and $\alpha_\ell=e_\ell$ if $i=\ell$. We persist in writing here e_{-i} instead of $-e_i$ to conform with the numbering of the base vectors and the coordinates of the $(2\ell+1)$-dimensional module V. The action of $T_{ij}(\xi)=x_\alpha(\xi)$ on a vector $v\in V$ has the following ef-fect. If $\alpha=e_i-e_j$ then it adds $\pm\xi v_j$ to v_i and $\mp\xi v_{-i}$ to v_{-j} (recall that in our notation $e_i+e_j=e_i-e_{-j}$). If $\alpha=e_i$ then it adds $\pm 2\xi v_{-i}$ to v_o and $\mp\xi v_o\mp\xi^2 v_{-i}$ to v_i. In both cases all the other coordinates remain unaltered.

3°. **The case** $\Phi=C_\ell$. Let's number the indices from 1 to 2ℓ as following $1,\dots,\ell,-\ell,\dots,-1$ and introduce on the free R-module $V=R^n$, $n=2\ell$, a symplectic form by

$$B(x,y)=(x_1y_{-1}-x_{-1}y_1)+\dots(x_\ell y_{-\ell}+x_{-\ell}y_\ell),$$

for $x=(x_1,\dots,x_{-1})$; $y=(y_1,\dots,y_{-1})\in V$. Then $G_{sc}(\Phi,R)$ is isomorphic to the *symplectic group* $Sp(2\ell,R)$, defined by this form and the group $G_{ad}(\Phi,R)$ to the corresponding *projective group* $PSp(2\ell,R)$. Recall the equations for the matrix entries of elements of $Sp(2\ell,R)$. For the even $n=2\ell$ define the matrix

$$F=\mathrm{sdiag}(1,\dots,1,-1,\dots,-1)$$

where the series of 1-s and (-1)-s consist of ℓ members each. Then $\Gamma=Sp(2\ell,R)$ consists of those matrices $a\in GL(n,R)$ for which $aF_{2\ell}a^t=F_{2\ell}$, A matrix $a=(a_{ij})$ is symplectic if and only if

$$\lambda a'_{ij}=\varepsilon_i\varepsilon_j a_{-j,-i},$$

where ε_i is the *sign* of i, i.e. $\varepsilon_i=1$, if $1\le i\le \ell$, and $\varepsilon_i=-1$, if $-\ell\le i\le -1$.

Recall that the root system of type C_ℓ is dual to that of type B_ℓ and consists of the roots $\pm e_i\pm e_j$ and $\pm 2e_i$ for $1\le i,j\le \ell$, where as before e_i is the canonical base of the ℓ-dimensional euclidean space $\mathbb{R}^\ell$ and $e_{-i}=-e_i$. The *elementary root unipotents* have the form

$$T_{i,-i}(\xi)=e+\xi e_{i,-i}$$

if $i=-j$ (these are the *long root unipotents* corresponding to the long roots $2e_i$), and the form

$$T_{ij}(\xi)=T_{-j,-i}(-\varepsilon_i\varepsilon_j\xi)=e+\xi e_{ij}-2\xi e_{-j,-i}$$

if $i\neq\pm j$ (these are the *short root unipotents* corresponding to the short root e_i-e_j).

Let $i\neq j$. All the root unipotents $T_{ij}(\xi)$ form the root subgroup $X_{ij}=\{T_{ij}(\xi),\ \xi\in R\}$, which is called a *root subgroup*. The subgroup of $Sp(2\ell,R)$ generated by all the root subgroups X_{ij} is called the *elementary symplectic subgroup* and denoted by $Ep(2\ell,R)$. Now

$$K_1(C_\ell,R)=KSp_1(2\ell,R)=Sp(2\ell,R)/Ep(2\ell,R)$$

is the ordinary *symplectic* K_1*-functor* [27,114].

The weight diagram of type $(C_\ell,\bar\omega_1)$ is a chain:

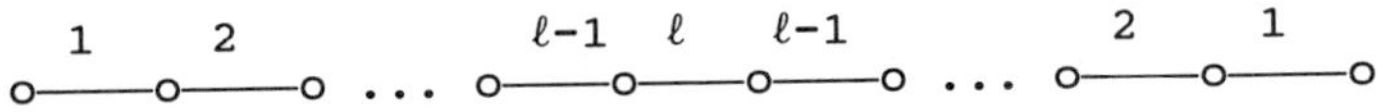

where bonds marked by i correspond to the simple root $\alpha_i=e_i-e_{i+1}$ if $1\le i\le\ell-1$ and $\alpha_\ell=2e_\ell$ if $i=\ell$. The action of $T_{ij}(\xi)=x_\alpha(\xi)$ on a vector $v\in V$ has the following effect. If $\alpha=e_i-e_j$ then it adds $\pm\xi v_j$ to v_i and $\mp\varepsilon_i\varepsilon_j\xi v_{-i}$ to v_{-j} (again in our notation $e_i+e_j=e_i-e_{-j}$). If $\alpha=2e_i$ then it just adds $\pm\xi v_{-i}$ to v_i. In both cases all the other coordinates remain unaltered.

4°. **The case $\Phi=D_\ell$.** Let's keep the same numbering of indices as for the case of $\Phi=C_\ell$ and introduce on the free R-module $V=R^n$, $n=2\ell$, a quadratic form by

$$Q(x_1,\dots,x_{-1})=x_1x_{-1}+\dots+x_\ell x_{-\ell}.$$

This form has the maximal Witt index and makes V an *Artin space.* Then $G_{sc}(\Phi,R)$ is again the *spinorial group* $Spin(n,R)$, defined by the form Q, and $G_{ad}(\Phi,R)=PSO(n,R)$. For one of the lattices P lying between $Q(\Phi)$ and $P(\Phi)$ (if ℓ is odd, there is just one such lattice,if ℓ is even three of them) we have $G_P(\Phi,R)=SO(n,R)$. Recall the equations for this latter group. Define for an even $n=2\ell$ the matrix $f=f_n$ as $f=\mathrm{sdiag}(1,\dots,1)$, in other words the entries of $f=(f_{ij})$ are defined by $f_{ij}=\delta_{i,-j}$. Now $SO(2\ell,R)$ consists of those matrices $a\in GL(2\ell,R)$, for which $\det(a)=1$ and $af_na^t=f_n$. In terms of the matrix entries the orthogonality condition for a matrix a is expressed by the following equations $a'_{ij}=a_{-j,-i}$. To be a column of an orthogonal matrix a vector $x=(x_1,\dots,x_{-1})$ has to satisfy the quadratic equation $x_1x_{-1}+\dots+x_\ell x_{-\ell}=0$.

Recall that the root system of type D_ℓ consists of the long roots of a system of type B_ℓ. In the usual realization D_ℓ has the roots $\pm e_i \pm e_j$, $1\le i,j\le \ell$. The *elementary root unipotents* are just the *long root unipotents* and thus have the form

$$T_{ij}(\xi)=T_{-j,-i}(-\xi)=e+\xi e_{ij}-\xi e_{-j,-i},$$

for $i\ne\pm j$. They generate the subgroup $EO(2\ell,R)$ of $SO(2\ell,R)$ which is called the *elementary orthogonal group*. All the comments from the case $\Phi=B_\ell$ apply here. Again the functor

$$K_1(D_\ell,R)=Spin(2\ell,R)/Epin(2\ell,R)$$

does not in general coincide with the Bass functor $KO_1(n,R)$.

The weight diagram of type $(D_\ell,\bar\omega_1)$ is *not* a chain:

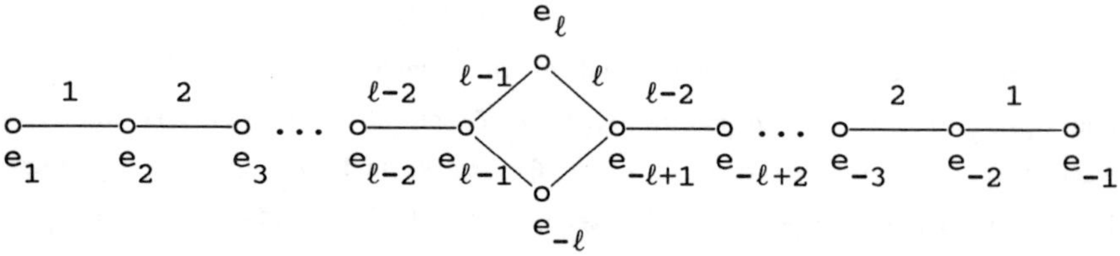

where bonds marked by i correspond to the simple root $\alpha_i= =e_i-e_{i+1}$ if $1\le i\le\ell-1$ and $\alpha_\ell=e_{\ell-1}+e_\ell$ if $i=\ell$. The action of $T_{ij}(\xi)=x_\alpha(\xi)$ on a vector $v\in V$ has the same effect as in the case $\Phi=B_\ell$. Namely if $\alpha=e_i-e_j$ then $x_\alpha(\xi)$ adds $\pm\xi v_j$ to v_i and $\mp\xi v_{-i}$ to v_{-j}, all the other coordinates being unaltered.

The identifications for the classical cases go back to the works of R.Ree [180] and J.Dieudonné [81] (where the case of the orthogonal group over a non-perfect field of characteristic 2 was treated). Look at J.-I.Hée [118] for a modern exposition. For the exceptional cases identifications with the groups stabilizing certain systems of tensors are far less commonly known. Here we describe the case of the group of type G_2 also studied by Ree. In fact the group of type G_2 is in many respects extremely close to the classical groups and should be considered one - or else the groups of types D_4 and B_3 should be thought of as being exceptional. We discuss this subject in the next two sections.

It is already much more difficult to describe explicitly the equations for the spinorial groups. R.Brown [45] has shown that the *split* spinorial groups may be characterized as the stabilizers of a bilinear form B *and* a tensor of valency four (actually he has considered a linear map $V\otimes V$ $V\otimes V$ but it can be transformed to a four-linear form using the pairing induced by B).

5°. **The case $\Phi=G_2$.** Let $V=R^7$ be a free R-module with the coordinates of the vectors being numbered by 1,2,3,0,-3,-2,-1. Introduce the following pair of forms on V: the same quadratic form Q as for the B_3 case:

$$Q(x)=-2x_0^2+x_1x_{-1}+x_2x_{-2}+x_3x_{-3},$$

where

$$x=(x_1,x_2,x_3,x_0,x_{-3},x_{-2},x_{-1})\in V,$$

and an alternating trilinear form F(x,y,z) defined by monomials

$$x_0x_1x_{-1}+x_0x_2x_{-2}+x_0x_3x_{-3}+x_1x_2x_3+x_{-1}x_{-2}x_{-3}.$$

Of course one has to polarize these monomials in such a way as to generate an alternating form. Thus, say the first monomial $x_0x_1x_{-1}$ gives rise to the following six summands in F(x,y,z):

$$x_0y_1z_{-1}+y_0z_1x_{-1}+z_0x_1y_{-1}-x_0z_1y_{-1}-y_0x_1z_{-1}-z_0y_1x_{-1}.$$

The form F is usually referred to as the *Dickson form* or, sometimes as the *alternating Dickson form* (see next section).

Then the group Isom(Q,F,R) consists of all the transformations from $GL(V)=GL(7,R)$, for which one has both $Q(gx)=Q(x)$ and $F(gx)=F(x)$ for all $x\in V$ and it is proven in [180] that actually $G(G_2,R)=\mathrm{Isom}(Q,F,R)$. Of course in [180] only the case of a field R=K was treated but it is well known that two affine group schemes are isomorphic if the corresponding groups of points over any (algebraically closed) field are isomorphic.

The weight diagram of type $(G_2,\bar{\omega}_1)$ is a chain:

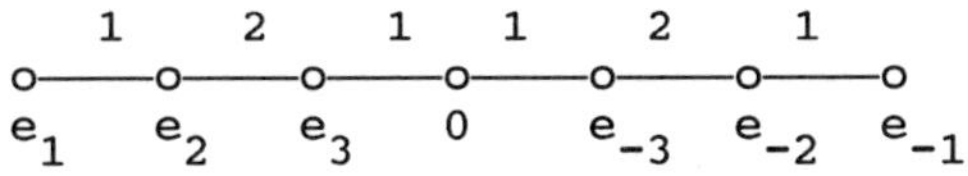

where α_1 is short and α_2 is long. The action of any $x_\alpha(\xi)$ on the vectors of the 7-dimensional module can be read off the diagram (apart from the signs).

Thus, say, to determine the action of $x_\alpha(\xi)$ for $\alpha=2\alpha_1+\alpha_2$ we have to find all the paths at the diagram which have marks 1,1 and 2 on their bonds (in any order) *and* their composite paths. So multiplication by $x_\alpha(\xi)$ in this case adds $\pm\xi v_0$ to v_1, $\pm\xi v_{-3}$ to v_2, $\pm\xi v_{-2}$ to v_3, $\pm 2\xi v_{-1}$ to v_0 *and* $\pm\xi^2 v_{-1}$ to v_1 (no attempt to fix signs here has been made, see [126]).

In fact the cases $(A_\ell,\bar\omega_1),(B_\ell,\bar\omega_1),(C_\ell,\bar\omega_1),(G_2,\bar\omega_1)$ considered above are *the only* cases when the weight diagram is a chain. For all the other cases any sort of a *complete order* on the coordinates of a vector from the representation (and thus on the columns and rows of a matrix expressing an endomorphism of this representation) is of no real value and thus may be misleading or even obnoxious. It is very convenient though to think of the coordinates, the columns and the rows as being *partially ordered*. Even in the case $(D_\ell,\bar\omega_1)$ it is an offence to the nature of things to claim that the ℓ-th coordinate precedes the $(-\ell)$-th coordinate.

§6. The cubic form for E_6

In this section we describe a realization of the simply connected Chevalley group $G=G_{sc}(E_6,R)$ of type E_6 as the isometry group of a trilinear form on a 27-dimensional vector space which gives full control of the equations satisfied by the entries of matrices from G. This is our crucial example and we discuss it in some detail. It is already much more sophisticated than the classical cases but still the underlying structure is much more transparent and easy to check up than for the senior cases E_7 and E_8 - or even for the subordinate case F_4. We sketch analogous constructions for other exceptional groups in the next section.

1°. **The 27-dimensional module.** It is very well known that the dimension of the module (V,π) of the group $G=G(E_6,R)$ with the highest weight $\omega=\bar\omega_1$ equals 27. The dual module V^* has highest weight $\bar\omega_6$. The weight diagram for the case $(E_6,\bar\omega_1)$ reproduced below gives already a more realistic image of what the weight diagrams actually look like.

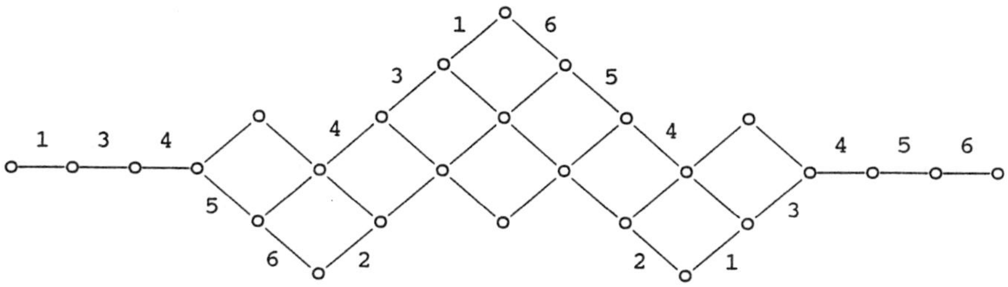

Here as always the nodes correspond to the weights $\lambda\in\Lambda=\Lambda(\pi)$ of the representation π (all of them have multiplicity 1 in this case since they form a single Weyl-orbit) and the numeration of the fundamental roots of E_6 follows that of N.Bourbaki [41]:

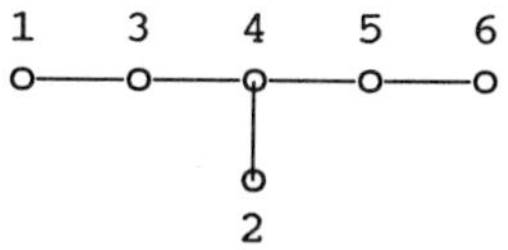

This diagram will be our critical example in what follows. In fact this is just the diagram of the cosets $W(E_6)/W(D_5)$, two cosets being joint by a bond marked i if and only if one passes from one to another by the fundamental reflection w_i (see [72]). One readily sees that by observing that the stabilizer of the highest weight in the Weyl group is generated by all of the fundamental reflections apart from w_1.

Let us recall that we think of the vectors $v\in V$ as the weighted diagrams with the element v_λ of the ring R put into node corresponding to $\lambda\in\Lambda(\pi)$. The action of the root unipotents $x_\alpha(\xi)$ on these vectors can be easily recovered from the diagram. Thus if we want to know the action of an element $x_\alpha(\xi)$ we have to decompose α into the sum of fundamental roots $\alpha=\alpha_{i_1}+\ldots+\alpha_{i_s}$ and then find all the paths in the diagram which have marks $i_1,\ldots,i_s$ (in any possible order) on their bonds. There are 6 such paths for any root α and if such a path starts at λ and finishes at ν then $x_\alpha(\xi)$ adds $\pm\xi v_\lambda$ to v_ν. For example if $\alpha=\alpha_1+\alpha_3+\alpha_4$ then $x_\alpha(1)$ adds/subtracts the coordinates in the nodes marked by crosses to/from the coordinates marked by squares as follows:

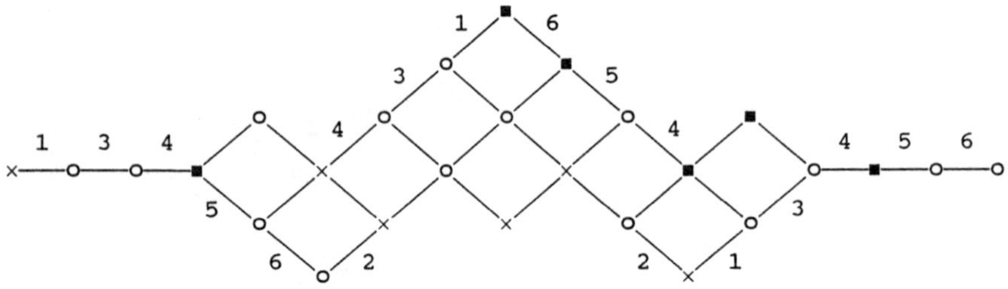

An explicit description of the signs of the action is a different story. Of course since the representation appear as a sub-representation for the adjoint module for the group of type E_7 the signs can be obtained from the structure constants $N_{\alpha\beta}$ for the Lie algebra of type E_7, thus, for example, from [165] or [101] and a table will be reproduced in a paper by the author and E.B.Plotkin. A more promising description is via the Frenkel-Kac cocycle [97,198,98].

2°. **The cubic form**. The first realization of the simply connected Chevalley group of type E_6 as the isometry group of a trilinear form on a 27-dimensional space was obtained by L.E.Dickson in 1905 and the form is usually referred to as the *Dickson form* (or sometimes, to distinguish it from the alternating Dickson form which appeared in the construction of G_2, as the *symmetric Dickson form*). In 1950 C.Chevalley and R.D.Schafer discovered a construction of E_6 and F_4 in terms of the 27-dimensional *exceptional Jordan algebra* - a certain algebra of hermitian matrices over the *Cayley-Dickson (octonion) algebra* [60]. Finally in 1952 H.Freudenthal gave an extremely beautiful construction which we reproduce below. Dickson's original construction of the form displays symmetry with respect to A_5+A_1. Analogous realizations were later used by C.Chevalley [56], E.Shult and M.Aschbacher [13]. Constructions in terms of the Cayley-Dickson algebra are related to D_4 and F_4 while in the Freudenthal's construction symmetry with respect to $3A_2$ is apparent. Of course H.Freudenthal himself has considered only the classical case of a characteristic 0 field. These realizations were generalized to fields of characteristics different from 2 and 3 in the subsequent works of T.Springer, J.Tits, G.B.Seligman, N.Jacobson, F.Veldkamp, A.Cohen, B.Cooperstein and others, see, for example, [130,189,194,196,200,266,62,91] and the references therefrom.

T.Springer [195] characterized the cubic form in question axiomatically. Of course it is just the norm form of the corresponding exceptional Jordan algebra.

Let M(3,R) be the full matrix ring of degree 3 over R, and V= $=\{(x,y,z)\,|\,x,y,z\in M(3,R)\}$ the free M(3,R)-module of rank 3 considered as a free R-module of rank 27. Define on V a cubic form F by the following formula:

$$F(x,y,z)=\det(x)+\det(y)+\det(z)-\mathrm{tr}(xyz).$$

Then $G_{sc}(E_6,R)$ may be identified with the group Isom(F,R), consisting of all the transformations $g\in GL(V)\cong GL(27,R)$, such that $F(g(x,y,z))=F(x,y,z)$. Of course H.Freudenthal considered only the classical fields and a priori the fields of characteristics 2 and 3 might constitute a problem here. Actually they do not. The extant proofs for the identifications use very deep geometric results: either the classification of semisimple algebraic groups or results of J.Tits on classification and local characterization of buildings. Below we outline one more construction of the form which displays symmetry with respect to the whole Weyl group $W(E_6)$ and a proof of the identification which uses only elementary linear algebra. This proof is very much along the lines of a proof of the Ree–Dieudonné theorem as exposed in the talk [118] of J.-Y.Hée.

Actually as we shall see the form f emerges inevitably if we are looking for a nontrivial *degenerate* form invariant under the action of the extended Weyl group $\tilde{W}(E_6)$ on the 27-dimensional space. The semisimple classical groups arise as the isometry groups of non-degenerate bilinear forms, so why does one have to look for a degenerate form here? Because a classical theorem due to C.Jordan - S.Lie - W.Burnside - A.Hurwitz - R.Bott - J.Tate - H.Matsumura - P.Monski - D.Mumford - J.Ax - C.H.Sah - P.Orlik - L.Solomon - H.Suzuki - ... - ... (see for example [169]) asserts that over a field of characteristic 0 the isometry group of a *non-degenerate* form of degree $m\geq 3$ is finite. So for the exceptional cases the situation is strictly opposite to the classical ones, only the exceptional forms are of interest. Precisely the fact that the form F is very degenerate guarantees that the isometry group of the trilinear form coincides with the isometry group of the corresponding cubic form so that there are no difficulties for the characteristics 2 and 3.

At this point it might be beneficial to make it clear that for a given algebraically closed field K it's not a problem to find *some* tensor whose stabilizer coincides with a given semisimple group. Virtually *any* tensor stabilized by the group will go. Why? Just because up to some explicitly listed exceptions the irreducible tensor-indecomposable semisimple subgroups are maximal among the closed connected subgroups of the classical groups in which they are contained (for the characteristic 0 case this is a classical Dynkin theory [84,85] while for positive characteristics it was established quite recently in the outstanding works of G.Seitz, I.D.Suprunenko and D.Testerman, see [187,188,229] for the detailed statement and further references). So it suffices to pick up any tensor which is not stabilized by the classical group. The real problem is to construct such an invariant which does not depend on the characteristic and gives simple and convenient equations.

3°. The exceptional Jordan algebra. In terms of the Cayley-Dickson algebra C the form is described as follows. Let $\mathrm{char}K\neq 2,3$. Let C be the *split octonion algebra* over the field K corresponding to a quadratic form N. This is the unique *composition algebra* (i.e. $N(xy)=N(x)N(y)$ for any $x,y\in C$) of dimension 8 over K with zero-divisors (in [91] one may find its matrix realization and further references). Denote by $x\longmapsto x^*$ the canonical involution on C and by $T(x)=x+x^*$ the trace. Now choose scalars $\gamma_1,\gamma_2,\gamma_3\in K$ and define the involution $a\longmapsto g^{-1}(a^*)^tg$, where $g=\mathrm{diag}(\gamma_1,\gamma_2,\gamma_3)$, on the ring $M(3,C)$. Let J be the set of hermitian matrices with respect to this involution. In other words J consists of all the matrices of the form

$$a=\begin{pmatrix} \alpha_1 & x_3 & \gamma_1^{-1}\gamma_3x_2^* \\ \gamma_2^{-1}\gamma_1x_3^* & \alpha_2 & x_1 \\ x_2 & \gamma_3^{-1}\gamma_2x_1^* & \alpha_3 \end{pmatrix}$$

Then J is closed under the multiplication $ab=\frac{1}{2}(a.b+b.a)$ where a.b is the usual matrix multiplication in $M(3,C)$. Under this multiplication J is an exceptional Jordan algebra (see [91]). For future reference we reproduce a picture which illustrates how the three scalar components and the three octonion components are fitted in the 27-dimensional space V:

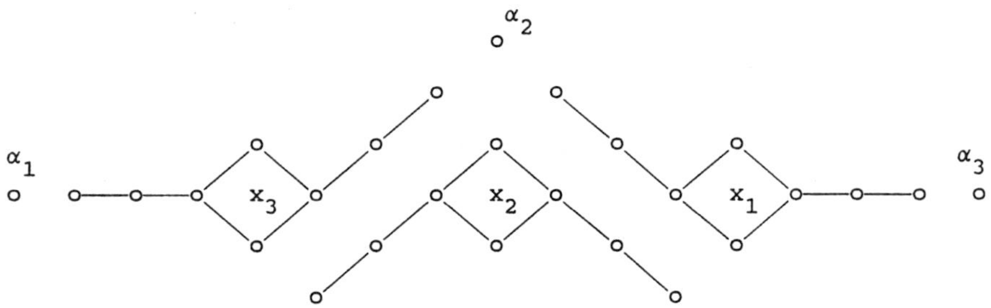

Now the usual formula for the determinant

$$\det(a)=\alpha_1\alpha_2\alpha_3+T(x_1x_2x_3)-\alpha_1\gamma_3^{-1}\gamma_2N(x_1)-\alpha_2\gamma_1^{-1}\gamma_3N(x_2)-\alpha_3\gamma_2^{-1}\gamma_1N(x_3)$$

defines a cubic form on J and this is it. It is usually more convenient to polarize det to a trilinear form f on J such that $f(x,x,x)=\det(x)$.

Actually using the form one can introduce a further very useful binary operation on J. One can define a nondegenerate quadratic form Q on J setting

$$Q(a)=\frac{1}{2}(\alpha_1^2+\alpha_2^2+\alpha_3^2)+\gamma_3^{-1}\gamma_2N(x_1)+\gamma_1^{-1}\gamma_3N(x_2)+\gamma_2^{-1}\gamma_1N(x_3).$$

Let (,) denote the polarization of Q. Now traditionally one defines the *Freudenthal cross product* $x\times y$ of the elements $x,y\in J$ by the following formula: $(x\times y,z)=3f(x,y,z)$. The cross product $x\times y$ differs from the usual Jordan product xy by some linear combination of x,y and e. But for characteristic 3 at this point one feels extremely unhappy. Since the Jordan structure is much more than one actually needs to recover the Chevalley group G of type E_6 and all the necessary formulas involve cross product rather than the Jordan one, we suppress the factor 3 in the above formula. Also we do not have a G-invariant quadratic form on our module - actually this form is an extra structure which comes from F_4. So we have to carefully distinguish V and V^*. Recall that we think of elements from V as columns and the elements from V^* as being rows so that the natural pairing $V^*\times V\longrightarrow K$ is given by multiplication. As soon as we have a trilinear form f on V we can define our cross product $V\times V\longrightarrow V^*$ by the formula $(x\times y)z=f(x,y,z)$. So the cross product of two columns is a row and analogously the cross product of two rows is a column. A further bit of extra structure for the the Jordan algebra setting is that the Jordan algebra has a distinguished element e - the identity matrix which is the sum $e=e_{11}+e_{22}+e_{33}$ of three primitive idempotents - compare the "distinguished" triple below.

As opposed to the D_4-symmetry shown at the previous picture the Dickson-Chevalley-Aschbacher's construction is based on the A_5+A_1-symmetry as shown at the following picture:

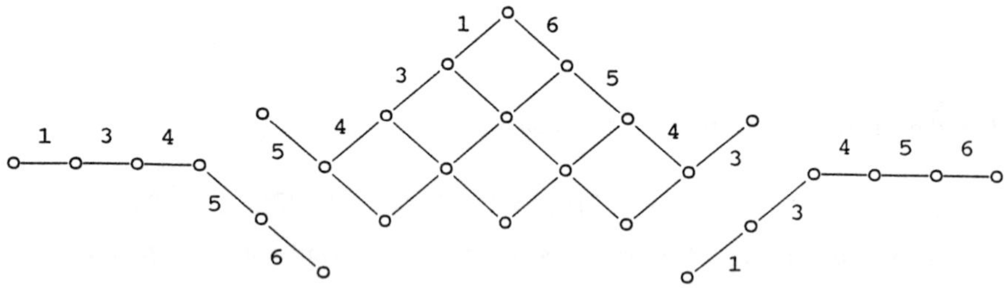

Here the space V is realized as $V=U\oplus\Lambda^2(U^*)\oplus U$ for the 6-dimensional natural representation space U of $SL(6,R)=G(A_5,R)$ (or else as $V=U\oplus\Lambda^4(U)\oplus U$) and the form is determined in terms of the natural pairing (or the intersection product). A detailed description of the form with an explicit table of signs may be found in [13].

5°. **One more construction**. Let us look at the *weight graph* of the representation $(E_6,\bar{\omega}_1)$ instead of its *weight diagram*: the nodes of the graph coincide with the nodes of the diagram and two nodes are linked with a bond if their difference is a root (not just a fundamental root as in the weight diagram). This is a remarkable *distance-transitive graph of girth* 2 and its structure is of course well known see, for example, [71]. Any node of the graph is adjacent to 16 further nodes and non-adjacent to the 10 remaining ones. This is illustrated by the following picture:

■ 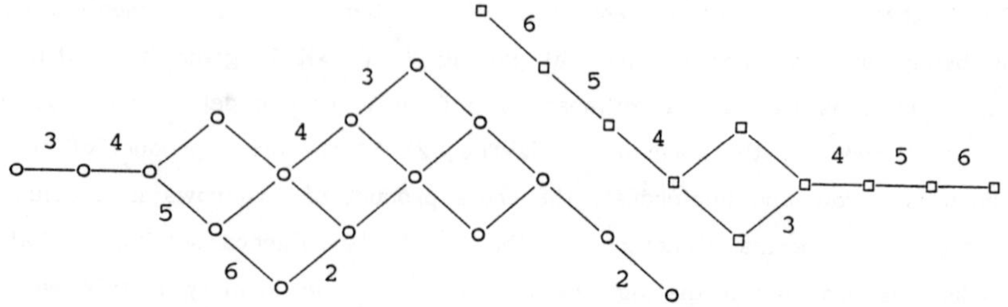

where the white squares represent nodes non-adjacent to the high weight ω denoted by the black square. It is crucial for what follows that these 10 weights form the familiar diagram of type $(D_5,\bar{\omega}_1)$. Of course the 16 weights form another very common diagram - that of the half-spinorial representation of D_5. Now we want our cubic form to be invariant under the action of the extended Weyl group $\tilde{W}(E_6)$ and have the least possible number of monomials. If the form $F(x,y,z)$ has a monomial $x_\lambda y_\mu z_\nu$ for some adjacent λ and μ then it has 15 further such monomials, but if λ and μ are non-adjacent - only 9. Because we want our form to be symmetric we are compelled to take ν non-adjacent to λ and μ. Actually looking at the preceding diagram one immediately discovers that for any two non-adjacent weights there is a unique third weight non-adjacent to both of them. Such a triple is shown at the following picture:

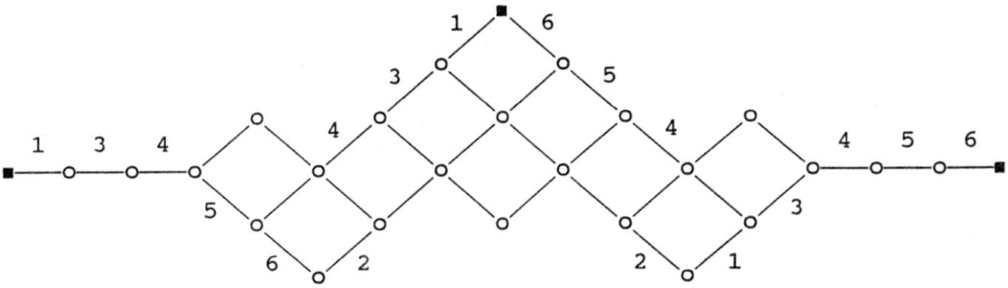

Let us denote the sets of all the ordered and unordered triples of pair-wise non-adjacent weights by Θ and Θ_o respectively:

$$\Theta=\{(\lambda,\mu,\nu)\,|\,\lambda,\mu,\nu\in\Lambda(\pi),\ \lambda-\mu,\lambda-\nu,\mu-\nu\notin\Phi\cup\{0\}\},$$

$$\Theta_o=\{\{\lambda,\mu,\nu\}\,|\,\lambda,\mu,\nu\in\Lambda(\pi),\ \lambda-\mu,\lambda-\nu,\mu-\nu\notin\Phi\cup\{0\}\}.$$

As we've just noticed any two elements of a triple determine the third element in a unique fashion and the Weyl group $W(E_6)$ acts transitively on Θ and Θ_o. Since the stabilizer of the highest weight and the largest weight non-adjacent to the highest weight is generated by all the fundamental reflections apart from w_1 and w_6, we see that $\Theta=W(E_6)/W(D_4)$. Thus the order of Θ equals 270 and the order of Θ_o equals 45. If we look at Λ as the 27 lines on a cubic hypersurface then Θ represents the triples of pair-wise non-intersecting lines. The triple shown at the preceding diagram is maximal with respect to the lexicographic order on Θ and we refer to it as the *distinguished triple* and denote it by (λ_o,μ_o,ν_o). Of course $\lambda_o=\omega$.

So let us define the cubic form F(x) and the corresponding trilinear form f(x,y,z) on the space V with base v^λ, $\lambda\in\Lambda$, by the following formulae:

$$F(x)=\sum \pm x_\lambda x_\mu x_\nu,\quad \{\lambda,\mu,\nu\}\in\Theta_o,$$

$$f(x,y,z)=\sum \pm x_\lambda y_\mu z_\nu,\quad (\lambda,\mu,\nu)\in\Theta,$$

where the signs are determined in a unique way by the fact that the form is invariant under the action of the extended Weyl group $\tilde{W}(E_6)$. To see this one just has to observe that multiplication by any element of $T(E_6,\mathbb{Z})$ always changes an *even* number of signs of the base vectors v^λ,v^μ,v^ν for any triple $(\lambda,\mu,\nu)\in\Theta$. Let us check this for the distinguished triple of base vectors. In fact, this group is generated by the semisimple root elements $h_\alpha(-1)$ and it is well known that a semisimple root element $h_\alpha(\varepsilon)$ multiplies 6 of the base vectors by ε and 6 by ε^{-1} as illustrated by the following picture which presents the case $\alpha=\alpha_1$:

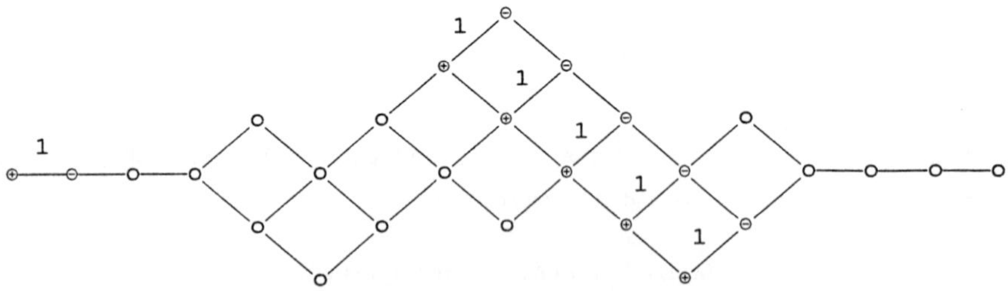

where the nodes with pluses represent the base vectors multiplied by ε while the nodes with minuses represent those multiplied by ε^{-1}. So multiplication by $h_{\alpha_1}(-1)$ changes 2 signs in our distinguished triple $v^{\lambda_o},v^{\mu_o},v^{\nu_o}$ as also does the multiplication by $h_{\alpha_6}(-1)$. All the other $h_\alpha(-1)$ for fundamental α's do not change the triple at all. This means that to determine the sign with which $x_\lambda y_\mu z_\nu$ appears in the form F(x) one just has to pick up any $n_w\in\tilde{W}(E_6)$ such that its image $w\in W(E_6)$ satisfies $w(\lambda_o,\mu_o,\nu_o)=(\lambda,\mu,\nu)$ and take the sign of $x_\lambda y_\mu z_\nu$ to be +1 if $n_w(v^{\lambda_o},v^{\mu_o},v^{\nu_o})$ is different from (v^λ,v^μ,v^ν) by an even number of signs and -1 otherwise. As we've mentioned it can be expressed somewhat less clumsily in terms of the Frenkel-Kac cocycles. The construction of the form is now complete. Let us note that even in the case when 2 and 3 are not invertible the form f(x,y,z) determines the form F(x) in a unique fashion - precisely because f(x,y,z) is very degenerate.

One may note that the weight elements mentioned in §2 may change an odd number of signs. It follows from the theorem 1 of [255] that the elements $h_{\bar{\omega}_1}(\varepsilon)$ have 1 eigenvalue ε, 10 eigenvalues ε^{-1} and 16 eigenvalues 1, as illustrated by the following picture:

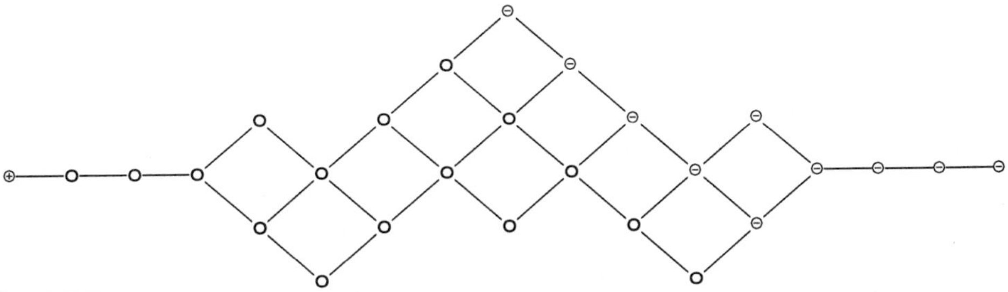

Here again the node with plus represents the base vector multiplied by ε while the nodes with minuses represent those multiplied by ε^{-1}. But the form f is not supposed to be invariant under the action of the extended Chevalley group $\bar{G}_{sc}(E_6,R)$ - it may be multiplied by a scalar factor!

Now we may define the cross product by the formula at the end of the previous section. This product may be described as follows. Take any two base vectors v^{λ} and v^{μ}. If λ and μ are adjacent, then $v^{\lambda}\times v^{\mu}=0$. If they are non-adjacent then $v^{\lambda}\times v^{\mu}=v_{\nu}$ where ν is the unique weight such that $(\lambda,\mu,\nu)\in\Theta$, and v_{ν} is the base vector of the dual module V^{*} corresponding to ν. Extend this product by linearity. This product is much better than the Jordan product though still somewhat upsetting in characteristic 2. The reason is that the coefficient 2 appears and $x\times x$ tends to be zero when we would prefer it not to be - and actually $x\times x$ is the thing which really matters in what follows. Of course the right solution is to introduce another partial polarization of the form f which has to take care of the characteristic 2 behaviour. This is precisely the Aschbacher's approach in [13,14]. Namely one has to introduce the form $\ell(x,y)$ which is *quadratic* in x and *linear* in y. Of course just setting $\ell(x,y)=f(x,x,y)$ would lead us nowhere and we have to define ℓ by the formula

$$\ell(x,y)=\sum \pm x_\lambda x_\mu y_\nu, \quad (\lambda,\mu,\nu)\in\Theta,$$

where the sum is taken over all pairs $(\{\lambda,\mu\},\nu)$ such that the corresponding triple (λ,μ,ν) belongs to Θ. It is now much safer to define the quadratic map $x^{\#}=x\times x$ by $(x\times x)y=\ell(x,y)$ and $x\times y=(x+y)^{\#}-x^{\#}-y^{\#}$. One has either to modify the proof given below in this spirit or think that $\mathrm{char}K\neq 2$. All the missing details are supplied in my paper "Structure of Chevalley groups over commutative rings. III. The equations".

6°. **The identification**. Here we sketch a proof of the following result.

Theorem 1. *Let* R *be any commutative ring with* 1, f *the form constructed in the preceding section. Then the simply-connected Chevalley group* $G_{sc}(E_6,R)$ *may be identified with* Isom(f).

Let us recall that the isometry group Isom(f) consists of all the transformations $g\in GL(V)$ such that $f(gx,gy,gz)=f(x,y,z)$ for all $x,y,z\in V$. The form f has integral coefficients and thus Isom(f) may be viewed as the group of points of an affine group scheme over $\mathbb{Z}$. First of all observe that one has to prove the theorem only for fields - which is of course a standard fact in the theory of affine group schemes (see, for example, [270]). So let R=K be a field.

The group $G_{sc}(E_6,K)=E_{sc}(E_6,K)$ is generated by $\widetilde{W}(E_6)$ and any elementary root subgroup X_α, say, for $\alpha=\alpha_1$. This means that to prove the inclusion $G_{sc}(E_6,K)\leq \mathrm{Isom}(f)$ one only has to verify that the form f is invariant under the action of X_α, or,what is the same, that the $x_\alpha(\xi)$ do not change the values $f(v^\lambda,v^\mu,v^\nu)$ for all $\lambda,\mu,\nu\in\Lambda$. This may be done by an easy direct calculation.

The proof of inclusion $\mathrm{Isom}(f)\leq G_{sc}(E_6,K)$ is a little bit more complicated. First of all recall generalities about the orbits of $G=G_{sc}(E_6,K)$ in this action. Let us call a non-zero vector $x\in V$ *white* (or *singular*) if $f(x,x,y)=o$ for any $y\in V$; *grey* if $f(x,x,x)=0$ but there exists $y\in V$ such that $f(x,x,y)\neq 0$; *black* if $f(x,x,x)\neq 0$ (of course if one wants to properly take care of the characteristics 2 and 3 here one has to write rather $\ell(x,y)=0$ or $F(x)=0$ but we omit these details here). A vector is white if $x\times x=0$. Typical examples of white, grey and black vectors are $v^\lambda o$, $v^\lambda o+v^\mu o$, $v^\lambda o+v^\mu o+\varepsilon v^\nu o$, $\varepsilon\neq 0$, respectively. In terms of the Cayley-Dickson algebra white, grey and black vectors correspond to matrices of ranks 1, 2 and 3 respectively. Then J.Mars [150] has proven that there is just one G-orbit of white vectors and just one G-orbit of grey vectors. Moreover he observed that the black vectors form one orbit with respect to the extended Chevalley group $\overline{G}=\overline{G}(E_6,K)$, though they may break into several G-orbits for a non algebraically closed field K. Of course we cannot refer to the results of [150] directly because we use a different definition of the form, but using the diagrammatic description of the G-action it is straightforward to check that the white vectors again form a single orbit with respect to G. So the white vectors are precisely those which lie in the orbit of $v^{\lambda}o$, or, in other words, the orbit of the highest weight vector.

Now our definition of the form f implies that every column of a matrix $g\in Isom(f)$ is white (since $f(v^{\lambda},v^{\lambda},V)=0$ for any $\lambda\in\Lambda$ forces $f(gv^{\lambda},gv^{\lambda},V)=0$). As there is just one G-orbit of white vectors this implies that there exists $a\in G$ such that the first column of ag (the one corresponding to the highest weight ω) coincides with the first column of the identity matrix, i.e.

ag= 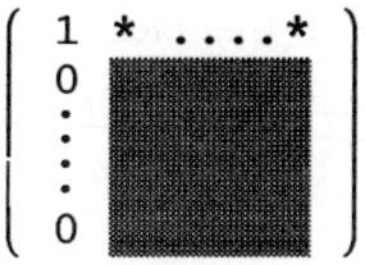

.

But we have already proven that $G\leq Isom(f)$ so that ag again belongs to Isom(f). Now since $f(v^{\lambda},v^{\mu},V)=0$ for any two adjacent $\lambda,\mu\in\Lambda$ this automatically implies that ag has a lot of further zeros. In fact all the matrix entries $(ag)_{\lambda\mu}$ such that μ is adjacent to ω while λ is not should be automatically zeros, since otherwise

$$f(v^{\omega},v^{\mu},V)=f(agv^{\omega},agv^{\mu},V)=f(v^{\omega},agv^{\mu},V)\neq 0$$

which is absurd. This means that actually ag has the following block triangular form:

ag= 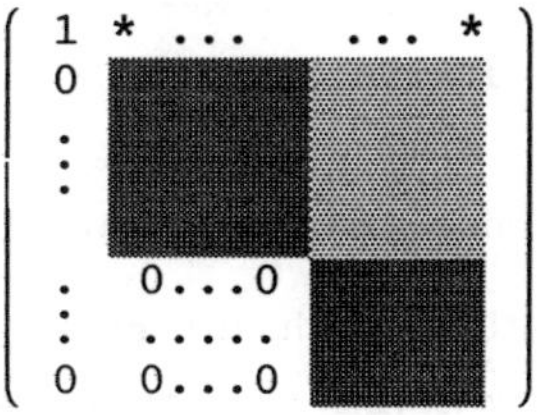

where the second diagonal block has degree 16 and the third one degree 10. Now we may apply the same argument to the first row of the matrix ag (which is an element of the dual module V^* bearing a trilinear form transferred from V via the isomorphism defined by the canonical base v^λ). Again there is one G-orbit of white vectors and thus there exists a $b\in G$ such that

$$agb=\begin{pmatrix} 1 & 0 \; \dots & \dots \; 0 \\ 0 & & 0\dots 0 \\ \vdots & \blacksquare & \dots\dots \\ \vdots & & 0\dots 0 \\ \vdots & 0\dots 0 & \\ \vdots & \dots\dots & \blacksquare \\ 0 & 0\dots 0 & \end{pmatrix}$$

where again the second diagonal block has degree 16 and the third one degree 10. Of course 10 is the dimension of the usual representation of the orthogonal group $G(D_5,K)$ while 16 is the dimension of its half-spinorial representation. Actually the fact that the the block of size 10 is included into a matrix from Isom(f) of this shape again forces certain equations on the block, since

$$f(v^\omega,agbv^\lambda,agbv^\mu)=f(agbv^\omega,agbv^\lambda,agbv^\mu)=f(v^\omega,v^\lambda,v^\mu)=0$$

if λ and μ are equal or adjacent. Look at the 10-dimensional subspace U of V generated by v^λ, λ non-adjacent to ω. Our trilinear form f defines a bilinear form B on U by $B(u,v)=f(v^\omega,u,v)$, $u,v\in U$. By the very definition of f this bilinear form B is non-degenerate and has maximal Witt index. But this signifies precisely that the matrix $(agb)_{\lambda\mu}$, λ,μ nonadjacent to ω, satisfies the equations

$$B(agbv^\lambda,agbv^\mu)=f(v^\omega,agbv^\lambda,agbv^\mu)=0$$

which define split orthogonal group of degree 10 as described in the previous section. It remains to refer to the Ree-Dieudonné theorem to identify the latter with $G(D_5,K)$.Proof of inclusion $Isom(f)\le G$ and the theorem is now complete.

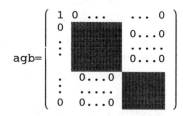

In the same way one can prove that the extended Chevalley group $\bar{G}=\bar{G}_{sc}(E_6,R)$ may be identified with the group Sim(f) consisting of all the transformations $g\in GL(V)$ preserving f up to a scalar factor i.e.such that there exists a $\varepsilon\in R^*$ such that $f(gx,gy,gz)=\varepsilon f(x,y,z)$ for all $x,y,z\in V$.

These identifications give a very practical way to control the equations satisfied by the entries of matrices from G and $\bar{G}$ and we will use this in the subsequent sections.

§7. A phoenix view of other exceptional groups

In his writings Chuang Chow compares a bird (let it be a phoenix), who soars so high that he cannot actually see the details of what is below and so has somewhat fuzzy picture of reality, with another one (let it be a snipe) who has never raised over his bog and thus has somewhat restricted picture of reality. He then draws a moral conclusion: "Wenn der Horizont verschieden ist sind es auch die Gedanken". In this section we give a phoenix-like view of what's on for other exceptional groups skipping all the details and restricting ourselves to pictures and some voodooing. A more balanced exposition has to appear in my paper "Structure of Chevalley groups over commutative rings. III. The equations".

1°. **The case $\Phi=F_4$.** The groups $G(F_4,R)$ appear as stabilizers of *appropriate* black vectors in the 27-dimensional representations of $G(E_6,R)$ described above. In other words they may be realized as stabilizers of restriction of the trilinear form f to certain hyperplanes.There are several orbits of black vectors which give rise to various forms of the group F_4 not all of which are split [233,185]. In terms of the split octonion algebra one gets the split F_4 if one takes the automorphisms of the exceptional Jordan algebra. To be an automorphism a norm-preserving map (i.e. an element from E_6) has to stabilize just one point - the identity element of the algebra [60,129,131,189,...]. So the Jordan algebra is the right object for F_4 rather than for E_6. The diagram of the minimal representation of F_4 looks as follows. Here as in N.Bourbaki [41] the roots α_1 and α_2 are long while the roots α_3 and α_4 are short. The dotted bonds linking the zero weight corresponding to α_3 with α_4 and $-\alpha_4$ and the zero weight corresponding to α_4 with α_3 and $-\alpha_3$ are not shown. As may be seen from the picture the elementary long root unipotents $x_\alpha(\xi)$ have 6 non-zero entries outside the principal diagonal,while the short root unipotents differ from the identity matrix in 12 or 11 positions (their residue is 10).

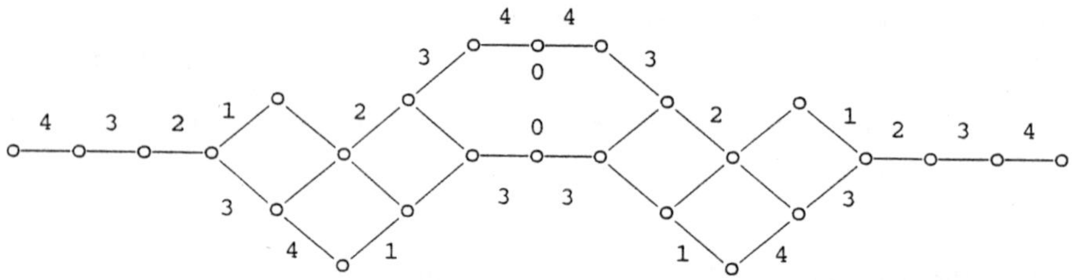

1°. **The case $\Phi=E_7$.** The minimal module (V,π) for the group $G_{sc}(E_7,R)$ has dimension 56. It is a microweight module with the highest weight $\omega=\bar{\omega}_7(E_7)$ isomorphic to its dual. In fact the representation π is *symplectic* and we fix a nondegenerate symplectic form h on V invariant under $G_{sc}(E_7,R)$. The diagram of π is shown at the following picture. The elementary root unipotents $x_\alpha(\xi)$ have 12 non-zero entries outside the principal diagonal. Quite recognizably its restriction to $G_{sc}(E_6,R)$ decomposes into two 1-dimensional summands and two contragredient 27-dimensional summands. This suggests that E_7 stands to E_6 more or less as E_6 stands to D_5. So one hopes to be able to define a *four-linear* form on the module V which would induce when one argument is frozen the familiar *trilinear* form on an appropriate 27-dimensional subspace. This is actually the case, though here construction of the form is somewhat more awkward.

For the classical case of a characteristic 0 field existence of a four-linear form f on V such that $G_{sc}(E_7,R)$ coincides with the isometry group of h,f was claimed by E.Cartan in 1894 though his construction of the form was an error. The first correct construction is again due to H.Freudenthal [93] and may be described as follows (see for example [17,115,198]. Let V be the space

$$V=\{(x,y)\mid x,y\in M(8,K),x^t=-x,y^t=-y\}$$

of pairs of 8×8 alternating matrices with entries in K (its dimension over K equals 56). Define a symplectic inner product h on the space V by

$$h((x_1,y_1),(x_2,y_2))=\frac{1}{2}(tr(x^ty)-tr(xy^t))$$

and a *quartic* form F by

$$F((x,y))=pf(x)+pf(y)-\frac{1}{4}tr((xy)^2)+\frac{1}{16}tr(xy)^2,$$

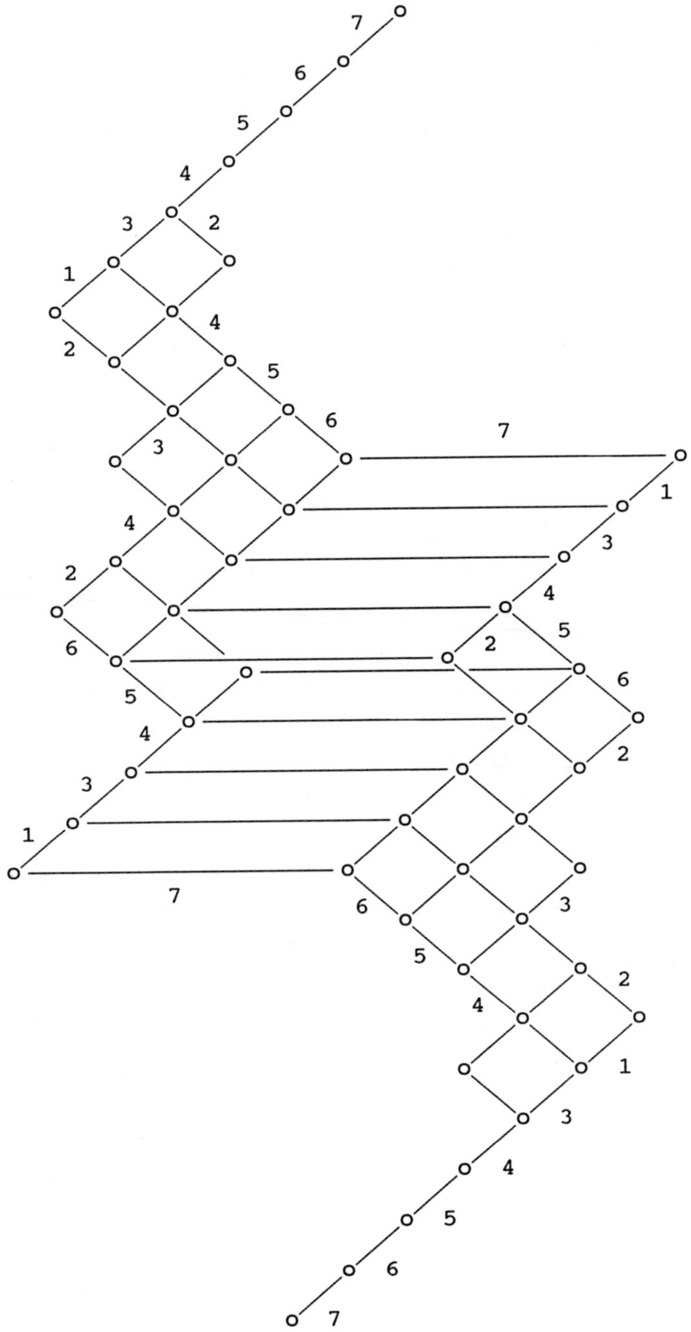
7
6
5
4
3
2
1
4
2
5
6
3
7
1
3
4
2
4
5
2
6
6
5
2
4
3
1
7
6
3
5
2
4
1
3
4
5
6
7

where pf(x) denotes the *pfaffian* of an alternating matrix x. Polarize F to a *symmetric* four-linear form f on V. Then Freudenthal proves that the isometry group of the pair of forms h,f coincides with $G_{sc}(E_7,R)$. In other words $G_{sc}(E_7,R)$ consists of those *symplectic* (with respect to h) matrices which preserve the form f or, what is the same for $charK\neq 2,3$, the form F, in the usual sense

$$G_{sc}(E_7,R)=\{g\in Sp(56,K)\,|\,f(g(x,y))=f((x,y))\}.$$

Freudenthal's original proofs apply to the case of a field of characteristic 0 and is restated the result in [95] in terms of the Cayley-DicksonJordan algebra. For the case of a field of characteristic different from 2 *and* 3 such a construction of $G_{sc}(E_7,K)$ was obtained by G.Seligman in 1962. The corresponding four-linear form f was characterized axiomatically (in the spirit similar to that of Springer characterization of the E_6 cubic form) by R.Brown in [44] and is sometimes referred to as the *Brown form*. He also used the E_6-construc- tion of the form, see also [17,115]. In fact the four-linear form is intrinsically related with the 56-dimensional exceptional *ternary* algebra (see [44,87, 88,91,92,162] for the definition and further references) which consists of 2×2-matrices over the exceptional Jordan algebra with scalar diagonal entries and some funny multiplication (as we know the dimension of the Jordan algebra is 27 and 56=1+27+27+1). Of course the *ternary* product leads to a *four-linear* form essentially in the same way as a *binary* product leads to a *trilinear* one.

Later M.Aschbacher [14] has given also another construction of the form based on A_6 rather then E_6. If U is the usual 7-dimensional module for $SL(7,K)=G_{sc}(A_6,K)$, then viewed as a SL(7,K)-module V is decomposed as follows: $V=U+\Lambda^2U+\Lambda^2U^*+U^*$ (so that 56=7+21+21+7). One may find further details of this construction as well as another proof for the identification of $G_{sc}(E_7,K)$ with the isometry group of this four-linear form in a preprint by B.Cooperstein [68]. Moreover M.Aschbacher was able to construct the form whose isometry group is isomorphic with $G_{sc}(E_7,K)$ for fields of characteristic 2 – the *Aschbacher form*. But this form is *not symmetric*! So the forms for odd characteristics and characteristic 2 are essentially different. The reason, why it is so is that now it is impossible to construct a form for which the extended Weyl group $\widetilde{W}(E_7)$ acts transitively on the monomials and which is preserved by a unipotent root subgroup. The monomials which come from E_6 constitute just one Weyl-orbit of monomials and to construct a form which is actually stabilized by some X_α one has to balance this form with another one coming from a different Weyl-orbit. At this moment coefficient 2 enters the stage. So actually in this case one gets a uniform identification of the Chevalley group $G_{sc}(E_7,R)$ with the isometry group of the form only if $2\in R^*$.

1°. **The case $\Phi=E_8$.** In a sense this is the general case. There are no microweight representations and the smallest representation of the group $G(E_8,R)$is the adjoint one which involves also the adjoint as well as the microweight representations of the groups of types E_6 and E_7. Since the diagram is a little bit too large to reproduce it here as a whole (though compare [165]) we present it by pieces. Let V=L be the Chevalley algebra of type E_8 over the ring R. It has dimension 248 over R. Viewed as the $G(E_7,R)$-module L decomposes to the direct sum of three trivial 1-dimensional representations (maximal root, its opposite and one of the zero-weights), two 56-dimensional ones and the adjoint one for E_7 which has dimension 133 (so 248=1+56+133+1+ +56+1). The 56-dimensional representation has just been described and it remains only to show how it is glued to the others. This is done on the first of the following pictures which presents the maximal root and the 56-dimensional representation, arrows show how it is glued to the adjoint representation while the double line leads to the isolated zero-weight. It remains to describe the adjoint representation of the group $G(E_7,R)$. When restricted to $G(E_6,R)$ it decomposes into one 1-dimensional representations (the isolated zero-weight), two contragredient 27-dimensional representations and the adjoint one which has dimension 78 (so 133=27+78+1+27). Now the 27-dimensional representations are familiar to us and they are glued to the adjoint by half-spinorial D_5 as shown below. Actually this is only one half of the picture. The double lines join it to the corresponding zero-weights which in turn are joined with another half. To finish with this we reproduce once more the part of this diagram describing the adjoint representation for E_6 in a more handy fashion.

Now the trilinear form f for E_8 is very easy to describe. It is what is usually called in the Russian secondary schools the "*mixed product*". Recall that our V=L bears the Lie bracket [,]: $V\times V$ V as well as the Killing form (,): $V\times V$ R. It remains to define an alternating form f by f(x,y,z)=

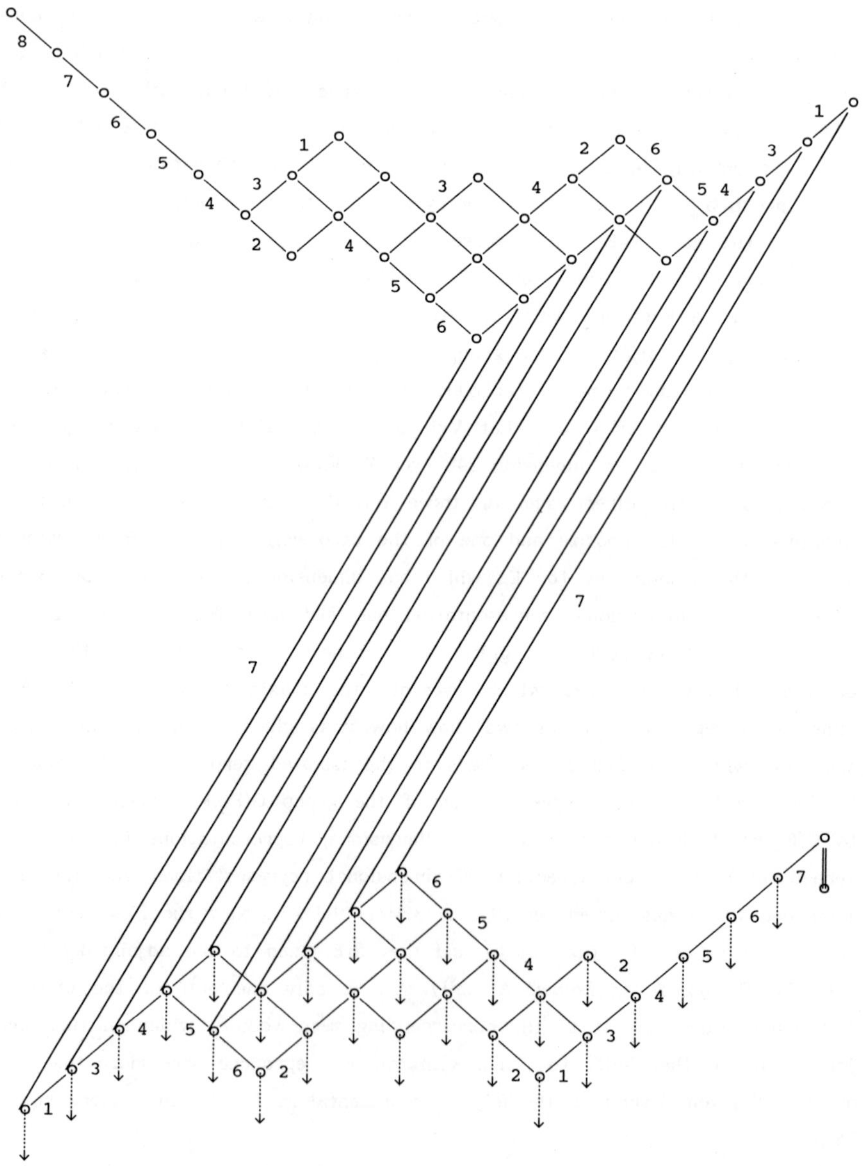
8
7
6
5
4
3
2
1
4
5
6
3
4
2
6
5 4
3
1
7
7
6
5
4
2
7
6
5
4
3
1
4
5
6
2
2
1
3

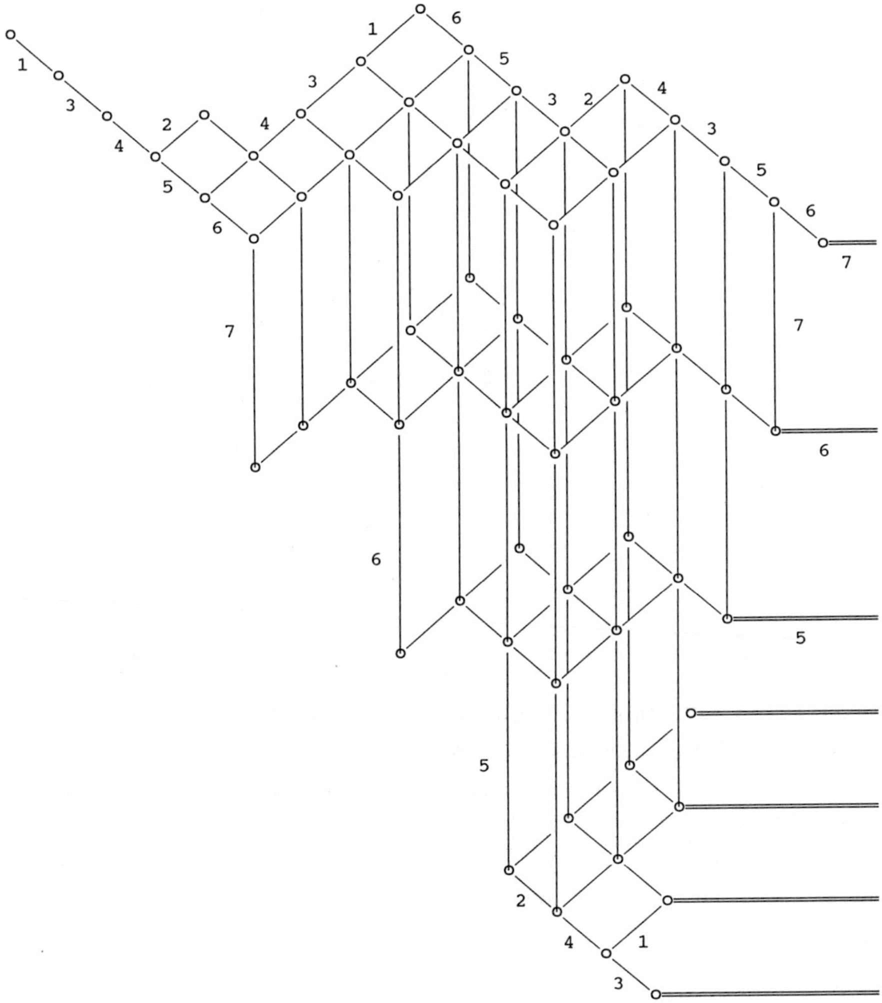
1
3
4
2
5
6
4
3
1
6
5
3
2
4
3
5
6
7
7
7
6
6
5
5
2
4
1
3

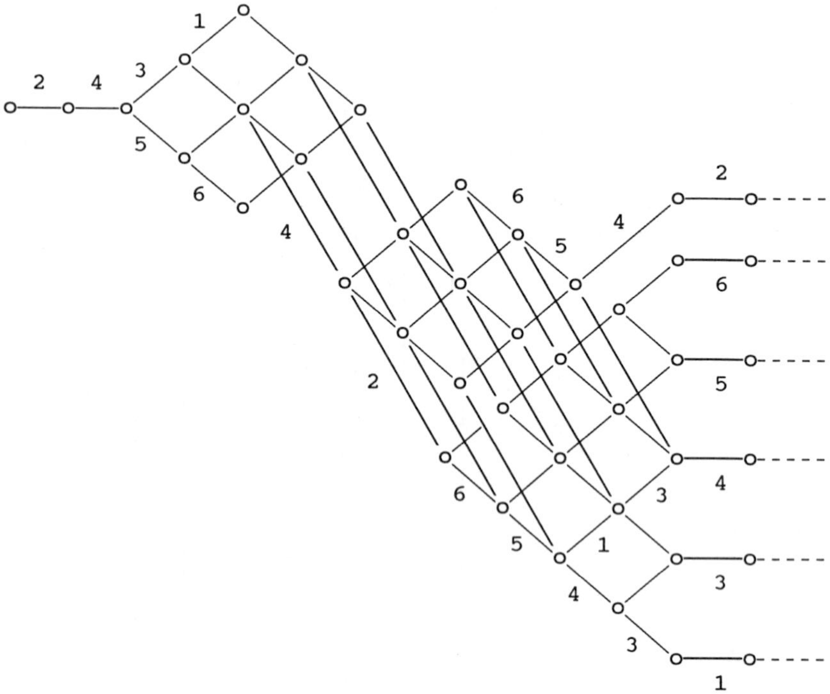

$=([x,y],z)$, $x,y,z\in V$. Then again $G(E_8,R)$ may be identified with the isometry group of the form which gives explicitly the equations on matrix entries. Of course for the characteristic 0 case this is again due to H.Freudenthal [94] (more generally this holds for any simple group of rank at least 2). To identify $G(E_8,R)$ one first has to identify with the corresponding isometry groups the groups $G_{ad}(E_7,R)$ and $G_{ad}(E_6,R)$ so in a sense the E_8-case is not simpler then the general case. To show that actually almost *any* invariant will go let us mention that in the same paper H.Freudenthal has proven that $G(E_8,R)$ may be equally well characterized as the isometry group of the four-linear form $f(x,y,z,u)=([x,y],[z,u])$.

Since the questions in which we are interested are essentially independent of a particular group in the given isogeny class one might think that it would have been natural to treat all the Chevalley groups simultaneously in the adjoint representations. But our present proofs in the following part are done case by case and of course the "natural" modules are usually so much nicer than the adjoint ones. I.Frenkel has suggested at this point that one might avoid case by case analysis by considering certain infinite-dimensional representations of the groups.

It seems that there is still a lot to be done to understand the geometry of these representations even for the case of a field. For example, the classification of the orbits of the classical groups on the Grassmanians is fairly trivial and most of their maximal subgroups act irreducibly in their minimal representations. To the contrary the results of D.Testerman [229] show that the maximal subgroups of the exceptional groups which act irreducibly on their minimal modules are "known" and the main difficulty is to classify the reducible ones. Major part of the Aschbacher's works on the maximal subgroups of $G(E_6,K)$ is to prove that most of the stabilizers of subspaces in V are not maximal by some general arguments without explicitly determining the orbits and the stabilizers. One has to look at the paper [62] where the orbits of $G(E_6,K)$ on the 2-subspaces are determined to realize the amount of work necessary to solve the problem completely.

CH.3. STRUCTURE THEOREMS

In this chapter we discuss the proofs of the main structure theorems for the Chevalley groups over commutative rings.

§9. The structure theorems

In this section we formulate the main structure theorems for the Chevalley groups over commutative rings. In the following sections we outline the existing approaches to their proofs and two new approaches.

1°. Commutator formulae. The first of the main structure theorems asserts that the elementary subgroups are normal in a Chevalley group $G=G(\Phi,R)$.

Theorem 2. *Let Φ be an irreducible root system of rank at least 2, I be an ideal of the ground ring R. then the following commutator formulae hold:*

$$[E(\Phi,R,I),G(\Phi,R)]\leq E(\Phi,R,I),$$

$$[E(\Phi,R),C(\Phi,R,I)]\leq E(\Phi,R,I),$$

Apart from the case when $\Phi=B_2$ or $\Phi=G_2$ and R has a residue field of 2 elements, these inclusions are in fact equalities.

Recall that $C(\Phi,R,I)$ is the full congruence subgroup of level I. Probably even for the general linear group over the ring $\mathbb{Z}$ this sort of phenomena was first recognized in the works of H.Bass [23,24] and was extended to other classical groups by A.Bak [18,19]. Of course both Bass and Bak always operated "at the stable level" i.e. when the rank of the group is large with respect to a certain dimension of the group. Usually they formulated the conditions in terms of the stable rank for the linear and symplectic cases and in terms of the Krull dimension or, rather the dimension of the *maximal spectrum* Max(R) of the ring R. Say for the general linear group H.Bass has shown that these formulas hold whenever $n \geq sr(R)+1$.

A real breakthrough is due to A.A.Suslin [218] who has proven that the elementary subgroup $E(\Phi,R,I)$ is always normal in $GL(n,R)$ when R is commutative and $n \geq 3$. We will refer to this result as a *Suslin's theorem*. Later A.A.Suslin and V.I.Kopeiko [220,141] generalized this result to the even-dimensional split orthogonal group and symplectic group. The symplectic case was independently rediscovered by G.Taddei [225] and the odd-dimensional split orthogonal group was considered by the author in 1983 (see references in [254,262]). These proofs were based on direct matrix calculations (see below).

For the Chevalley groups M.R.Stein [205,206] has considered the formulas at the stable level in terms of the absolute stable rank or dimMax(R) In fact the formulas follow immediately from the surjective stability of the K_1-functor and some further results in that line are obtained in the works of E.B.Plotkin and the author [171,173–175,259].

For the classical groups the second formula was discovered independently by L.N.Vaserstein and Z.I.Borewicz and the author [244,40,248,249,256,262], the proofs being quite different. One might note that normality of $E(\Phi,R,I)$ in the whole group was used in both cases.

Now the proofs coming from the original Suslin's proof were not easy to generalize to other exceptional groups. In fact in [244] L.N.Vaserstein has noticed that one can give quite different proof based on some "localization and patching techniques" coming from the Quillen–Suslin's solution of the Serre's problem. And the first general proof for the normality of the absolute elementary subgroup $E(\Phi,R)$ was obtained by G.Taddei [226,227] using this method. Then L.N.Vaserstein [247] has shown how to deduce the commutator formulae from this (as we've seen in §5 the relative case follows easily from the absolute one).

In 1987 my Ph.D. student A.V.Stepanov has discovered a remarkably elementary proof of the Suslin's normality theorem. Analyzing it I have noticed that unlike the original Suslin's proof it might be generalized to the other Chevalley groups with a reasonable amount of effort and have very rapidly produced analogous proofs for other classical groups and the group of type E_6. Then the remaining cases were considered in the same style in a joint work with E.B.Plotkin. This proof has been sketched in [VPS] and we give somewhat more details here.

The formulae from theorem 2 are sometimes referred to as the *standard commutator formulae*. Let us mention briefly the following two possible generalizations. It follows from the Bass-Vaserstein K_1-stability theorem for the general linear group that

$$[GL(n,R,I),GL(n,R)]\le E(n,R,I),$$

whenever $n\ge 3, sr(R)+1$. For the general Chevalley groups this would have been something like

$$[G(\Phi,R,I),G(\Phi,R)]\le E(\Phi,R,I).$$

Although it is of course still true at the stable level, there is absolutely no way to prove this sort of commutator formulae independently of the dimension of the ground ring. Actually what this last formula says is that $G(\Phi,R,I)/E(\Phi,R,I)$ lies in the centre of $G(\Phi,R)/E(\Phi,R,I)$, and there are no reasons for this to be true in general.

Another generalization which was considered for the general linear group by A.W.Mason and W.W.Stothers [155,151,152] is to look at the commutators of the relative subgroups. Let A and B are two ideals of R. By our methods one very easily gets the following formula

$$[E(\Phi,R,A),G(\Phi,R,B)]=[E(\Phi,R,A),E(\Phi,R,B)]$$

whenever $rk\Phi\ge 2$. It is obvious that (with some minor exceptions in rank 2 for the first inequality) one has

$$E(\Phi,R,AB)\le[E(\Phi,R,A),E(\Phi,R,B)]\le G(\Phi,R,AB),$$

but explicit calculation of this group in is an extremely difficult problem (compare [155,151,152]).

2°. Normal subgroups. The second of the main structure theorems concerns what is traditionally called "the description of the normal subgroups", though what one usually does is description of subgroups normalized by the elementary subgroups (whether they are actually normalized by the whole Chevalley group is another story).

One is usually looking for the "standard description" expecting all the groups to lie close to the "natural" normal subgroups defined by the ideals of the ground ring. Namely one says that the *standard description* of the subgroups of a Chevalley group $G(\Phi,R)$ normalized by the elementary subgroup $E(\Phi,R)$ holds if for any such subgroup F there exists a unique ideal I of R such that

$$E(\Phi,R,I)\le F\le C(\Phi,R,I).$$

That there should be some minor restrictions for the standard descriptions is obvious since for some types it doesn't hold for some fields of small characteristic or cardinality. Still it is a very marginal phenomenon and the standard description holds in most cases.

Theorem 3. *Let* Φ *be an irreducible root system of rank at least* 2 *and* R *be a commutative ring with* 1. *Suppose that* $2\in R^*$ *if* $\Phi=A_\ell,B_\ell,F_4$ *and* $3\in R^*$ *for* $\Phi=G_2$. *Then the standard description of the subgroups in* $G(\Phi,R)$ *normalized by* $E(\Phi,R)$ *holds.*

It would impossible to talk here about the history of this theorem, which goes back to the late 19-th century. Of course for the fields (and skew-fields) description of the normal subgroups was the major part of the whole theory of the classical groups and probably up to 60% of all the publications in the field were dedicated to this problem. The study of the normal structure of some arithmetic groups (like $SL(2,\mathbb{Z})$ for example) goes back to the late 19-th century as well. We'll skip description of this period of history which is very well covered by the existing literature [81,167,168,114].

The problem over rings was first approached as such by J.Brenner and something like 20 years later by W.Klingenberg. But they as their numerous successors thought mostly of such classes of rings as local or semilocal rings or Dedekind domains. Of course it would be impossible to give here even a short description of this period as well and of the interrelations with arithmetic via the congruence subgroup problem etc.

If one thinks in terms of general rings here again the true story starts with the work of H.Bass [23] where he has proven that the standard description holds for GL(n,R) under assumption $n\geq 3, sr(R)+1$. For rings satisfying some stability conditions this was generalized by H.Bass himself and A.Bak to other split classical groups [25,18-20].

The true revelation came with the work of J.S.Wilson [271] where he has proven that the standard description holds for GL(n,R) whenever R is commutative and $n\geq 4$. He got also some partial results concerning the case n=3. The next year I.Z.Golubchik announced with an idea of a proof that the standard description holds for $n\geq 3$ for all the commutative rings as well as for some non-commutative ones. This theorem, usually referred to as the *Wilson-Golubchik theorem* initiated a different epoch in the theory of linear groups when one have started to think in terms of rings without any restrictions whatsoever (although a lot of old-fashioned works limited to "zero-dimensional" rings appears even now).

In [244] L.N.Vaserstein has given another proof of the Wilson-Golubchik theorem based on the "localization and patching" techniques while Z.I.Borewicz and the author have [40] published a proof similar in spirit to Stepanov's proof of Suslin's normality theorem. I.Z.Golubchik himself has generalized it to other split classical groups [103-105] as well as to a very wide class of "unitary" groups over non-commutative rings.

The first work in which the problem was approached in the context of Chevalley groups was that of E.Abe [1] where he considered the case of local rings. Then E.Abe and K.Suzuki [10] have proven results which gave a very good approximation to the standard description. In fact, as was noted in [252] an easy corollary of their results shows that the standard description of normal subgroups holds in the elementary group $E(\Phi,R)$ for any commutative ring subject to usual minor restrictions.

But to pass from the normal subgroups of $E(\Phi,R)$ to the subgroups of $G(\Phi,R)$ normalized by $E(\Phi,R)$ one has to know that $E(\Phi,R)$ is normal in $G(\Phi,R)$ and as we know this was not proven until 1986. In the full generality theorem 3 was first proven by L.N.Vaserstein [247]. Actually he has given a more precise result which says that the standard description holds if any element $a\in R$ belongs to the ideal generated by 2a and a^2 if $\Phi=A_\ell,B_\ell,F_4$ and to the ideal generated by 3a and a^3 for $\Phi=G_2$. If these conditions are not satisfied there are nonstandard subgroups normalized by $E(\Phi,R)$. But actually if one modifies the notion of standardness to take into account the different behaviour of the long roots and the short roots in the exceptional cases (see [1,10] for the precise definitions), then E.Abe [6,7] has shown that there are almost no restrictions at all for such a parastandard description. The last touch has been made by D.L.Costa and G.E.Keller [70] who removed the remaining restrictions for the symplectic case.

Here again one may note several immediate generalizations for the standard description. In his paper [271] J.S.Wilson has noticed that one might as well describe subnormal subgroups of $GL(n,R)$. Recall that a subgroup F of a group G is called *subnormal of depth* (alias *defect*) d if there exists a normal series $G=G_o \triangleright G_1 \triangleright \ldots \triangleright G_{d-1} \triangleright G_d=F$ (where $G_i \triangleright G_{i+1}$ means that G_{i+1} is a normal subgroup of G_i. This problem is very closely related with the description of subgroups in $GL(n,R)$ normalized by a relative elementary subgroup $E(n,R,I)$ for some ideal I of R which was independently proposed by A.Bak [19]. For the general linear group these problems stand as follows.

If F is a subgroup of $GL(n,R)$ (where as usual R is commutative and $n\geq 3$), normalized by the relative elementary subgroup $E(n,R,I)$, then there exists an ideal A in R such that

$$E(n,R,AI^m)\leq F\leq C(n,R,A).$$

where $m=4$. This is a recent result of L.N.Vaserstein [250]. Before this has been proven by J.S.Wilson [271] with $n=7$ for $n\geq 4$, by A.Bak for $m=24$ under additional assumption that $n\geq sr(R)+1$, by L.N.Vaserstein [246] with $m=6$, by Li Fuan and Liu Mulan [146] with $m=40$, and by the author [261] with $m=5$.

This result readily implies that if F is a subnormal subgroup of $GL(n,R)$ of depth d them there exists an ideal A of R such that

$$E(n,R,A^s)\leq F\leq C(n,R,A).$$

where $s=(m^d-1)/(m-1)$.

Of course using our method of proof it is straightforward to generalize these results to all the Chevalley groups and this is done in the work of the author and E.B.Plotkin "Structure of the Chevalley groups over commutative rings. VI. The main structure theorems". In the last section we mention briefly some further generalizations of the description of normal subgroups.

3°. **Basic reduction for the standard description.** After the work of H.Bass [23] it is a common understanding that in the presence of the standard commutator formulae for the ring R and all of its factor-rings the standard description of subgroups normalized by the elementary subgroups is equivalent to the fact that - again over any factor-ring of R - any non-central subgroup with this property contains a non-trivial elementary root unipotent. In turn to check this it is sufficient to find in any such a group a non-central element contained in a proper parabolic subgroup (see §2).

In fact, let F be a subgroup of the Chevalley group $G(\Phi,R)$ normalized by the elementary subgroup $E(\Phi,R)$. Let us denote by A the set of all $\xi\in R$ such that for some root β the elementary unipotent $x_\beta(\xi)$ belongs to F. Then A is an ideal of R and $E(\Phi,R,A)\le F$ (here we use restrictions on the characteristic, it follows from the works of E.Abe and his followers that without these restrictions this is not necessarily true and one has to modify the notion of standardness). In particular if $x_\alpha(\xi)\in F$ for some root α then $\xi\in A$.

Now look at the factor-ring R/A and consider the image $\overline{F}$ of F in $G(\Phi,R/A)$ under the reduction homomorphism modulo A. Then it is clear that $\overline{F}$ is normalized by $E(\Phi,R/A)$ and if it is non-central,then by the assumption there exists a non-trivial elementary unipotent $x_\alpha(\overline{\xi})\in\overline{F}$, where $\xi\notin A$. Thus $x_\alpha(\xi)=xy$, where $x\in F$, $y\in G(\Phi,R,A)$, and thus for any $g\in E(\Phi,R)$ one has $[g,x_\alpha(\xi)]=[g,xy]=[g,x]x[g,y]x^{-1}$. But $[g,x]\in F$ since F is normalized by $E(\Phi,R)$ and $[g,y]\in E(\Phi,R,A)$ by the second standard commutator formula and $x[g,y]x^{-1}\in E(\Phi,R,A)\le F$ by the first one. This means that $x_\alpha(\xi)\in F$ and thus $\xi\in F$ what is absurd. This contradiction shows that $\overline{F}$ is central in $G(\Phi,R/A)$ and thus $F\le C(\Phi,R,A)$.

Now it is extremely easy to pull out a transvection from a non-central element contained in a proper parabolic subgroup $G(P)$ (see §2). Let x be such an element. Then write x as $x=yz$, $y\in L(P)$, $z\in U(P)$. If already y is noncentral, then taking $g\in U(P)$ we may get a non-trivial element $[x,g]\in U(P)$. On the other hand if x is central then $z\ne e$ and one might take $g\in L(P)$ in such a way that $[x,g]\in U(P)$ is again non-trivial. It is of course straightforward to pull out a non-trivial elementary unipotent having a non-trivial elementary unipotent from U(P).

4°. **Rank 1 case**. It is known that all of the structure theorems fail completely for the rank 1 case, i.e. for the group SL(2,R). As we have already mentioned P.H.Cohn [64], R.Swan [224] (in the arithmetic case) and A.A.Suslin [217] (in the functional case) have shown that the group E(2,R) is not necessarily a normal subgroup of SL(2,R) even if R is a *Dedekind ring of arithmetic type* ("*Hasse domain*").

It is even worse with the description of normal subgroups. We cannot attach here positive results due to W.Klingenberg, N.Lacroix, J.-P.Serre, G.Keller, D.Costa, L.N.Vaserstein, P.Menal, A.Mason and many others which solve the problem of description of normal subgroups for SL_2 over some special classes of rings (like, say, semi-local or Hasse domains with infinite multiplicative groups). But for the general commutative rings the problem of description of normal subgroups in SL_2 will never be solved. Why?

Look at the apparently easiest case of the group $SL(2,\mathbb{Z})$. It is of course very well known that it has lots of normal subgroups - actually $2^{\aleph_0}$, see [153], so most of them should be far from being standard. In fact a classical result asserts that $PSL(2,\mathbb{Z})$ is isomorphic to the free product $\mathbb{Z}/2\mathbb{Z}*\mathbb{Z}/3\mathbb{Z}$. This amounts to saying that a group G may be realized as a factor-group of $PSL(2,\mathbb{Z})$ if and only if it may be generated by a pair of elements of orders 2 and 3 respectively. There are lots and lots of such groups. To give some idea of how difficult this problem is let us mention two recent results.

It has been proven in [Sc],[MP] that every countable group can be embedded into an infinite simple group with a pair of generators of orders 2 and 3 respectively whose product has infinite order. On the other side it is extremely plausible that almost every finite simple group is a factor-group of $SL(2,\mathbb{Z})$ (see [83] for a detailed survey). That all the alternating groups A_n, n=6,7,8, are (2,3)-generated was known to Miller in 1901, in 1989 A.Woldar has checked that all the sporadic groups apart from M_{11}, M_{22}, M_{23} and McL are (2,3)-generated too and the current works of Ch.Tamburini, L.Di Martino and the author show that for a finite group of Lie type the property of not being (2,3)-generated is a very rare phenomenon restricted to some very small groups in characteristics 2 and 3, like Suzuki groups and a few more eccentric groups like PSL(2,9) and PSU(3,9) which appear as exceptions more or less everywhere. This means that a classification of normal subgroups of $SL(2,\mathbb{Z})$ should imply the classification of finite simple groups as well as of huge classes of infinite simple groups - and in fact much more than that since there usually are many essentially different ways to pick up a (2,3)-generating pair in a group.

5°. Noncommutative case. Here we very briefly sketch the current situation for the non-commutative ground rings. Of course, for a non-commutative ring one has to restrict oneself to some classical groups. There are no natural analogues of the exceptional groups for non-commutative coefficients. Even symplectic groups do not exist (this is not a speculation but a theorem proven independently by Li Shangzhi and the author, see the forthcoming survey of the author "Generation in Chevalley groups" for the precise statement). Let us concentrate on the case of the general linear group GL(n,R). The situation for the unitary groups is more or less analogous.

As was already mentioned for the general linear group the results on the stable level do not depend on the commutativity of the ground ring. In his paper [102] I.Z.Golubchik has established the normal description for some classes of non-commutative rings including, for example, von Neumann regular rings (this result was some 10 years later rediscovered by L.N.Vaserstein). During the next 8 years he has generalized the standard description to a huge class of rings. In technical terms these are the rings which satisfy the following two conditions: all of their primitive factor-rings satisfy the Ore condition and they have finite "localizational dimension" (see [107]). To hint how huge this class really is let us mention that it contains the class of PI-rings (i.e. rings satisfying a polynomial identity) as its small part.

Any two elements a,b of a commutative ring are linearly dependent: ba-ab=0. The Ore condition is the most general form of the idea of "linear dependence". On the other hand I.Z.Golubchik had to prove the standard description in the absence of the standard commutator formulae. Even now there is no proof for the standard commutator formulae in such a generality and even for the PI-rings they have been proven something like 10 years after the standard description (the previous results of I.Z.Golubchik and A.V.Mikhalev [108,109] say only that the elementary subgroup is *subnormal*). So the condition on "localizational dimension" was used to pull out a transvection in several steps.

This result was first announced by I.Z.Golubchik in 1978 and was present in his Ph.D.thesis [104] (I was a member of jury for this thesis and can certify that the proof certainly was there not later then in 1981). It is a mystery why this proof was never properly published (the publication [Go5] is not easily accessible even in Russia). I myself and others in my presence have several times asked I.Z.Golubchik himself and his Ph.D. advisor A.V.Mikhalev about that and never could get a satisfactory answer. Even now after something like 13 years this result was not superseded by numerous subsequent publications and remains by far the most general result in terms of the class of rings considered. The reason is that the usual "localization and patching" method comes from commutative rings and, unlike the Golubchik's method, operates essentially in terms of the centre of a ring rather than of the ring itself.

In a series of papers L.N.Vaserstein and others have developed methods which allowed to prove analogous results for some further classes of infinite-dimensional rings of topological and analytical origin (like, say, Banach algebras), but we cannot list a complete bibliography here.

Now on the other side there are counterexamples to both the standard commutator formulae and the standard description for the general rings. The first example is due to V.N.Gerasimov [100] and is as follows. Take two general matrices $x=(x_{ij})$ and $y=(y_{ij})$ of degree m over a field K. This means that x_{ij}, y_{ij}, $1\leq i,j\leq m$, are non-commutative variables which commute with the coefficients but do not satisfy any polynomial identity. Then the factor-ring $R=K\langle x_{ij}, y_{ij}\rangle/$ $/(xy=e=yx)$ is the counter-example. Around 1980 several people have speculated that rings without IBN property (such rings are also called *non-dimensional*) could be counter-examples too, but no definitive proof ever appeared.

The problem to find the precise class of rings for which the standard commutator formulae and the standard description hold is an extremely attractive one but probably a hopeless one too. A conjecture of Z.I.Borewicz and the author said that it might be the class of absolutely weakly finite rings but it was disproven by A.V.Stepanov [214] and S.G.Khlebutin [138-140] Now there is even no plausible guess what the class of rings might have been. One may note though that what really breaks down in the general case is the normality of $E(n,R)$ in $GL(n,R)$. As for the description of normal subgroups in $E(n,R)$ there is some evidence provided by A.V.Stepanov that this is standard for any ring whenever $n\geq 3$.

§10. The proofs for A_ℓ

In this section we discuss different proofs for the structure theorems in the easiest case of the special linear group. A standing assumption for this section is that the ring R is commutative and $n\geq 3$.

1°. **Linear transvections**. Let us denote by ${}^nR$ the module of *rows of length* n as opposed to the module R^n of *columns of height* n. As we know $V^*={}^nR$ is the dual module for $V=R^n$ and the pairing is given by the usual product ${}^nR\times R^n\longrightarrow R$ which sends a pair consisting of a row $v=(v_1,\dots,v_n)\in{}^nR$ and a column $u=(u_1,\dots,u_n)^t\in R^n$ to $vu=v_1u_1+\dots+v_nu_n\in R$. Let us denote by e_i and f_i respectively the i-th column and the i-th row of the identity matrix e.

A *transvection* (or if we want to stress the fact that we are talking about the case A_ℓ a *linear transvection*) is a matrix of the form $x=t_{uv}(\xi)=e+u\xi v$ where $u\in R^n$, $v\in{}^nR$, $\xi\in R$ are subject to the condition $vu=0$. Then $\det(x)=1$ and thus $x\in SL(n,R)=G_{sc}(A_{n-1},R)$ where we adhere to the identification described in the preceding section. The set of all the transvections with the given *centre* and *axis* will be referred to as the corresponding *unipotent root type subgroup*

$$X_{uv}=\{t_{uv}(\xi),\ \xi\in R\}.$$

In this case a root unipotent is an element conjugated to an elementary transvection $t_{ij}(\xi)$. Since a conjugate of a transvection is again a transvection

$$gt_{uv}(\xi)g^{-1}=t_{gu,vg^{-1}}(\xi),\ g\in GL(n,R),$$

and any elementary transvection $t_{ij}(\xi)$ is a linear transvection corresponding to the column e_i and the row f_j one concludes that an element of root type is necessarily a transvection. If R=K is a field the converse is also true: any transvection is conjugated to an elementary transvection. This is not true in general however: denote by T(n,R) the subgroup of SL(n,R) generated by all the transvections $t_{uv}(\xi)$, $u\in R^n$, $v\in{}^nR$, $\xi\in R$, $vu=0$. Then T(n,R) may be strictly larger than E(n,R).

Why? In what follows the crucial role is played by the fact that transvections are additive in their arguments. Take a row $v=(v_1,\dots,v_n)\in{}^nR$ and a *two* columns $u=(u_1,\dots,u_n)^t, w=(w_1,\dots,w_n)^t\in\ \in R^n$ subject to the condition $vu=vw=0$. Then

$$t_{(u+w),v}(\xi)=e+(u+w)\xi v=(e+u\xi v)(e+w\xi v)=t_{uv}(\xi)t_{wv}(\xi)$$

since $vw=0$. Analogous equality holds with respect to the row v. Of course the "theoretical" explanation is that the angle between the root subgroups X_{uv} and X_{wv} equals $\pi/3$. In particular this shows that for any three pairwise distinct indices i,j,k the matrix $t=t_{ij}(\xi)t_{ik}(\zeta)$ is a transvection. But only very rarely this matrix is a root element in the traditional sense. In fact an obvious *necessary* condition for this matrix to be conjugated to an elementary transvection $t_{12}(\theta),\theta\in R$, is that ξ and ζ generate a principal ideal (since θ should generate the same ideal as ξ and ζ do). For $n\geq 4$ this condition is also *sufficient* but for SL_3 one has to impose even stronger restrictions then just being a *Bezout ring* on R if one wants any such t to be a root unipotent. In fact our new proofs are based on the calculations with the elements of root type which are products of two root unipotents but not root unipotents themselves.

2°. Whitehead-Vaserstein lemma. One of the key ingredients in any normality proof for the elementary subgroups is the following easy but very important fact to which or some of its immediate corollaries we refer as the *Whitehead-Vaserstein lemma*. Let I be an ideal in R, and $r,s\in\mathbb{N}$. Let further $x\in M(r\times s,R)$ be a matrix of size $r\times s$ with entries from R, and $y\in M(s\times r,I)$ be a matrix of size $s\times r$ with entries from I. Suppose $e+xy$ is invertible (the order of the identity matrix e should be clear from the context, thus it is r here and s in the next sentence). Then $e+yx$ is invertible too and

$$\begin{pmatrix} e+xy & 0 \\ 0 & e+yx \end{pmatrix}\in E(r+s,R,I).$$

Of course the classical Whitehead lemma pertains to the case $r=s$ (see [24,163]) that the calculation carries over to the case $r\neq s$ was noted by L.N.Vaserstein [240] (see also [241,251]). Now this lemma may be viewed as the first approximation towards normality result. This lemma readily implies that if $x=t_{uv}(\xi)$, $u\in R^n$, $v\in {}^nR$, $\xi\in I$, $vu=0$, is a transvection *and* one of the components of the row v (or the column u) equals 0 then $x\in E(n,R,I)$.

Indeed, let one of the components of v be zero. Since the components are sociologically equal we may think that $u_n=0$. Then write u and v as $u=(\hat{u},u_n)^t$ and $v=(\hat{v},0)$ where $\hat{u}\in R^{n-1}$ is a column of height n-1 and $\hat{v}\in {}^{n-1}R$ is a row of length n-1. Then $\hat{v}\hat{u}=0$ and one has

$$x=\begin{pmatrix} e+\hat{u}\xi\hat{v} & 0 \\ u_n\xi\hat{v} & 1 \end{pmatrix}=\begin{pmatrix} e+\hat{u}\xi\hat{v} & 0 \\ 0 & 1 \end{pmatrix}\begin{pmatrix} e & 0 \\ u_n\xi\hat{v} & 1 \end{pmatrix}.$$

Both factors on the right hand side are transvections belonging to E(n,R,I): for the second factor this is obvious while for the first follows from the

Whitehead-Vaserstein lemma.

An explanation of the lemma in terms of the Chevalley commutator formula as well as a recipe how to write analogous formulas for other Chevalley groups is given in [259]. The idea goes back to [27]. Let us explain it on this simplest example which appears already in [244]. Take $\xi\in R$ and look at the transvection $x=gt_{ij}(\xi)g^{-1}$, where $1\le i,j\le n-1$, and $g\in GL(n-1,R)$. Why it should belong to the elementary group $E(n,R,I)$ under the usual embedding

$$g\mapsto\begin{pmatrix} g & 0 \\ 0 & 1 \end{pmatrix}?$$

All the transvections being sociologically equal we may think that $(i,j)=$ $=(1,2)$. Then we may write $t_{12}(\xi)$ in the following form: $t_{12}(\xi)=[t_{1n}(1),t_{n2}(\xi)]$ (Chevalley commutator formula). Now

$$gt_{12}(\xi)g^{-1}=[gt_{1n}(1)g^{-1},gt_{n2}(\xi)g^{-1}]$$

where

$$gt_{1n}(1)g^{-1}=\begin{pmatrix} e & ge_1 \\ 0 & 1 \end{pmatrix},\ gt_{n2}(\xi)g^{-1}=\begin{pmatrix} e & 0 \\ \xi f_2 g^{-1} & 1 \end{pmatrix}$$

where of course ge_1 is the first column of the matrix g while $\xi f_2 g^{-1}$ is the second row of the matrix g^{-1} multiplied by ξ. Now this suggests that for any $u\in R^{n-1}$ and $v\in {}^{n-1}R$ such that $vu=0$ one has

$$\begin{pmatrix} e+u\xi v & 0 \\ 0 & 1 \end{pmatrix}=\left[\begin{pmatrix} e & 0 \\ -v & 1 \end{pmatrix},\begin{pmatrix} e & \xi u \\ 0 & 1 \end{pmatrix}\right]$$

Of course as soon as such a formula is written it is readily verified by a direct calculation.

As one sees all the ingredients of this approach have natural counterparts for arbitrary Chevalley groups. What one gets is a decomposition of an element of root type which is contained in the Levi subgroup of a proper parabolic subgroup into a product of elements of root type from the unipotent radicals of this parabolic subgroup and the opposite one with the same Levi subgroup. Of course since the right hand side of the Chevalley commutator formula might consist of more than one elementary unipotent in the general case there are usually more than 4 factors (see [259] for details).

3°. Proofs using stability. The first proof which covered fairly large classes of rings is due to Bass and is based on the stability conditions. In fact the starting point here is what is called the *surjective stability for* K_1-*functor* which says that under appropriate restrictions on the ring R ("stability conditions") depending on the embedding $\Delta \subset \Phi$ of the root systems the natural embedding $G(\Delta,R) \longrightarrow G(\Phi,R)$ of the Chevalley groups induces *surjective* map of the corresponding K_1-functors $K_1(\Delta,R) \longrightarrow K_1(\Phi,R)$ or, what is the same, that

$$G(\Phi,R) = E(\Phi,R)G(\Delta,R).$$

There is a huge literature on the stability of K-functors of which we quote just some samples [23-25,27,240-242,203-206,217-219,239,171-175] (a detailed account of stability of $K_1(\Phi,R)$ was presented in the talk of E.B.Plotkin at the same conference).

Now for the general linear group the surjective stability amounts to the fact that if $n \geq sr(R)+1$ then

$$GL(n,R) = E(n,R)GL(n-1,R).$$

This goes back to the original work of H.Bass [23,27] and is very easy to prove. In fact take an arbitrary matrix $g \in GL(n,R)$ and look at its last row:

$$g = \begin{pmatrix} \rule[-1.5em]{0pt}{3em}\qquad\qquad\qquad \\ a_1 \cdots a_{n-1} \ a_n \end{pmatrix}$$

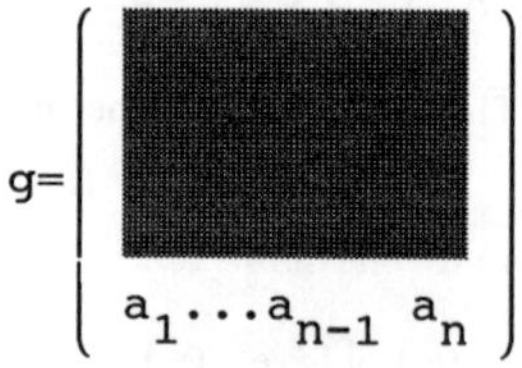

Recall that $SL(n,R) = G(A_\ell,R)$, $GL(n,R) = \overline{G}(A_\ell,R)$, where $n=\ell+1$, and we are thinking of the row as follows:

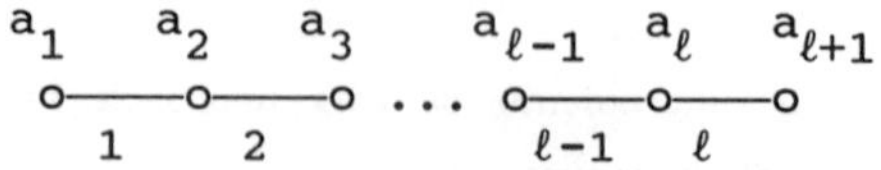

$$
\begin{array}{cccccc}
a_1 & a_2 & a_3 & a_{\ell-1} & a_\ell & a_{\ell+1} \\
\circ\!\!-\!\!\!&\!\!-\!\!\circ\!\!-\!\!\!&\!\!-\!\!\circ & \cdots \circ\!\!-\!\!\!&\!\!-\!\!\circ\!\!-\!\!\!&\!\!-\!\!\circ \\
1 & 2 & & \ell-1 & \ell &
\end{array}
$$

Since $(a_1,...,a_{\ell+1})$ is unimodular the condition $sr(R) \leq n$ implies that there exist $b_1,...b_\ell \in R$ such that $(a_1+a_{\ell+1}b_1,...,a_\ell+a_{\ell+1}b_\ell)$ is unimodular. Now set

$$x = t_{\ell+1,1}(b_1)...t_{\ell+1,\ell}(b_\ell).$$

Then the last row of gx is $(a_1+a_{\ell+1}b_1,...,a_\ell+a_{\ell+1}b_\ell,a_{\ell+1})$ and since the first ℓ components of the row generate R as a (right) ideal, there exist $c_1,...,c_\ell \in R$ such that

$$a_{\ell+1}-1=(a_1+a_{\ell+1}b_1)c_1+\ldots+(a_\ell+a_{\ell+1}b_\ell)c_\ell.$$

Now set

$$y=t_{1,\ell+1}(c_1)\ldots t_{\ell,\ell+1}(c_\ell).$$

Then the element of $z=gxy$ in the SE-corner is 1 and we can apply the Chevalley-Matsumoto decomposition theorem to conclude that z - and hence g - belongs to $G(A_{\ell-1},R)E(A_\ell,R)$.

Now it follows immediately from the the Whitehead-Vaserstein lemma that $GL(n-1,R)$ normalizes $E(n,R,I)$ (see, for example, [23,244]). So, of course, does $E(n,R)$ by the very definition of the relative groups. Thus $E(n,R,I)$ is normal in $GL(n,R)$ whenever $n\geq sr(R)+1$. One may note that actually commutativity of the ground ring R was never used in this proof.

It is possible to imitate this proof also for other Chevalley groups. Of course in most cases one can add a multiple of a component only to some of the remaining components and not to all of them. Thus one is forced either to work with shorter rows which are not necessarily unimodular or to take into account the equations among the components. The first approach was used by M.Stein in his stability papers and this is how the absolute stable rank enters the stage. For some of the classical cases L.N.Vaserstein has introduced some ad hoc stability conditions involving a quadratic/hermitian equation [242]. An extremely plausible conjecture about the general form of such stability conditions for all the Chevalley groups has been recently formulated by E.B.Plotkin.

4°. Suslin's proof of Suslin's theorem. Suslin's original proof of his normality theorem is based on the fact that any solution of a single homogeneous linear equation over a commutative ring may be written as a linear combination of some obvious "two-term" solutions. Let R be an arbitrary commutative ring, $(v_1,\ldots,v_n)\in {}^nR$ - a unimodular row of elements of R. Consider the homogeneous linear equation

$$v_1x_1+\ldots v_nx_n=0.$$

Then commutativity implies that the vectors $v_je_i-v_ie_j$, $1\leq i,j\leq n$, satisfy this equation. We will call these solutions *obvious*. The starting point in the normality proof is the following observation (which is sometimes referred to as the Suslin's lemma): *every solution of a homogeneous linear equation over a commutative ring is a linear combination of the obvious ones.*

In fact recall that the row $(v_1,\dots,v_n)$ is unimodular. This means that there exists a column $(w_1,\dots,w_n)^t \in R^n$ such that $v_1w_1+\dots+v_nw_n=1$. Let now $(u_1,\dots,u_n)^t \in R^n$ be any solution of the equation. Then a direct calculation using commutativity of R shows that

$$(u_1,\dots,u_n)^t=\sum(w_ju_i-w_iu_j)(v_je_i-v_ie_j)$$

where the sum is taken over $1\le i,j\le n$. This proves the lemma. Denote the (i,j)-th summand in this formula by w_{ij}. Then w_{ij} has at most two non-zero components.

Now the normality proof runs as follows. Look at a transvection $t_{uv}(\xi)=e+u\xi v$ where $u\in {}^nR$, $v\in R^n$, $\xi\in I$, are subject to the condition $vu=0$ *and* moreover v is *unimodular*. Then as we have just proven $u=\sum w_{ij}$, $1\le i,j\le n$, and thus, by the additivity of $t_{uv}(\xi)$ in the first argument,

$$t_{uv}(\xi)=\prod t_{w_{ij},v}(\xi).$$

Here each factor $t_{w_{ij},v}(\xi)$ is again a transvection but now w_{ij} has at most two non-zero components. Since we assume that $n\ge 3$ at least one component of w_{ij} equals zero and thus all $t_{w_{ij},v}(\xi)$ belong to E(n,R,I) by the Whitehead-Vaserstein lemma. But then their product $t_{uv}(\xi)$ belongs to E(n,R,I) as well. Now if $t_{rs}(\xi), r\ne s, \xi\in I$, is any elementary transvection of level I and $g\in GL(n,R)$, then

$$gt_{rs}(\xi)g^{-1}=t_{u,v}(\xi),$$

where $u=ge_r$ and $v=f_sg^{-1}$ (so that u is the r-th column of the matrix g, while v is the s-th row of the matrix g^{-1}). Since any row of an invertible matrix is unimodular, the preceding argument shows that $gt_{rs}(\xi)g^{-1}\in E(n,R,I)$ and the proof of the theorem is now complete.

As one immediately discovers realization of this idea for all the other Chevalley groups - apart from $Sp_{2\ell}$ of course - is much more tricky since in these cases our columns and rows have to satisfy some quadratic equations. Thus to decompose an element of root type in such a way one has to solve *systems of linear and quadratic* equations. One is not particularly happy doing this over an arbitrary commutative ring. For the cases one actually needs this can be done, but the arising formulas are by no means obvious and the calculations somewhat awkward. What is remarkable, there is no need for all this as we see in a moment.

5°. Stepanov's proof of Suslin's theorem. This proof is based on an apparently absurd idea: if it is difficult to decompose the large thing, let's decompose the small one with which we have started. Let again $t_{rs}(\xi)$, $r\neq s$, $\xi\in I$, be any elementary transvection of level I and $g\in GL(n,R)$. We want to show that $gt_{rs}(\xi)g^{-1}\in E(n,R,I)$. Let us try to find x_i, $1\leq i\leq n$, subject to the following conditions:

i) $t_{rs}(\xi)=\prod x_i$, $1\leq i\leq n$;

ii) x_i, $1\leq i\leq n$, are transvections of level I;

iii) $x_i g^{-1}e_i=g^{-1}e_i$, $1\leq i\leq n$.

Of course the last condition says that multiplication by x_i on the left does not change the i-th column of the matrix g^{-1}.

Now suppose we have found such x_i's. Then

$$gt_{rs}(\xi)g^{-1}=\prod gx_ig^{-1},\ 1\leq i\leq n;$$

where each of the matrices gx_ig^{-1} is a transvection whose i-th column coincides with e_i (since the i-th column of x_ig^{-1} coincides with the i-th column of g^{-1}). Such a transvection belongs to E(n,R,I) by the Whitehead-Vaserstein lemma and hence $gt_{rs}(\xi)g^{-1}\in E(n,R,I)$.

All the transvections being sociologically equal we may take $(r,s)=(1,2)$ for simplicity of notation. Let's look for x_i in the following form:

$$x_i=t_{12}(\textit{something})t_{13}(\textit{another something}).$$

As we now, such a matrix is *always* a transvection so the condition ii is satisfied. Try to satisfy iii. Look at the i-th column of a matrix:

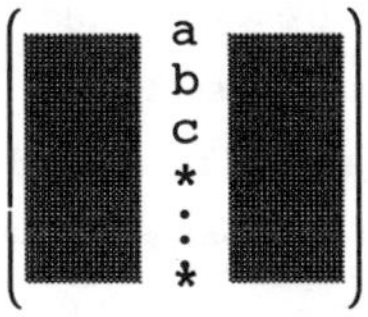

Then our transformation x_i multiplies b by *something* and adds to a and multiplies c by *another something* and adds to a too. Now since our ring is commutative cb-bc=0. This means that if *something*=c and *another something*=-b the column remains intact. In our case $b=g'_{2i}$ and $c=g'_{3i}$ and thus we may set $x_i=t_{12}(g'_{2i})t_{13}(-g'_{3i})$ as the first approximation to our x_i. But we want x_i have level I. Of course simultaneous multiplication of *something* and *another something* by the same thing does not affect the property iii. So we may take $x_i=t_{12}(\xi g'_{2i})t_{13}(-\xi g'_{3i})$ as our second approximation which satisfies both properties ii and iii. We still have the property i to reconcile with. When one multiplies x_i's one just adds their entries in the position (1,2) to get the corresponding entry of the product, the same holds for the (1,3)-entry. What we want at the end is ξ in the position (1,2) and 0 in the position (1,3). So we have to sum g'_{3i} out whit the same coefficients which sum g'_{2i} to 1. But this is of course what one would expect of rows of a matrix and columns of its inverse. So we finally set

$$x_i=t_{12}(g_{i2}\xi g'_{2i})t_{13}(-g_{i2}\xi g'_{3i}).$$

Again this is a common factor so that the property iii is not affected and at the same time (since R is commutative!) these new x_i's enjoy property i. The construction of x_i's and the proof of Suslin's theorem is now complete.

Now this proof is quite different. Looking at the proof one immediately discovers that calculations with matrices were never used here. Indeed to verify conditions i and ii one uses only elementary calculations while condition iii is precisely what the stable calculations are about. In fact in his thesis A.V.Stepanov has given a lot of further variations of this proof and many generalizations [213]. For example, he was able to rewrite in the same style also the proofs in the stable range. As we know we can very well produce both elementary and stable calculations for arbitrary Chevalley groups. That's the idea.

6°. Borewicz-Vavilov's proof of Wilson-Golubchik's theorem. This idea became apparent only when Stepanov has invented this remarkable trick. But in a less explicit form it occurred already in the simplified proof of the Wilson-Golubchik normal subgroup theorem presented in [40].

Let us recall that to establish the standard description of normal subgroups it suffices to prove that any subgroup F which is normalized by E(n,R) and is not contained in the centre of GL(n,R) contains a non-trivial elementary transvection $t_{ij}(\xi)$, $i\neq j$, $\xi\in R$, $\xi\neq 0$. As we've already mentioned it is obvious and well-known (see [23]) that to do this one has only to exhibit a non-central matrix $z\in F$ which is contained in a proper parabolic subgroup, say a non-central matrix of the form

$$z=\begin{pmatrix} \xi & u \\ 0 & c \end{pmatrix}\in F$$

where $\xi\in R^*$, $u\in R^{n-1}$ and $c\in GL(n-1,R)$.

Take a non-central matrix $g=(g_{ij})\in F$ with the inverse $g^{-1}=(g'_{ij})$ and form the same matrices $x_r=t_{12}(g'_{2r})t_{13}(-g'_{3r})$ as the first approximation towards x_r in the Stepanov's proof. As we know the r-th column of x_rg^{-1} coincides with the r-th column of g^{-1} and thus the r-th column of $g_r=gx_rg^{-1}$ coincides with e_r. Now look at the matrices

$$z_r=[g,x_r]=g(x_rg^{-1}x_r^{-1})=g_rx_r^{-1}\in F.$$

Clearly one has

$$z_r=(e+\sum g_{i1}(g'_{2r}g'_{3j}-g'_{3r}g'_{2j})e_{ij})u^{-1},\ 1\le i\le n,\ 1\le j\le n.$$

Thus the first column of z_1 coincides with e_1 and if z_1 is non-central we may take $z=z_1$. Since z_1 has 1 in the NW corner it may be central only if it $z_1=e$, which means that

$$g_{i1}(g'_{21}g'_{3j}-g'_{31}g'_{2j})=0,\ 2\le i\le n,\ 1\le j\le n.$$

But this guarantees that all the matrices z_r, $2\le r\le n$, have the necessary block-triangular form. If at least one of the matrices z_i is non-central, we are done. If all of them are central alias scalar, then

$$g_{i1}(g'_{2r}g'_{3j}-g'_{3r}g'_{2j})=0,\ 2\le i\le n,\ 1\le j,r\le n,\ i\ne j.$$

But these products vanish also when j=i (when j=r tautologically and when j≠r because of the symmetry in j and r). Thus

$$g_{i1}(g'_{2r}g'_{3j}-g'_{3r}g'_{2j})=0,\ 2\le i\le n,\ 1\le j,r\le n.$$

By the Laplace theorem all the minors taken from the second and the third row of the matrix g^{-1} generate the identity ideal. Hence all the entries g_{i1}, $2\le i\le n$, are zeros and thus the matrix g itself has the necessary block-triangular form. Since g is non-central this finishes the proof.

Now this proof is somewhat less transparent than the Stepanov's proof because the additional factor x_r^{-1} in the expression for z_r disturbs the perfect symmetry of the columns. Still if one looks at the proof a bit closer one discovers that here again the general calculations have never been used and everything can be expressed as a combination of elementary steps and calculations involving a single column at a time.

7°. Localization and patching proofs. Yet another proof of the normality theorem and the normal subgroup theorem uses the localization technique. Its starting point is the following theorem from [218] which Suslin himself calls "Quillen's theorem" because of the similarity with the Quillen's solution [179] of the Serre's problem. Of course the Suslin's solution of the Serre's problem was based on the same idea too [216]. The idea is now commonly referred to as the *Quillen-Suslin localization principle.*

The "Quillen's theorem" of Suslin runs as follows: *Let* $n\geq 3$ *and* g *be a matrix from the principal congruence subgroup* $GL(n,R[x],xR[x])$. *Then the matrix* g *belongs to* $E(n,R[x])$ *if and only if for any maximal ideal* $\mathfrak{m}\in MaxR$ *the image of the matrix* g *in* $GL(n,R_{\mathfrak{m}}[x])$ *belongs to* $E(n,R_{\mathfrak{m}}[x])$. The main idea of the proof is roughly speaking as follows. Let $a\in R$ and R_a be the quotient ring of R with respect to the multiplicative system $\{a^n\}$, $n\in\mathbb{N}$. If a is not a zero-divisor then R may be identified with a subring of R_a. Now consider the set I(g) of all $a\in R$ such that $g\in E(n,R_a[x])$. Then the Quillen-Suslin localization principle says that I(g) is an ideal. By the very condition of the theorem I(g) is not contained in any maximal ideal and thus coincides with R. In particular $1\in I(g)$.

Then various modifications of this idea were developed by A.A.Suslin himself, V.I.Kopeiko, E.Abe, T.Vorst, D.Costa and others to study Chevalley groups over polynomial rings [220,141,4,268,69]. L.N.Vaserstein [244] has noticed that this idea may be used to prove the main structure theorems for commutative (or, more generally, almost commutative rings, or, still more generally for such rings that $S^{-1}R$ satisfies $sr(S^{-1}R)\leq d$ for many multiplicative subsystems S of R). This idea was then used by G.Taddei [226,227] to prove that $E(\Phi,R)$ is normal in $G(\Phi,R)$ and by L.N.Vaserstein and E.Abe [247,6,7] to describe the normal subgroups of $G(\Phi,R)$.Later it was developed in many publications of L.N.Vaserstein [248,249], etc., to cover large classes of groups far from being split.

8°. Winnie-the-Pooh comments. At this point another major taoist thinker Winnie-the-Pooh comments "different proofs would prove different things, otherwise it had been the same proof". This is indeed the case and it seems rather odd that this was not explicitly stated in the literature. Not only the scope of applicability of the four proofs is different, but they actually prove quite different things.

First look at the Whitehead-Vaserstein lemma. It proves that $T(n,R)\le$ $\le E(n+1,R)$. Now recall that the *injective stability for* K_1*-functor* says that under appropriate restrictions on the ring R (close to but slightly different from the conditions for the surjective stability) embedding $\Delta\subset\Phi$ of the root systems induces *injective* map of the corresponding K_1-functors $K_1(\Delta,R)$ $K_1(\Phi,R)$ or, what is the same,

$$E(\Delta,R)=E(\Phi,R)\cap G(\Delta,R).$$

The Bass-Vaserstein theorem states that for the linear case injective stability starts not later than at $n=sr(R)+1$ (see [23,24,27,252,253,221]), or, in other words,

$$E(n,R)=E(n+1,R)\cap GL(n,R)$$

for any $n\ge sr(R)+1$. This means that for such an n one has $T(n,R)=E(n,R)$

Let's introduce the following provisional notation. Denote by $E'(n,R)$ the normal closure of $E(n,R)$ and by $E''(n,R)$ the subgroup generated by all the transvections $t_{uv}(\xi)=e+u\xi v$ where $u\in {}^nR$, $v\in R^n$, $\xi\in I$, are subject to the condition $vu=0$ *and* moreover v is *unimodular*. It follows from the formula $gt_{ij}(\xi)g^{-1}=$ $=t_{ge_i,e_jg^{-1}}(\xi)$ that $E'(n,R)$ is generated by all the transvections $t_{uv}(\xi)$ where v is a row of an invertible matrix. Since every row of an invertible matrix is unimodular one has $E'(n,R)\le E''(n,R)$ but the converse is not obvious.

Now the Suslin's proof shows that if R is commutative and $n\ge 3$ then $E''(n,R)=E(n,R)$ so somewhat more than just the normality of the elementary subgroup while the Stepanov's proof shows only that $E'(n,R)=E(n,R)$ so precisely the normality. It is no wonder that a weaker statement requires a simpler proof. Now it follows immediately from the definition that for a commutative ring any unimodular row of length 2 is the first row of some invertible matrix (see [24]) or, in other words, $E'(2,R)=E''(2,R)$.

The localization proofs show an entirely different thing from what we originally wanted, namely that under some conditions a matrix which is elementary "locally" is elementary "globally". But as soon as $E'(n,R)$ or $E''(n,R)$ or $T(n,R)$ or whatever else happens to be locally elementary (what is often the case) it would be globally elementary as well.

§11. The proofs for B_ℓ and D_ℓ

In this section we describe how to modify for the orthogonal groups the approach outlined in the previous section for the linear case. Actually we give some details only for the normality of the elementary subgroup, but it should be clear from the previous section that then the description of normal subgroups may be gained by a bit of extra work. We do not reproduce the preceding proofs for normality, but in this case they are already much longer than ours.

1°. Orthogonal transvections. Start with a geometric definition. Let again $V=R^n$ be a free R-module of rank n with a non-degenerate symmetric bilinear form B. As in §§3,6 we identify every vector u from V with its column of coordinates and an endomorphism of the module V with its matrix in the Witt base $e_1,\dots,e_{-1}$. Recall that a vector u is called *isotropic* if $(u,u)=0$. In other words this means that $u\in u^\perp$, where $u^\perp$ is the set of all vectors from V orthogonal to u, i.e. $u^\perp=\{v\in V \mid (u,v)=0\}$. Let now u is an isotropic vector, $v\in u^\perp$ is a vector orthogonal to u, and $\xi\in R$ is a scalar. We define a linear transformation $T_{u,v}(\xi)$ of the module V by the formula

$$T_{u,v}(\xi)x=x-\xi(u,x)v+\xi(v,x)u-\frac{1}{2}\xi^2(v,v)(u,x)u,$$

where $x\in V$. It is clear that this transformation is invertible and its inverse is $T_{u,v}(-\xi)$. It is classically known that for any $u,v\in V$, $\xi\in R$ such that $(u,u)=(u,v)=0$, the transformation $T_{u,v}(\xi)$ belongs to $G=SO(n,R)$. The transformations $T_{u,v}(\xi)$ are sometimes called *orthogonal transvections.* Many authors use the terms *Eichler-Siegel-Dickson transvections* or *ESD-transvections.* The subgroups $TO(n,R)$ of $SO(n,R)$ generated by all the ESD-transvections $T_{u,v}(\xi)$, $\xi\in R$, $(u,u)=(u,v)=0$, is called the *Eichler subgroup.* For a field (and on the stable level) the group $TO(n,R)$ coincides with the elementary subgroup $EO(n,R)$, but in general it may be strictly larger. The ESD-transvections are precisely what we call the elements of root type in the natural representation of orthogonal groups. The transformation $T_{u,v}(\xi)$ is called an element of *long root type* if the vector v is also isotropic and of *short root type* otherwise.

Conjugation by an element $g\in SO(n,R)$ has the following effect on $T_{u,v}(\xi)$

$$gT_{u,v}(\xi)g^{-1}=T_{gu,gv}(\xi).$$

This implies that the group TO(n,R) is normal in G=GO(n,R).

The elementary orthogonal transvections are obtained in the following way. Take $u=e_i, v=e_{-j}$. Then a straightforward calculation shows that $T_{u,v}(\xi)=T_{ij}(\xi)$.

Return to the matrix notation. For a column $u\in R^n$ we denote by $\tilde{u}\in {}^nR=(R^n)^*$ the row $(fu)^t$, where, $f=f_n$, if $n=2\ell$ is even and $f=F_n$, if $n=2\ell+1$ is odd. This means that the i-th component $\tilde{u}_i$ of the row $\tilde{u}$ is equal to the (-i)-component u_{-i} of the column u if $i\neq 0$, while $\tilde{u}_0=2u_0$. Comparing this notation with the definition of the inner product on $V=R^n$ we see that $(u,v)=\tilde{v}u$. In particular a vector $u\in V$ is isotropic if and only if $\tilde{u}u=0$. Now with this notation the matrix of $T_{u,v}(\xi)$ looks as follows. Let $\xi\in R$ and $u,v\in R^n$ are such that $\tilde{u}u=\tilde{v}u=0$. Then in the Witt base $e_1,\dots,e_{-1}$ the ESD-transvection $T_{u,v}(\xi)$ is given by the matrix

$$T_{u,v}(\xi)=e+u\xi\tilde{v}-v\xi\tilde{u}-\frac{1}{2}u\xi^2\tilde{v}v\tilde{u}.$$

In the special case of a long root type element one has $\tilde{u}u=\tilde{v}u=\tilde{v}v=0$ and thus

$$T_{u,v}(\xi)=e+u\xi\tilde{v}-v\xi\tilde{u}.$$

The following formula immediately follows from the definition of $T_{u,v}(\xi)$. Let u,v,w be three isotropic vectors such that $\tilde{v}u=\tilde{w}u=0$ and $\xi,\zeta\in R$. Then

$$T_{u,v}(\xi)T_{u,w}()=T_{u,\xi v+\zeta w}.$$

Recall that a submodule U of V is called *totally isotropic* if every vector $u\in U$ is isotropic. If $2\in R^*$ this is equivalent to the usual definition which requires that $(u,v)=0$ for any $u,v\in U$. Now we construct certain special elements of root type which play the crucial role in our calculations. Let $u,v,w\in V$ generate a totally isotropic submodule. Then for any $\eta,\zeta,\vartheta\in R$ the product

$$t=T_{u,v}(\eta)T_{u,w}(\zeta)T_{v,w}(\vartheta)$$

is a long root type element. In fact a straightforward calculation using the previous formula shows that $t=T_{x,y}(1)$, where $x=u+v+w$, and

$$y=((\eta+\zeta-\vartheta)u+(\eta-\zeta+\vartheta)v+(-\eta+\zeta+\vartheta)w)/2.$$

One gets a very important special case of this formula taking $u=e_i$, $v=e_j$, $w=e_h$, where all the six indices $\pm i,\pm j,\pm h$ are pair-wise distinct. Then the element $t=T_{i,-j}(\eta)T_{i,-h}(\zeta)T_{j,-h}(\vartheta)$ is a long root type element for arbitrary $\eta,\zeta,\vartheta\in R$. Take for example i=1, j=2, h=3. Then the matrix of t looks as follows:

$$t=\begin{pmatrix} 1 & 0 & 0 & \zeta & \eta & 0 \\ & 1 & 0 & \vartheta & 0 & -\eta \\ & & 1 & 0 & -\vartheta & -\zeta \\ & & & 1 & 0 & 0 \\ & & & & 1 & 0 \\ & & & & & 1 \end{pmatrix}.$$

As we shall see in a while for any column v one may pick up $\eta,\zeta,\vartheta\in R$ in such a way that t stabilizes v. This is an example of what was called a fake root unipotent in §5. It is not generally speaking a root unipotent i.e. is not conjugated to an element of the form $x_\alpha(\xi)$ - certainly not if the ideal generated by η,ζ,ϑ is not principal. Still it satisfies all the polynomial equations satisfied by the long root elements and in particular the Whitehead-Vaserstein type formula which expresses its conjugate in a proper parabolic subgroup as a product of elementary factors.

2°. Normality. Now let's try to imitate Stepanov's proof for the orthogonal groups. Let $T_{rs}(\xi)$, $r\neq\pm s$, $\xi\in I$, be an elementary orthogonal transvection of level I. We want to show that $gT_{rs}(\xi)g^{-1}\in EO(n,R,I)$ for any $g\in SO(n,R)$. Since EO(n,R,I) is generated by *long* root type elements we may without any loss of generality suppose that $i,j\neq 0$. Again we try to find x_i's subject to the same conditions i–iii as in the Stepanov's proof but of course, now we want x_i's to be *orthogonal* transvections, actually elements of *long* root type in SO(n,R).

Again since all the elementary long root elements are sociologically equal, we may without loss of generality take (r,s)=(1,-2). We look for x_i in the following form:

$$x_i=T_{1,-2}(\eta)T_{1,-3}(\zeta)T_{2,-3}(\vartheta)$$

As we know, such a matrix is *always* a long root type element, so that ii is satisfied. Now try to satisfy iii. Look at the i-th column of the matrix g^{-1}, which we represent by a diagram as follows:

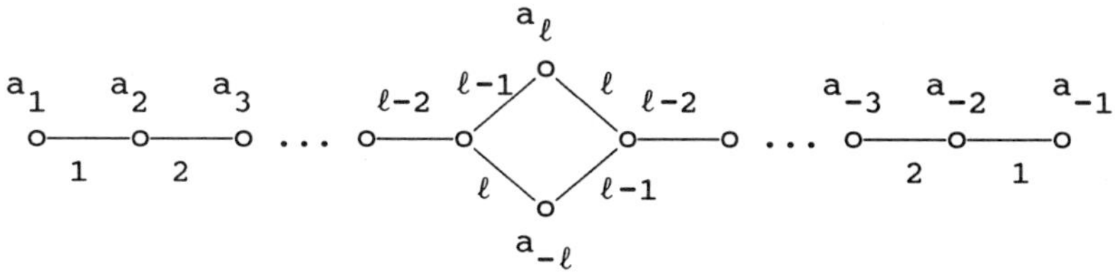

Then our transformation x_i multiplies a_{-1} by $-\zeta$ and adds to a_3 and multiplies a_{-2} by $-\vartheta$ and adds to a_3 too. If we want a_3 to remain unaltered we may take $\zeta=a_{-2}$ and $\vartheta=-a_{-1}$. In that case we have added $a_{-2}a_{-3}$ to a_1 and $-a_{-1}a_{-3}$ to a_2. Now we may choose $\eta=-a_{-3}$. Then it multiplies a_{-2} by $-a_{-3}$ and adds to a_1 and multiplies a_{-1} by a_{-3} and adds to a_2. By the commutativity of the ground ring all the new terms cancel and the column remains unaltered. In our case $a_{-1}=g'_{-1i}$, $a_{-2}=g'_{-2i}$ and $a_{-3}=g'_{-3i}$ so that we may take

$$x_i=T_{1,-2}(-g'_{-3i})T_{1,-3}(g'_{-2i})T_{2,-3}(-g'_{-1i})$$

as the first approximation to our x_i. Again simultaneous multiplication of η,ζ and ϑ by a common factor does not affect the property iii. Thus we may take

$$x_i=T_{1,-2}(-g_{i,-3}\xi g'_{-3i})T_{1,-3}(g_{i,-3}\xi g'_{-2i})T_{2,-3}(-g_{i,-3}\xi g'_{-1i}).$$

Then these x_i's satisfy all the three required properties.

Now one finishes the proof of normality of EO(n,R,I) in SO(n,R) in precisely the same way as for the linear case. Namely one has

$$gT_{rs}(\xi)g^{-1}=\prod gx_ig^{-1},\ 1\le i\le n;$$

where each of the matrices gx_ig^{-1} is an orthogonal transvection of level I whose i-th column coincides with e_i. An appropriate analogue of the Whitehead-Vaserstein lemma shows that such a transvection belongs to EO(n,R,I) and, finally, $gT_{rs}(\xi)g^{-1}\in$ $\in$EO(n,R,I).

3°. And more. But now another miracle happened - indeed two of them. First, when we've constructed the fake root unipotent stabilizing a column we never used the fact that this was a column of an orthogonal matrix! In other words we never used the quadratic equation. This means that there are enough unipotents in the usual matrix representation of the orthogonal group SO(n,R) to stabilize *any* column of *any* matrix from GL(n,R). This means that we are able not just to rewrite the description of normal subgroups of SO(n,R) in this style. We are able to prove that any non-central subgroup of GL(n,R) which is normalized by EO(n,R) contains an element from a proper parabolic subgroup. Then it is of course an easy exercise with the Chevalley commutator formula to pull out linear or orthogonal elementary root unipotents and to establish "standard" description for the subgroups of GL(n,R) normalized by EO(n,R) in terms of *pairs* of ideals of the ground ring. To show how powerful this description is let us mention that it is a common generalization of the following three facts: the Wilson-Golubchik theorem of normal subgroups of GL(n,R), the Golubchik theorem on normal subgroups of SO(n,R) and the classification of overgroups of SO(n,R) in GL(n,R). Of course before this last classification has been known only in the case of fields (by R.Dye, O.King and Li Shangzhi, some cases, like say R=ℝ, were treated in early 60's by R.Brauer, W.Noll and others). The result was announced in the abstracts of the International Conference in Algebra in Novosibirsk, 1989, the details are to be found in a forthcoming paper "Overgroups of the classical groups over commutative rings".

The second remarkable feature of the proof was that we've never used the whole stock of unipotents of the split orthogonal group. The proof relied solely on the fact that there was a "singular plane" in the geometry of the long root subgroups what is certainly true if the underlying module V has a three-dimensional totally isotropic subspace. This means that in all our proofs we may take any orthogonal group which is "isotropic enough" instead of the split one. Thus we get a new proof for normality of the elementary subgroups, normal subgroup theorem and more for any orthogonal group defined by a form of the Witt index at least *three* (or, by a slightly different choice of fake unipotent, *two* in most cases, see the next section). A recent paper of L.N.Vaserstein [248] covers even a larger class of groups, defined in terms of the*local* Witt indices, so that he can prove analogous results whenever the form is globally isotropic, i.e. has Witt index at least *one* and the corresponding module has a large enough rank. By the very nature of our proof we have to have enough unipotents *globally* and so our results apply to a smaller class of groups but the proofs are straightforward and short and some of the results more precise. This will appear in my joint paper with W.Hołubowski "Structure of isotropic orthogonal groups over commutative rings".

§12. The proofs for C_ℓ

In this section we outline the new proofs for the symplectic case. Again we restrict ourselves to the proof of normality,the proof for the standard description of normal subgroups follows the same lines. Since $B_2=C_2$ these proofs apply to the odd-dimensional orthogonal groups as well and sometimes give better results than the ones described in the previous section. But since the geometry of long root subgroups in the symplectic group is extremely poor (two distinct long root subgroups are either orthogonal or opposite) the proofs become perceptibly more tricky than for the linear and orthogonal cases.

1°. **Symplectic transvections.** As in the orthogonal case we start with a geometric definition. Let now $V=R^n$ be a free R-module of rank $n=2\ell$ with a non-degenerate symplectic bilinear form B with the same matrix $F=\mathrm{sdiag}(1,\dots,1,-1,\dots,-1)$ as in §6. We denote by $e_1,\dots,e_{-1}$ the standard base of V which is by definition of B a Witt base of V. Let $u\in V$ is a vector and $\xi\in R$ is a scalar. We define a linear transformation $T_u(\xi)$ of the module V by the formula $T_u(\xi)x=x+(u,x)\xi u$. If $u,v\in V$ are two orthogonal vectors we define $T_{u,v}(\xi)$ by the formula

$$T_{u,v}(\xi)x=x+\xi(u,x)v+\xi(v,x)u.$$

When $2\in R^*$ one has $T_u(\xi)=T_{u,u}(\xi/2)$. It is classically known that for any $u,v\in V$, $\xi\in R$ such that $(u,v)=0$, the transformations $T_u(\xi)$ and $T_{u,v}(\xi)$ belong to $G=Sp(n,R)$. The transformations $T_u(\xi)$ are called *symplectic transvections*. They are just the usual (linear) transvections contained in the symplectic group. The transformations $T_{u,v}(\xi)$ are often called *symplectic transvections* too, but more commonly *Eichler-Siegel-Dickson transvections* or *ESD-transvections*. The transformations $T_u(\xi)$ are precisely what we call elements of long root type in the natural representation of orthogonal groups while $T_{u,v}(\xi)$ for linearly independent u and v are the elements of short root type in this representation.

Conjugation by $g\in Sp(n,R)$ has the usual effect on $T_u(\xi)$ and on $T_{u,v}(\xi)$:

$$gT_u(\xi)g^{-1}=T_{gu}(\xi),\ gT_{u,v}(\xi)g^{-1}=T_{gu,gv}(\xi).$$

The elementary symplectic transvections are obtained in the following way. Take $u=e_i,v=e_{-j}$. Then $T_{u,v}(\xi)=T_{ij}(\xi)$ and in particular $T_u(\xi)=T_{i,-i}(\xi)$.

For a column $u\in R^n$ we denote by $\tilde{u}\in {}^nR=(R^n)^*$ the row $(Fu)^t$. In other words if $u=(u_1,\dots,u_\ell,u_{-\ell},\dots,u_{-1})^t$ then $\tilde{u}=(u_{-1},\dots,u_{-\ell}, -u_\ell,\dots,-u_1)$. Comparing this notation with the definition of the inner product on $V=R^n$ we see that $(u,v)=\tilde{v}u$. With this notation the matrices of $T_u(\xi)$ and $T_{u,v}(\xi)$ in the Witt base $e_1,\dots,e_{-1}$ look as follows.

$$T_u(\xi)=e+u\xi\tilde{u},\ T_{u,v}(\xi)=e+u\xi\tilde{v}+v\xi\tilde{u}.$$

and one has the same sort of formulae for the ESD-transvections in the symplectic case as one has in the orthogonal case.

2°. Long root approach. First we construct some special elements of *long* root type. Take $u,v\in U$ such that $(u,v)=0$. Then for any $\eta,\vartheta\in R$ the product

$$t=T_{u,v}(\eta\vartheta)T_u(\eta^2)T_v(\vartheta^2)$$

is a long root type element. In fact a straightforward calculation shows that $t=T_{\eta u+\vartheta v}(1)$.

One gets a very important special case of this formula taking $u=e_i$, $v=e_j$, where all the four indices $\pm i,\pm j$ are pair-wise distinct. Then the element $t=T_{i,-j}(\eta\vartheta)T_{i,-i}(\eta^2)T_{j,-j}(\vartheta^2)$ is a long root type element for arbitrary $\eta,\vartheta\in R$. Take for example $i=1$, $j=2$. Then the matrix of t looks as follows:

$$t=\begin{pmatrix} 1 & 0 & \vartheta\eta & \eta^2 \\ & 1 & \vartheta^2 & \vartheta\eta \\ & & 1 & 0 \\ & & & 1 \end{pmatrix}.$$

As we shall see in a while again for absolutely any column v one may pick up η, $\vartheta\in R$ in such a way that t stabilizes v. All the remarks from the orthogonal case apply. This is again an example of what was called a fake root unipotent in §5.

Now we start imitating Stepanov's proof for the symplectic case. Let $T_{rs}(\xi)$, $r\neq s$, $\xi\in I$, be an elementary symplectic transvection of level I. We want to show that $gT_{rs}(\xi)g^{-1}\in$ $\in Ep(n,R,I)$ for any $g\in Sp(n,R)$. Since $Ep(n,R,I)$ is generated by *long* root type elements we may (changing g by an element from $Sp(n,R)$ conjugating a given long root unipotent with an elementary one) take $i=-j$ here. Again we are looking for the elements x_i's subject to the same conditions i–iii as in the Stepanov's proof but now we want them to be *symplectic* transvections, actually elements of *long* root type in Sp(n,R).

We may without loss of generality take (r,s)=(1,-2). We look for x_i in the following form:

$$x_i=T_{1,-2}(\eta\vartheta)T_{1,-1}(\eta^2)T_{2,-2}(\vartheta^2)$$

As we have just checked such a matrix is *always* a long root type element, so that ii is satisfied. Now to satisfy iii we look at the i-th column of the matrix g^{-1}, which is represented by a diagram as follows:

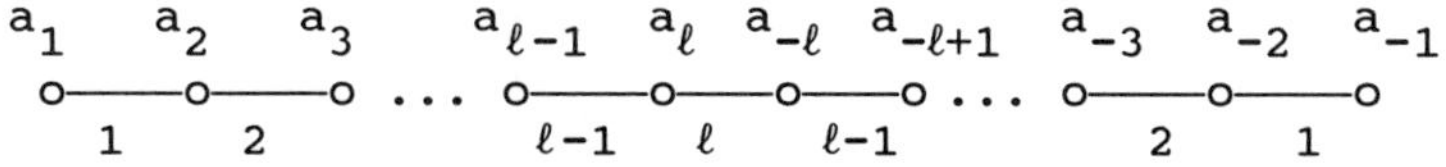

Then our transformation x_i multiplies a_{-1} by $\eta\vartheta$ and adds to a_2 and multiplies a_{-2} by ϑ^2 and adds to a_2 too. If we want a_2 to remain unaltered we may take $\eta=a_{-2}$ and $\vartheta=-a_{-1}$. By the commutativity of the ground ring the new terms which we have added to a_1 cancel and the column remains unaltered. In our case $a_{-1}=g'_{-1i}$ and $a_{-2}=g'_{-2i}$ so that we may take

$$x_i=T_{1,-2}(-g'_{-1i}g'_{-2i})T_{1,-1}((g'_{-2i})^2)T_{2,-2}((g'_{-1i})^2)$$

as the first approximation to our x_i. Again simultaneous multiplication of η and ϑ by a common factor $\xi\in I$ does not affect the property iii.Thus we may take

$$x_i=T_{1,-2}(-\xi g'_{-1i}g'_{-2i})T_{1,-1}(\xi(g'_{-2i})^2)T_{2,-2}(\xi(g'_{-1i})^2)$$

as the second approximation. But the naive approach collapses at this point since it is not obvious how to sum out $-g'_{-1i}g'_{-2i}$ and $(g'_{-1i})^2$. The reason, why it is not obvious is the same as for most of the things which are not obvious – it is impossible. Now the proof still can be written but one has to check that for a fixed matrix $g\in Sp(n,R)$ the elements x_i of this form actually generate Ep(n,R,I). Of course here one has to vary not just i but 1 and 2 as well. This can be done (even in a much more complicated setting of "net subgroups", see [256] and [262] for the statements and [254] for a proof, the details will appear in my paper "Subgroups of symplectic groups over commutative rings"). But this is not straightforward and we better describe another approach.

3°. Short root approach. Now we construct certain *short* root type elements. It is immediate to check that for arbitrary $\zeta_j\in R$, $j\neq\pm i$, and $\eta\in R$ the product

$$t=T_{i,-i}(2\eta)\prod T_{ij}(\zeta_j),\ j\neq\pm i,$$

is an element of short root type. If $2\in R^*$ then one can even replace 2η by η here (anyhow in general the elements of short root type do not generate the whole elementary subgroup, so that a combination of the two approaches is in order). In particular for n=4 and i=1 one has short root type elements

$$t=T_{12}(\zeta)T_{1,-2}(\eta)T_{1,-1}(2\vartheta-\zeta\eta)$$

with matrices

$$t=\begin{pmatrix} 1 & \zeta & \eta & 2\vartheta \\ & 1 & 0 & \eta \\ & & 1 & -\zeta \\ & & & 1 \end{pmatrix}.$$

Now a suitable modification of the Stepanov's original approach is in order and we return to the case of the special linear group. Having a product of two elementary transvections allowed us to keep one column invariant. But for any system of m linear homogeneous equations there are obvious (m+1)-term solutions whose components are expressed as minors of order m of the matrix of coefficients (see [253]). Thus if we have m+1 elementary transvections we may stabilize m columns of g^{-1}. As a pattern take $n\geq4$ and look at the product $z=t_{12}(\zeta)t_{13}(\eta)t_{14}(\vartheta)$ which is always a transvection. Then given any *two* columns one may choose ξ,η,ϑ here in such a way that both columns are fixed by the action of z. Indeed, let $u=(u_1,\dots,u_n)^t$ and $v=(v_1,\dots,v_n)^t$ be the columns. Then take $\zeta=u_3v_4-u_4v_3$, $\eta=u_4v_2-u_2v_4$, $\vartheta=u_2v_3-u_3v_2$. By the commutativity of the ground ring z leaves u and v unchanged. In particular take a matrix $g\in GL(n,R)$ choose a pair of indices j<h and look at the j-th and h-th columns of g^{-1}. Then

$$z=t_{12}(g'_{3j}g'_{4h}-g'_{4j}g'_{3h})t_{13}(g'_{4j}g'_{2h}-g'_{2j}g'_{4h})t_{14}(g'_{2j}g'_{3h}-g'_{3j}g'_{2h})$$

leaves the j-th and the h-th columns of g^{-1} unaltered. Of course we can easily modify this and find an element which stabilizes any number of columns $\leq n-2$.

THis is precisely what we'll do for the symplectic group. Decompose the element t as above into product as follows:

$$t=\begin{pmatrix}1 & \zeta & \eta & 2\vartheta \\ & 1 & 0 & \eta \\ & & 1 & -\zeta \\ & & & 1\end{pmatrix}=\begin{pmatrix}1 & 0 & 0 & \vartheta \\ & 1 & 0 & \eta \\ & & 1 & -\zeta \\ & & & 1\end{pmatrix}\begin{pmatrix}1 & \zeta & \eta & \vartheta \\ & 1 & 0 & 0 \\ & & 1 & 0 \\ & & & 1\end{pmatrix}.$$

Then we have just checked that one can pick up ζ,η and ϑ in such a way that multiplication by the second factor does not change the j-th and the h-th column of g^{-1} (see the formula for z above). But now if the matrix $g\in Sp(4,R)$ is symplectic then the equations guarantee that multiplication by the first factor does not alter the (-j)-th and the (-h)-th rows of g.

This means that the submatrix of the short root type element gtg^{-1} standing on the intersection of the j-th and h-th columns and the (-j)-th and the (-h)-th rows coincides with the corresponding submatrix of the identity matrix. Now there are two essentially different cases: $j=-h$ and $j\neq-h$. If $j\neq-h$ then the element gtg^{-1} is contained in a proper parabolic subgroup and thus belongs to Ep(4,R) by the Whitehead-Vaserstein lemma. On the other hand if $j=-h$, then gtg^{-1} belongs to Ep(4,R) by one of the first lemmas of [141] which says that $e+u\xi\tilde{v}+v\xi\tilde{u}\in Ep(n,R,I)$ if $\xi\in I$ and $v_1=v_{-1}=0$.

Now the proof goes as follows. We write $T_{12}(\xi)$ as a product of factors x_{jh}, $j<h$, which are short root type elements of the form

$$x_{jh}=t_{12}(\kappa_{jh}\xi((g'_{-2,j}g'_{-1,h}-g'_{-1,j}g'_{-2,h}))$$

$$t_{1,-2}(\kappa_{jh}\xi(g'_{-1,j}g'_{2h}-g'_{2j}g'_{-1,h}))t_{1,-1}(\kappa_{jh}\xi(g'_{2j}g'_{-2,h}-g'_{-2,j}g'_{2h})),$$

where the common factor $\kappa_{jh}\xi$ does not affect the property iii while the minors may be summed out by putting $\kappa_{jh}=g_{j,-2}g_{h,-1}-g_{j,-1}g_{h,-2}$, so that the property i is satisfied and the product of x_i equals $T_{12}(\xi)$. Of course if one actually wants to make the factors belong to proper parabolic subgroups for $n\geq6$ one has to look at the larger short root type elements defined in terms of minors of order ℓ. The necessary changes in the proof are obvious.

§13. The proofs for E_6

Here we describe the proof for the group of type E_6. In principle the proof here follows very closely those for A_ℓ and D_ℓ and is much more straightforward than for C_ℓ but here another complication occurs. We cannot construct a unipotent stabilizing an arbitrary column and since we are interested only in the columns of the matrices from $G(E_6,R)$ we construct such a unipotent only for a *white* column. In other words in this case we have to use the quadratic equations.

1°. **Freudenthal's transvections.** The elements of root type in the usual 27-dimensional representation (V,π) of the group $G=G(E_6,R)$ with the highest weight $\bar{\omega}_1$ are well known in the Jordan algebra theory - they are the so called Freudenthal's transvections which were first introduced by H.Freudenthal and later used by N.Jacobson, T.Springer, F.Veldkamp, R.Bix and many others. We do not need such transformations in general form to prove the structure theorems since everything can be done by a class of "fake root unipotents which we consider below. But to give the reader some idea of what it is we define them in the simplest situation by the following formula. Take a column $u\in V$, a row $v\in V^*$ and a scalar $\xi\in R$. Then we may define a linear transformation of V by the following formula:

$$T_{u,v}(\xi)x=x+\xi(x,v)u+\xi v\times(u\times x),\ \ x\in V.$$

Here the cross product of the columns u and x is a row and the cross product of the rows $u\times x$ and v is a column. If $(u,v)=0$ and either the column u or the row v is white, then this transformation preservs the cubic form i.e. belongs to the group $G=G(E_6,R)$. These are our "elements of root type". We are interested only in the case when *both* the column u and the row v are white - these are elements of *long* root type - although E_6 does not have short roots the corresponding group has short root subgroups (we've encountered this phenomenon already for the group of type D_ℓ). Somethimes even a more general class of transformations is considered uder the name of Freudenthal's transvections, defined by *three* columns and rows rather than by two (see [Sp1],[Sp3]). Of course $T_{u,v}(\xi)$ satisfies the usual formulae like, say,

$$gT_{u,v}(\xi)g^{-1}=T_{gu,vg^{-1}}(\xi).$$

We will not expound here the theory of these transformations in detail and restrict ourselves to showing that the usual elementary unipotents (and hence by the previous formula all of their conjugates) may be easily interpreted as a particular case of those. Look at the diagram representing the coordinates x_λ of a vector $x \in V$ which are numbered as follows (see 2° below):

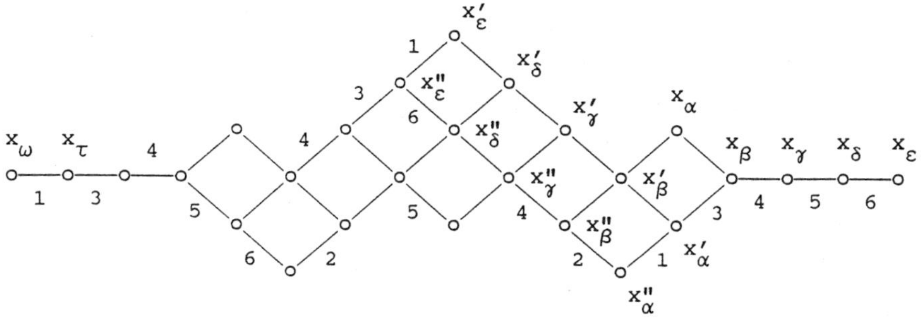

Then $x_{\alpha_1}(\xi)$ adds $\pm\xi x_\tau$ to the coordinate x_ω and adds $\pm\xi x'_\rho$ to x''_ρ where $\rho\in\{\alpha,\beta,\gamma,\delta,\varepsilon\}$. Now take $u=v^\omega \in V$ and $v=(v^\tau)^* \in V^*$. Then $(x,(v^\tau)^*)=x_\tau$ and the formula for $T_{u,v}(\xi)$ says presicely that we have to add $\xi x_\tau v^\omega$ to x. On the other side $v^\omega \times v^\lambda$ is not zero only when λ is a weight corresponding to the coordinates $x'_\rho, x_\rho, \rho\in\{\alpha,\beta,\gamma,\delta,\varepsilon\}$. Thus taking the cross product with v^ω has the following effect on x: it interchanges the coordinates x_ρ and x'_ρ (possibly changing their signs too) while making all the other coordinates zero - we shouldn't forget that the resulting vector $y=v^\omega \times x$ will be a row not a column. Now the cross product $(v^\tau)^* \times (v^\lambda)^*$ is not zero only when λ is a weight corresponding to one of the coordinates y''_ρ, y_ρ, $\rho\in\{\alpha,\beta,\gamma,\delta,\varepsilon\}$. Again taking the cross product with $(v^\tau)^*$ interchanges y_ρ with y''_ρ (possibly changing their signs. Thus the only non-zero coordinates of the column $z=(v^\tau)^* \times y$ are $z''_\rho = \pm x'_\rho$, $\rho\in\{\alpha,\beta,\gamma,\delta,\varepsilon\}$ and the formula for $T_{u,v}(\xi)$ says that we have to add $\pm\xi x'_\rho$ to x''_ρ. Finally this means that $T_{u,v}(\xi)$ has the same effect on any column x as $x_{\alpha_1}(\xi)$ does (if we take for granted that the signs coincide which is actually the case).

2°. Stabilizing a column. To stabilize a column in this case we will use root elements corresponding to a maximal "singular subspace" in the geometry of root subgroups. Let us recall that in E_6 the maximal number of roots every two of which form the angle $\pi/3$ is five. Now we fix such a set which is maximal with respect to the chosen order of positive roots. We take the following roots

$$\alpha = {}^{12321}_{\ \ 2},\ \beta = {}^{12321}_{\ \ 1},\ \gamma = {}^{12221}_{\ \ 1},\ \delta = {}^{12211}_{\ \ 1},\ \varepsilon = {}^{12210}_{\ \ 1}.$$

Since all of their differences are roots too they form such a set. Now the products

$$z=x_{\alpha}(z_{\alpha})x_{\beta}(z_{\beta})x_{\gamma}(z_{\gamma})x_{\delta}(z_{\delta})x_{\varepsilon}(z_{\varepsilon})$$

are elements of long root type for all values of $z_{\alpha},\ldots,z_{\varepsilon}\in R$. Our proof of normality in this case will be based on the decomposition of $x_{\alpha}(\xi)$ into a product of 27 elements of this form each one of which stabilizes one column of the conjugating matrix g^{-1}. It is critical here that the columns of a matrix from $G(E_6,R)$ are *white* - without that everything would collapse. The action of the elements $x_{\alpha}(*),\ldots,x_{\varepsilon}(*)$ on the 27-dimensional module V with the highest weight $\overline{\omega}_1$ is shown at the following picture:

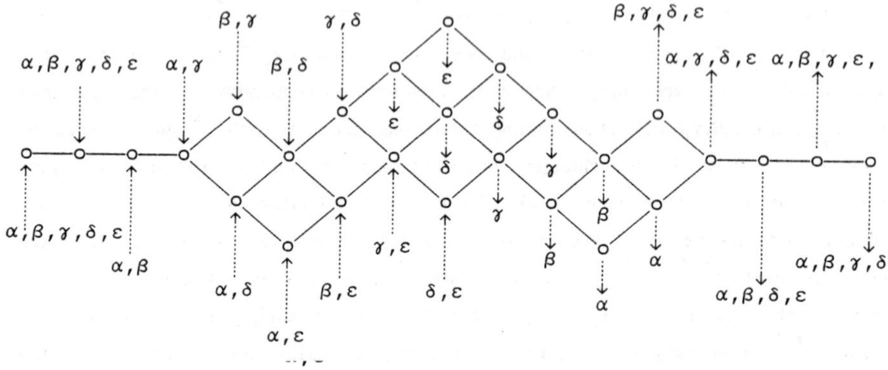

Here of course a dotted arrow starting off a weight and finishing at one of the symbols $\alpha,\ldots,\varepsilon$ shows that this coordinate is multiplied by ξ and then added to or subtracted from some other coordinate by the corresponding transformation $x_{\alpha}(\xi),\ldots,x_{\varepsilon}(\xi)$. Quite analogously a dotted arrow starting off one of the symbols $\alpha,\ldots,\varepsilon$ and finishing at a weight shows that something is added to or subtracted from this coordinate by the corresponding transformation $x_{\alpha}(\xi),\ldots,\ x_{\varepsilon}(\xi)$.

Now we take a vector $u=(u_{\lambda})\in V$ and try to choose the coefficients $z_{\alpha},\ldots,z_{\varepsilon}$ in the expression for z so that z stabilizes u. Recall that we are thinking of the coordinates u as the marks at the corresponding nodes of the weight diagram and for the purposes of this proof we temporarily christen them as shown at the following picture (as opposed to their regular numbering by the weights of the 27-dimensional representations, the reasons are that the number of currently used greek letters is less than that of the weights and our new numbering is more suggestive for the calculation).

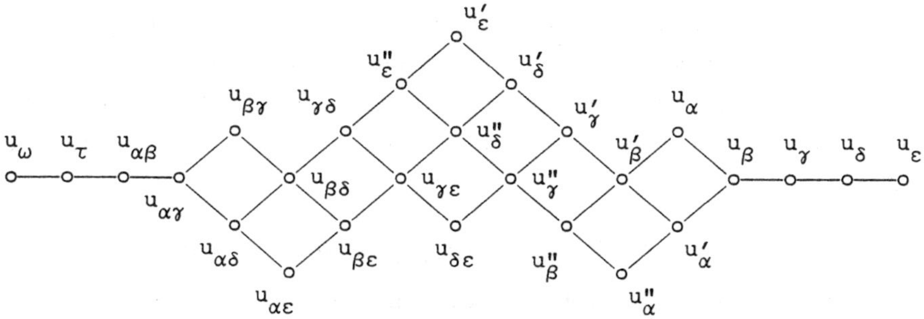

Now there is little doubt that if we want z to stabilize u we have to set $z_\alpha=\pm\xi u_\alpha,\dots,z_\varepsilon=\pm\xi u_\varepsilon$. Now if $\rho,\sigma\in\{\alpha,\beta,\gamma,\delta,\varepsilon\}$ then $x_\rho(z_\sigma)$ adds $\pm\xi u_\sigma u_\rho$ to $u_{\rho\sigma}$, while $x_\sigma(z_\rho)$ adds $\pm\xi u_\rho u_\sigma$ to $u_{\rho\sigma}$. Now the explicit knowledge of signs from [GS] to which we referred in §3 (or the calculations with the Frenkel-Kac cocycle if we are a bit more sophisticated) show that the signs of $z_\alpha,\dots,z_\varepsilon$ may be chosen in such a way that the summands $\pm\xi u_\sigma u_\rho$ and $\pm\xi u_\rho u_\sigma$ cancel, so that all the coordinates $u_{\rho\sigma}$, $\rho,\sigma\in\{\alpha,\beta,\gamma,\delta,\varepsilon\}$, remain unchanged. No we have no freedom left since z has been already uniquely determined up to a common factor ξ and we still have to check that u_ω and u_τ remain unaltered. In fact we have added to u_ω and u_τ the sums $\pm u_\alpha u''_\alpha\pm\dots\pm u_\varepsilon u''_\varepsilon$ and $\pm u_\alpha u'_\alpha\pm\dots\pm u_\varepsilon u'_\varepsilon$ respectively. Now for an arbitrary column there is absolutely no reason why these sums should be zero. But our column is *white*! This means precisely that they are zero in our case (again after a careful checking of signs in either way). This means that the column u is fixed under the action of z

3°. **Normality.** Now that we've found an element of long root type, stabilizing a white row it is straightforward to modify Stepanov's proof for our case.

Let $x_\alpha(\xi)$, $\xi\in I$, be an elementary orthogonal transvection of level I. We want to show that $gx_\alpha(\xi)g^{-1}\in E(E_6,R,I)$ for any $g\in G(E_6,R)$. We want to find long root type elements x_λ's, where $\lambda\in\Lambda(\pi)$, subject to the same conditions i–iii as in the Stepanov's proof. But now it is more than clear how to do this. Take first

$$x_\lambda = x_\alpha(\pm\xi g'_{\alpha\lambda}) x_\beta(\pm\xi g'_{\beta\lambda}) x_\gamma(\pm\xi g'_{\gamma\lambda}) x_\delta(\pm\xi g'_{\delta\lambda}) x_\varepsilon(\pm\xi g'_{\varepsilon\lambda}).$$

Here we persist in our recent ill habit of denoting the five last coordinates of the λ-th column $(g'_{*\lambda})$ of the matrix g^{-1} by the roots $\alpha,\dots,\varepsilon$ rather than by their true names corresponding to the weights as we've done throughout the whole paper. Now it is very easy to sum out all the components of x_λ's but the first one. To do this we have only to multiply all the components of x_λ by the common factor $g_{\lambda\alpha}$ (where of course $(g_{*\alpha})$ is the column of g "with number α" in the same sense as above - the fifth column from below). Thus our final x_λ equals

$$x_\lambda = x_\alpha(\pm\xi g_{\lambda\alpha} g'_{\alpha\lambda}) x_\beta(\pm\xi g_{\lambda\alpha} g'_{\beta\lambda}) x_\gamma(\pm\xi g_{\lambda\alpha} g'_{\gamma\lambda}) x_\delta(\pm\xi g_{\lambda\alpha} g'_{\delta\lambda}) x_\varepsilon(\pm\xi g_{\lambda\alpha} g'_{\varepsilon\lambda}).$$

These x_λ's satisfy all of the conditions i-iii. Now $gx_\lambda g^{-1}$ is contained in a proper parabolic subgroup and thus belongs to $E(E_6,R,I)$ by an appropriate analogue of the Whitehead-Vaserstein lemma. The proof is complete.

§14. Concluding remarks

Here we outline briefly the situation for the other exceprtional groups, some further developments and prospects.

1°. Some landmarks for other exceptional groups. The proof described above for E_6 is a very simple one. Actually it is too simple to carry over to other exceptional cases (apart from E_7 over a ring R such that $2\in R^*$) without some further complications. First usually one does not have a "singular subspace" in the geometry of root subgroups large enough to stabilize any column. So one has develop a D_ℓ-proof rather then the A_ℓ-proof described above (the proof for D_ℓ was an A_3 proof and the proof for E_6 was an A_5-proof). For E_7 in the 56-dimensional representation we decompose $x_\alpha(\xi)$ into a product of 56 root type elemenmts each of which is in turn a product of 7 unipotents this is an A_7-proof). For the adjoint representations of E_6, E_7 and E_8 we decompose $x_\alpha(\xi)$ into products of elements which are in turn products of 8, 12 or 14 elementary unipotents (so these are D_5,D_7 and D_8-proofs respectively). For F_4 it is even worse since when we present an elementary unipotent as a product of factors which in turn are products of 11 elementary unipotents, the components of these unipotents depend quadratically on the coordinates of the column which we want to stabilize (actually they have the form $\pm x_{\lambda_1} x_{\mu_1} + \ldots + x_{\lambda_m} x_{\mu_m}$ where $\lambda_1, \ldots, \mu_m$ are some weights and m=1,3. Then it is quite an exercise to show that the elements of such form from the stabilizers of columns of a given matrix $g \in G(F_4, R)$ actually generate the whole of $E(F_4, R)$.

These calculations were carried through jointly with E.B.Plotkin and the details will appear as the V-th and VI-th papers from the series "Structure of Chevalley groups over commutative rings" under the subtitles "Stabilizers of columns" and "The main structure theorems".

2°. Groups normalized by some elementary matrices. One can extend the standard description of normal subgroups to some classes of subgroups normalized by large groups of elementary matrices. Some of such possibilities, like the description of subgroups normalized by a relative elementary group, were already mentioned. There are many more and they have close connection with the study of subgroup structure and provide new results even in the case of fields. Thus we' ve already mentioned the classification of subgroups in GL(n,R) normalized by an elementary classical group.

In fact much more can be done. In a series of papers Z.I.Borewicz and the author studied subgroups of the general group which contain a subgroup of block diagonal matrices (see, for example, [40,256,262] and the references therefrom). Led by analogy on our works on subgroups normalized by a split maximal torus I.Z.Golubchik has then noticed that this descriptiion may be refined to the description of subgroups normalized such a subgroup [105]. Analogous problems for other classical groups have been studied by the author (see [256,262] for statement of the results and further references). In the general setting it goes here about description of subgroups in $G(\Phi, R)$ normalized by an elementary subgroup $E(\Delta, R)$, where Δ is a subsystem of Φ which is not too small. Of course one could think even of a much more general situation on an arbitrary embedding $\pi: G(\Delta, R) \longrightarrow G(\Phi, R)$, not necessarily coming from an embedding of root systems. In such a form this problem very closely touches the description of maximal subgroups of algebraic groups and finite groups of Lie type (see [187, 188] and references therein) and is not solved even in the case of fields although there is a lot of partial results for particular embeddings (see my forthcoming survey "Subgroups of Chevalley groups containing a semisimple subgroup).

There is a room for further development. Thus in several papers N.S.Romanovskii, Z.I.Borewicz, R.A.Schmidt, A.V.Stepanov, A.E.Zalesskii, the author and others have studied subgroups of G(Φ,R) normalized by E(Φ,S), where S is a subring of R (typically R was the ring of fractions of S, but some other situations have been considered as well). So a person with a passion to generalizations could now consider for example subgroups of G(Φ,R) normalized by E(Δ,S,I) where I is an ideal of S, etc.

3°. **Steinberg groups and K-functors.** As one easily observes we have described in this paper all the machinery necessary to imitate for arbitrary groups the van der Kallen - Tulenbaev result on the centrality of $K_2(n,R)$ for $n \geq 4$. Of course for the classical groups one has to use the results on the presentations of these groups over fields in terms of the small-dimensionall transformations (transvections, ESD-transvections, etc.) by S.Böge, E.Ellers, U.Spengler and many others and for the exceptional groups one has still to develop such presentations himself, but we stronly believe that all the tools are already here.

The techniques described in this paper should prove useful in further study of stability problems for the lower K-functors modeled on the Chevalley groups (it was how the large part of the underlying ideas originally emerged in the fundamental works of H.Matsumoto [157] and M.R.Stein [206]). The stability of $K_1(\Phi,R)$ was described in the talk of E.B.Plotkin at the same conference and we feel that now most of the results can be generalized to $K_2(\Phi,R)$ as well.

4°. **Non-split groups.** Most of the contents of this paper may be generalized to the groups which are "isotropic enough" but not necessarily split. For the orthogonal groups this has been already mentioned in §11. The first construction of some "real" forms of Chevalley groups as the "twisted groups" was introduced by R.Steinberg [207] (see also [50,52,209]). The Steinberg construction has been generalized and modified in several directions. Thus in [211,212] E.Stensholt has used it to construct embeddings among the Chevalley groups. In 2 E.Abe introduced the twisted groups over commutative rings (later this has been used in [222,215] to treat the normal subgroups of the twisted groups in the same style as it has been done before for the groups of normal types. Recently Cheng Chon Hu (see [54,55] many further works have been published in Chinese journals) has generalized the Steinberg's construction to a huge class of semi-simple groups including for example all the non-compact real forms (not just the few considered by R.Steinberg).

In fact all our proofs depend only on the existence of a certain configuration of unipotent root subgroups and not on the fact that the group in question is split. Apart from some recent geometric works of F.Veldkamp and J.Ferrar where only some very special classes of rings (basically the rings of stable rank 1) are considered we are not aware of any research on the structure of non-quasi-split groups over rings has ever been done. We propose to study the structure of some of these groups in our subsequent publications.

5°. **Infinite-dimensional groups.** Another field were our techniques might prove useful is the study of infinite dimensional analogues of the Chevalley groups. Many such infinite dimensional generalizations have been introduced and studied by E.Abe, A.Bak, H.Garland, J.-Y.Hée, V.Kac, R.Moody, J.Morita, D.H.Peterson, M.Takeuchi, K.L.Teo, J.Tits and others. It seems extremely plausible that large parts of the contents of the present paper may be generalized to these infinite-dimensional groups.

Acknowledgements

This survey is based on the notes of the lecture courses "Classical groups and Chevalley groups" (24 lectures, Fall 1989) and "Recent results on classical groups and Chevalley groups" (11 lectures, Spring 1990) delivered by the author at the Univ. of Crete and the Univ. of Notre Dame respectively as also on a series of talks given in the years 1989-1991 at some Universities in Poland, U.S.A., Japan, Italy, France and U.K. (including Univ. of Wrocław, Penn. State Univ., Northwestern Univ., Univ.of Oregon, Caltech, Yale Univ., Univ.of Virginia, Univ.of Tsukuba, Univ.of Kumamoto, Univ.degli Studi di Milano, Univ.Roma II, Univ.Paris VII, Univ.of Warwick, Cambridge Univ. and Univ.of Newcastle upon Tyne). The author expresses his most sincere thanks to D.I.Deriziotis, A.J.Hahn and O.T.O'Meara for organization of his stays in Greece and the U.S. during the 1989/90 academic year. Actually the Cretan course contained lots of details skipped here as well as a good deal of proofs and the notes taken by I.Antoniadis and A.Fakiolas were of great help.

In the 80's the theory of classical groups over rings has become to a large extend a Russian science and over years I benefited from numerous discussions with Z.I.Borewicz, V.N.Gerasimov, I.Z.Golubchik, V.I.Kopeiko, A.V.Mikhalev, E.B.Vinberg, A.E.Zalesskii, E.I.Zelmanov and others. Especially important for the development of my views on the subject was the constant contact with A.A.Suslin. I am very grateful to A.V.Stepanov who permitted me to develop his idea long before it was published in his own works. Much of the work was done in the long-standing cooperation with E.B.Plotkin. For many years I was not able to see L.N.Vaserstein often but we kept in touch. During last years I have discussed the subject with many mathematicians. Conversations and correspondence with M.Aschbacher, B.N.Cooperstein and I.Frenkel on the realizations and with A.Bak, A.J.Hahn, W.van der Kallen and M.R.Stein on the structure theorems have been particularly useful.

Very special thanks are due to E.Abe who profoundly influenced this work in many respects, first in absentia via his works and later by his deep understanding of the subject, unfailing interest and encouraging. He has organized my stay in Japan during August-September 1990. I greatly appreciate the hospitality of my Japanese colleagues, especially of J.Morita and K.Suzuki, and the excellent organization of the Hiroshima conference by K.Yamaguti and N.Kawamoto. N.Kamiya supported me in the dark January 1991 in a very unexpected way.

The final text was prepared in the Spring 1991 at the Univ. of Warwick under a research grant from the Science and Engineering Research Council of the United Kingdom. I very much appreciate the friendly and relaxed atmosphere of the Maths Department at Warwick and gratefully acknowledge help of my colleagues here, especially R.W.Carter. B.Westbury carefully read the final text and his suggestions helped to make it less ambiguous at many places. He corrected most of queer phrases. The remaining ones as well as some other discrepancies with the English English usage are left on purpose. M.Beynon and J.Rhodes helped me to solve the problems resulting from the incompatibility of computer systems. E.Shiels has been extremely considerate and made my life here easier in a number of ways.

I have been half way through with the text when the news came that an overwhelming majority of my fellow-citizens have voted for the true name of my home city. I dedicate this work to the city of Saint Peter which is, despite everything the spirits of evil have done with it, the best city in the world.

References

1. Abe E. Chevalley groups over local rings. *Tôhoku Math.J.*, 1969, V.**21**, N.3, P.474-494.
2. Abe E. Coverings of twisted Chevalley groups over commutative rings. *Sci. Repts.Tokyo Kyoiku Daigaku*, 1977, V.**13**, P.194-218.
3. Abe E. *Hopf algebras*. Cambridge Univ. Press, 1980. 284P.
4. Abe E. Whitehead groups of Chevalley groups over polynomial rings. *Comm. Algebra*, 1983, V.**11**, N.12, P.1271-1308.
5. Abe E. Whitehead groups of Chevalley groups over Laurent polynomial rings. *Preprint Univ.Tsukuba,* 1988, 14P.
6. Abe E. Chevalley groups over commutative rings. *Proc. Conf.Radical Theory, (Sendai, 1988).* Tokyo: Uchida Rokakuho Publ.Comp., 1989, P.1-23.
7. Abe E. Normal subgroups of Chevalley groups over commutative rings. *Contemp. Math.*, 1989, V.**83**, P.1-17.
8. Abe E., Hurley J. Centers of Chevalley groups over commutative rings. *Comm. Algebra*, 1988, V.**16**, N.1, P.57-74.
9. Abe E., Morita J. Some Tits systems with affine Weyl groups in Chevalley groups over Dedekind domains. *J.Algebra*, 1988, V.**115**, N.2, P.450-465.
10. Abe E., Suzuki K. On normal subgroups of Chevalley groups over commutative rings. *Tôhoku Math.J.*, 1976, V.**28**, N.1, P.185 -198.
11. Artin E. *Geometric algebra*. Wiley: N.Y. - London, 1957, 214P.
12. Aschbacher M. The 27-dimensional module for E_6. I. *Inv.Math.*, 1987, V.**89**, N.1, P.159-195.
13. Aschbacher M. Chevalley groups of type G_2 as the groups of a trilinear form. *J.Algebra*, 1987, V.**109**, N.1, P.193-259.
14. Aschbacher M. Some multilinear forms with large isometry groups. *Geom. dedic.*, 1988, V.**25**, N.1-3, P.417-465.
15. Azad H. Structure constants of algebraic groups. *J.Algebra*, 1982, V.**75**, N.1, P.209-222.
16. Azad H. The Jacobi identity. *Punjab Univ.J.Math.*, 1983, V.**16**,P.9-29.
17. Baily W.L. An exceptional arithmetic group and its Eisenstein series. - *Ann. Math.*, 1970, V.**91**, N.1-3, P. 512-549.
18. Bak A. On modules with quadratic forms. *Lecture Notes Math.*, 1969, V.**108**, P.55-66.

19. Bak A. Subgroups of the general linear group normalized by relative elementary groups. *Lecture Notes Math.*, 1982, V.**967**, P.1–22.

20. Bak A. *K-theory of forms.* Princeton Univ.Press: Princeton N.J. 1981. Ann. of Math.Stud., V.**98**, 268P.

21. Bak A. The nilpotent structure of $GL_{n\geq 3}$. *Preprint Univ.Bielefeld*, 1989, 42P.

22. Bartolone C., Bartolozzi F. Topics in geometric algebra over rings. *Rings and geometry.*, 1985, P.353–389.

23. Bass H. K-theory and stable algebra. *Publ.Math.Inst.Hautes Et.Sci.*, 1964. N.**22**, P.5–60

24. Bass H. *Algebraic K-theory.* Benjamin: N.Y. Amsterdam. 1968. 762P.

25. Bass H. Unitary algebraic K-theory. *Lecture Notes Math.*, 1973, V.**343**, P.57–265.

26. Bass H. Clifford algebras and spinor norms over a commutative ring. *Amer. J.Math.*, 1974, V.**96**, N.1, P.156–206.

27. Bass H., Milnor J., Serre J.-P. Solution of the congruence subgroup problem for SL_n ($n\geq 3$) and Sp_{2n} ($n\geq 2$). *Publ.Math.Inst.Hautes Et.Sci.*, 1967, N.**33**, P.59–137.

28. Bass H., Tate J. The Milnor ring of a global field. *Lecture Notes Math.*, 1973, V.**342**, P.349–446.

29. Baston R.J., Eastwood M.G. *The Penrose transform, its interactions with representation theory.* Oxford.Science Public.,Claredon Press, 1984, 213P.

30. Bayer-Fluckiger E. Principe de Hasse faible pour les systèmes de formes quadratiques. *J.reine angew.Math.*, 1987, Bd.**378**, N.1, S.53–59.

31. Berman S., Moody R. Extensions of Chevalley groups. *Israel J.Math.*, 1975, V.**22**, N.1, P.42–51.

32. Bix R. Octonion planes over local rings. *Trans.Amer.Math.Soc.*, 1980. V.**261**, N.1, P.417–438.

33. Bix R. Isomorphism theorems for Octonion planes over local rings. *Trans. Amer.Math.Soc.*, 1981, V.**266**, N.2, P.423–439.

34. Björner A. Orderings of Coxeter groups. *Contemp.Math.*,1984, V.**34**,P.175–195.

35. Björner A., Wachs M. Bruhat order of Coxeter groups and shellability. *Adv. Math.*, 1983, V.**43**, N.1, P.87–100.

36. Björner A., Wachs M. Generalized quotients in Coxeter groups. *Trans.Amer.*

Math.Soc., 1988, V.**308**, N.1, P.1–37.

37. Boe B.D., Collingwood D.H. A comparison theory for the structure of induced representations. I,II. *J.Algebra*, 1985, V.**94**,N.3, P.511–545; *Math.Z.*, 1985, Bd.**190**, N.1, S.1–11.

38. Borel A. *Linear algebraic groups*. Benjamin: N.Y. – Amsterdam, 1969, 398P.

39. Borel A. Properties and linear representations of Chevalley groups. *Lecture Notes Math.*, 1970, V.**131**, P.1–55.

40. Borewicz Z.I., Vavilov N.A. The distribution of subgroups in the general linear group over a commutative ring. *Proc.Steklov.Inst.Math*, 1985, N.3, P.27–46.

41. Bourbaki N. *Groupes et algèbres de Lie. Ch.4–6.* Hermann:Paris,1968, 288P.

42. Bourbaki N. *Groupes et algèbres de Lie. Ch.7,8.* Hermann:Paris,1975, 271P.

43. Brown R.B. A minimal representationm for the Lie algebra $\mathfrak{E}_7$. *Ill.J. Math.*, 1968, V.**12**, N.1, P.190–200.

44. Brown R.B. Groups of type E_7. *J.reine angew.Math.*, 1969, Bd.**236**, N.1, S.79–102.

45. Brown R.B. A characterization of spin representations. *Canad.J.Math.*, 1971, Bd.**23**, N.5, P.896–906.

46. Burgoyne N., Williamson C. Some computations involving simple Lie algebras – In: *Proc. 2nd Symp.Symbolic and algebraic manipulation*. New York: Ass. Comp. Mach., 1971.

47. Bürgstein H., Hesselink W.H. Algorithmic orbit classification for some Borel group actions. *Compos.Math.*, V.**61**, N.1, P.3–41.

48. Carter D., Keller G.E. Bounded elementary generation of $SL_n(\mathcal{O})$. *Amer.J. Math.*, 1983, V.**105**, N.3, P.673–687.

49. Carter D., Keller G.E. Elementary expressions for unimodular matrices. – *Comm.Algebra*, 1984, V.**12**, N.3–4, P.379–389.

50. Carter R.W. Simple groups and simple Lie algebras. *J.London Math. Soc.*, 1965, V.**40**, N.1, P.163–240.

51. Carter R.W. Conjugacy classes in the Weyl group. *Compos.Math.*, 1972, V.**25**, N.1, P.1–59.

52. Carter R.W. *Simple groups of Lie type*. Wiley: London et al., 1972, 331P.

53. Carter R.W. *Finite groups of Lie type: conjugacy classes and complex representations*. Wiley: London et al., 1985, 544P.

54. Cheng Chon Hu A family of simple groups associated with the Satake dia-

grams. *J.Austral.Math.Soc.*, ser.A, 1986, V.**41**, N.1, P.13–43.

55. Cheng Chon Hu Classical groups and generalized simple groups of Lie type. *J.Austral.Math.Soc.*, ser.A, 1989, V.**47**, N.1, P.53–70.

56. Chevalley C. Sur le groupe exceptionnel (E_6). *C.R.Acad.Sci.Paris,* 1951, V.**232**, P.1991–1993.

57. Chevalley C. Sur certains groupes simples. *Tôhoku Math.J.*, 1955, V.**7**, N.1, P.14–66.

58. Chevalley C. *Classification des groupes de Lie algebriques*, V.**2**. Paris: Secretariat Mathématique, 1956–1958, 240P.

59. Chevalley C. Certain schemas des groupes semi-simples. *Sem. Bourbaki,* – 1960–1961, Exp.219, P.1–16.

60. Chevalley C., Schafer R.D. The exceptional simple Lie algebras of types F_4 and E_6. *Proc.Nat.Acad.Sci.U.S.A.*, 1950, V.**36**, P.137–141.

61. Cline E., Parshall B., Scott L. Cohomology of finite groups of Lie type I,II. *Publ.Math.Inst.Hautes Et.Sci.*, 1975, N.**45**, P.169–191; *J.Algebra*, 1977, V.**45**, N.1, P.182–198.

62. Cohen A.M., Cooperstein B.N. The 2-spaces of the standard $E_6(q)$–module – *Geom.dedic.*, 1988, V.**25**, N.1–3, P.467–480.

63. Cohen A.M., Helminck A.G. Trilinear alternating forms on a vector space of dimension 7. *Comm.Algebra,* 1988, V.**16**, N.1, P.1–25.

64. Cohn P. On the structure of the GL_2 of a ring *Publ.Math.Inst.Hautes Et. Sci.*, 1966, N.**30**, P.365–413.

65. Cooperstein B.N. The geometry of root subgroups in exceptional groups. – *Geom.dedic.*, 1978, V.**8**, N.3, P.317–381; 1983, V.**15**, N.1, P.1–45.

66. Cooperstein B.N. Geometry of long root subgroups in groups of Lie type. – *Proc.Symp.Pure Math.*, 1980, V.**37**, P.243–248.

67. Cooperstein B.N. A note on the Weyl group of type E_7. *Preprint Univ. of California at Santa Cruz,* 1988, 10P.

68. Cooperstein B.N. The 56–dimensional module for E_7. I. *Preprint Univ. California at Santa Cruz*, 1988, 52P.

69. Costa D.L. Zero–dimensionality and the GE_2 of polynomial rings. *J.Pure Appl. Algebra*, 1988, V.**50**, P.223–229.

70. Costa D.L., Keller G.E. Radix redux: normal subgroups of symplectic groups. *Preprint Univ.Virginia*, 1991, 88P.

71. Coxeter H.S.M. *Regular polytopes*. 2nd ed. Macmillan Co: N.Y., 1963, 321P.

72. Curtis C.W., Iwahori N., Kilmoyer R. Hecke algebras and characters of parabolic type of finite groups with (B,N)-pairs. *Publ.Math.Inst. Hautes Et. Sci.*, 1971, N.**40**, P.81–116.

73. Demazure M. Schémas en groupes réductifs. *Bull.Soc.Math.France*, 1965, V.**93**, P.369–413.

74. Demazure M., Gabriel P. *Groupes algèbriques. I.* North Holland: Amsterdam et al., 1970, 770P.

75. Demazure M., Gabriel P. *Introduction to algebraic geometry and algebraic groups.* North Holland: Amsterdam et al., 1980, 357P.

76. Demazure M., Grothendieck A. *Schémas en groupes. I, II, III. Lecture Notes Math.*, 1971, V.**151**, 564P.; V.**152**, 654P.; V.**153**, 529P.

77. Dennis R.K., Vaserstein L.N. On a question of M.Newman on the number of commutators. *J.Algebra*, 1988, V.**116**, N.1, P.150–161.

78. Deodhar V.V. Some characterisations of Bruhat ordering on a Coxeter group and determination of the relative Möbius function. *Invent. Math.*, 1977, V.**39**, N.2, P.187–198.

79. Deodhar V.V. On Bruhat ordering and weight lattice ordering for a Weyl group. *Proc.Nederl.Akad.Wetensch.*, Ser.A, 1978, V.**40**, N.5, P.423–435.

80. Deodhar V.V. A splitting criterion for the Bruhat orderings on Coxeter groups. *Comm.Algebra*, 1987, V.**15**, N.9, P.1889–1894.

81. Dieudonné J. On simple groups of type B_n. *Amer.J.Math.*, 1957, V.**79**, N.5, P.922–923.

82. Dieudonné J. *La géométrie des groupes classiques*, 3rd ed. Springer: Berlin et al., 1971, 129P.

83. Di Martino L., Tamburini M.Ch. 2-generation of the finite simple groups and related problems. *Proc.Conf.Generations and Relations Alg.Systems (Lucca, 1990)*, 52P.

84. Dynkin E.B. Semi-simple subalgebras of semi-simple Lie algebra.*Amer.Math. Soc.Transl.Ser.*, 1957, V.**6**, P.111–244.

85. Dynkin E.B. The maximal subgroups of the classical groups. *Amer.Math.Soc. Transl.Ser.*, 1957, V.**6**, P.245–378.

86. Estes D., Ohm J. Stable range in commutative rings. *J.Algebra*, 1967, V.**7**, N.3, P.343–362.

87. Faulkner J.R. Orbits of the automorphism group of the exceptional Jordan algebra. *Trans.Amer.Math.Soc.*, 1970, V.**151**, N.2, P.433–441.

88. Faulkner J.R. Octonion planes defined by quadratic Jordan algebras. Mem. *Amer.Math.Soc.*, 1970, N.104, 71P.

89. Faulkner J.R. A construction of Lie algebras from a class of ternary algebras. *Trans.Amer.Math.Soc.*, 1971, V.**155**, N.2, P.397–408.

90. Faulkner J.R. A geometry for E_7. *Trans.Amer.Math.Soc.*, 1972, V.**167**, N.1, P.49–58.

91. Faulkner J.N, Ferrar J.C. Exceptional Lie algebras and related algebraic and geometric structures. *Bull.London Math.Soc.*, 1977, V.**9**, P.1–35.

92. Ferrar J.C. Strictly regular elements in Freudenthal triple systems. – *Trans.Amer.Math.Soc.*, 1972, V.**174**, N.2, P.313–331.

93. Freudenthal H. Sur le groupe exceptionnel E_7. *Proc.Nederl.Akad.Wetensch.*, Ser.A, 1953, V.**56**, P.81–89.

94. Freudenthal H. Sur les invariantes caractéristiques des groupes semisimples. *Proc.Nederl.Akad.Wetensch.*, Ser.A, 1953, V.**56**, P.81–89.

95. Freudenthal H. Beziehungen der E_7 und E_8 zur Oktavenebene. I–XI. *Proc. Nederl.Akad.Wetensch.*, Ser.A, 1954, V.**57**, P.218–230; P.363–368; 1955, V.**58**,P.151–157; P.277–285; 1959, V.**62**, P.165–201; P.447–474; 1963, V.**66**, P.457–487.

96. Freudenthal H. Die Beziehungen zwischen der E_7–Geometrie und die Konfiguration der 28 Doppeltangenten der Kurve vierter Ordnung. *Russian Math. Surveys*, 1976, V.**31**, N.5, P.148–152.

97. Frenkel I.B., Kac V. Basic representations of affine Lie algebras and dual resonance models. *Invent.Math.*, 1980, V.**62**, N.1, P.23–66.

98. Frenkel I.B., Lepowsky J., Meurman A. *Vertex operator algebras and the Monster*. Academic Press: N.Y. et al., 1988, 502P.

99. Gabriel P., Roiter A.V. *Representations of finite-dimensional algebras.* – Preprint Univ.Zürich, 1990, 264P. (Russian translation to appear as the V.**73** of "*Fundamental trends in Math.*"; English translation of the Russian translation to appear in Springer).

100. Gerasimov V.N. The group of units of a free product of rings. *Math. U.S.S.R.. Sbornik*, 1989, V.**62**, N.1, P.41–63.

101. Gilkey P., Seitz G. Some representations of exceptional Lie algebras – *Geom.dedic.*, 1988, V.**25**, N.1–3, P.407–416

102. Golubchik I.Z. On the general linear group over an associative ring. *Uspehi Mat.Nauk*, 1973, V.**28**, N.3, P.179–180 (in Russian).

103. Golubchik I.Z. On the normal subgroups of orthogonal group over an associative ring with involution. *Uspehi Mat.Nauk*, 1975, V.**30**, N.6, P.165 (in Russian).

104. Golubchik I.Z. *The normal subgroups of linear and unitary groups over rings.* Ph.D. Thesis, Moscow State Univ., 1981, 117P (in Russian).

105. Golubchik I.Z. On the subgroups of the general linear group $GL_n(R)$ over an associative ring R. *Russian Math.Surveys,* 1984, V.**39**, N.1, P.157–158.

106. Golubchik I.Z. On the normal subgroups of the linear and unitary groups over associative rings. In: *Spaces over algebras and some problems in the theory of nets.* Ufa, 1985, P.122–142 (in Russian).

107. Golubchik I.Z., Markov V.T. Localization dimension of PI-rings. *Trudy Sem. Petrovskogo*, 1981, N.6, P.39–46 (in Russian).

108. Golubchik I.Z., Mikhalev A.V. Elementary subgroup of a unitary group over a PI-ring. *Vestnik Mosk.Univ.*,ser.1,*Mat.,Mekh.*, 1985, N.1, P.30–36.

109. Golubchik I.Z., Mikhalev A.V. On the elementary group over a PI-ring. – In: *Studies in algebra*, Tbilisi, 1985, P.20–24. (in Russian).

110. Griffits P., Harris J. *Principles of algebraic geometry*. Wiley: N.Y., 1978, 813P.

111. Grünewald F., Mennicke J., Vaserstein L.N. On symplectic groups over polynomial rings. *Math.Z.*, 1991, Bd.**206**, N.1, P.35–56.

112. Hahn A.J. Algebraic K-theory, Morita theory and the classical groups. *Lecture Notes Math.*, 1986, V.**1185**, P.88–117.

113. Hahn A.J. The finite presentability of linear groups. *Contemp. Math.*, 1989, V.**82**, P.23–33.

114. Hahn A.J., O'Meara O.T. *The classical groups and K-theory.* Springer: Berlin et al., 1989, 576P.

115. Haris S.J. Some irreducible representations of exceptional algebraic groups. *Amer.J.Math.*, 1971, V.**93**, N.1, P.75–106.

116. Hartley B., Shute G. Monomorphisms and direct limits of finite groups of Lie type. *Quart.J.Math.*, 1984, V.**35**, N.1, P.49–71.

117. Hartshorn R. *Algebraic geometry*. Springer:N.Y. Heidelberg, 1977, 496P.

118. Hée J.-Y. Groupes de Chevalley et groupes classiques. *Publ.Math. Univ. Paris VII,* 1984, V.**17**, P.1–54.

119. Hesselink W. A classification of the nilpotent triangular matrices. *Compos. Math.*, 1985, V.**55**, N.1, P.89–133.

120. Hiller H. *Geometry of Coxeter groups.* Pitman: Boston and London, 1982, 213P.

121. Hiller H. Combinatorics and intersections of Schubert varieties. *Comment. Math.Helv.*, 1982, V.**57**, N.1, P.41–59.

122. Humphreys J.E. *Linear algebraic groups.* Springer: New York Heidelberg, 1975, 247P.

123. Humphreys J.E. *Ordinary and modular representations of Chevalley groups. Lecture Notes Math.*, 1976, V.**528**, 127P.

124. Humphreys J.E. *Introduction to Lie algebras and representation theory.* Springer: Berlin et al., 1980, 171P.

125. Hurley J.F. Some normal subgroups of elementary subgroups of Chevalley groups over rings. *Amer.J.Math.*, 1971, V.**93**, N.4, P.1059–1069.

126. Hurrelbrink J., Rehmann U. Eine endliche Präsentation der Gruppe $G_2(\mathbb{Z})$. *Math.Z.*, 1975, Bd.**141**, N.2, S.243–251.

127. Idowu A.J., Morris A.O. Some combinatorial results for Weyl groups. 1987, V.**101**, N.3, P.405–420.

128. Irving R.S. Projective modules in the category $\mathcal{O}_S$: self-duality. *Trans. Amer.Math.Soc.*, 1985, V.**291**, N.2, P.701–732.

129. Jacobson N. Some groups of transformations defined by Jordan algebras. II. Groups of type F_4. *J.reine angew. Math.*, 1960, Bd.**204**, N.1, S.74–98.

130. Jacobson N. Some groups of transformations defined by Jordan algebras. III. Groups of type E_{6I}. *J.reine angew. Math.*, 1961, Bd.**207**, N.1, P.61–85.

131. Jacobson N. *Exceptional Lie algebras.* Marcel Dekker: N.Y., 1971, 125P.

132. Jantzen J.C. *Representations of algebraic groups.* Academic Press: N.Y. et al., 1987, 443P.

133. Kac V. Some remarks on nilpotent orbits. *J.Algebra*, 1980, V.**64**, N.1, P.190–213.

134. van der Kallen W. *Infinitesimally central extensions of Chevalley groups. Lecture Notes Math.*, 1973, V.**356**, 147P.

135. van der Kallen W. Another presentation for the Steinberg group. *Proc. Nederl.Akad.Wetensch.*, Ser.A, 1977, V.**80**, P.304–312.

136. van der Kallen W. Generators and relations in algebraic geometry. *Proc. Intern.Congress Math. (Helsinki, 1978).* Acad.Sci.Fennica, Helsinki, 1980, P.305–310.

137. van der Kallen W. $SL_3(\mathbb{C}[x])$ does not have bounded word length. *Lecture Notes Math.*, 1982, V.**966**, P.357–361.

138. Khlebutin S.G. Sufficient conditions for the normality of the subgroup of elementary matrices. *Russian Math.Surveys*, 1984, V.**39**, N.3, P.245–246.

139. Khlebutin S.G. Some properties of the elementary subgroup. In: *Algebra, logic and number theory*, Moscow State Univ., 1986, P.86–90 (in Russian).

140. Khlebutin S.G. *Elementary subgroups of linear groups over rings.* Ph.D. Thesis, Moscow State Univ., 1987, 48P. (in Russian).

141. Kopeiko V.I. The stabilization of symplectic groups over a polynomial ring. *Math. U.S.S.R.. Sbornik*, 1978, V.**34**, P.655–669.

142. Kopeiko V.I. On a theorem of Suslin. *J.Sov.Math.*, 1983,V.**132**, P.119–121.

143. Kopeiko V.I. Unitary and orthogonal groups over rings with involution. In: *Algebra and discrete Math.*, Kalmytsk.Gos.Univ., Elista, 1985, P.3–14 (in Russian).

144. Kostant B. Groups over $\mathbb{Z}$. *Proc.Symp.Pure Math.*, 1966, V.**9**, P.90–98.

145. Li Fuan The structure of symplectic groups over arbitrary commutative rings. *Acta Math.Sinica*, 1987, V.**3**, N.3, P.247–255.

146. Li Fuan, Liu Mulan Generalized sandwich theorem. *K-theory*, 1987, V.1, P.171–184.

147. Lichtenstein W. A system of quadrics describing the orbit of the highest weight vector. *Proc.Amer.Math.Soc.*, 1982, V.**84**, N.4, P.605–608.

148. Liehl B. Beschränkte Wortlänge in SL_2. *Math.Z.*, 1984, Bd.**186**, S.509–524.

149. Manin Yu.I. *Cubic forms: algebra, geometry, arithmetic.* North Holland: Amsterdam–London, 1974, 292P.

150. Mars J.G.M. Les nombres de Tamagawa de certains groupes exceptionnels. *Bull.Soc.Math.France*, 1966, V.**94**, P.97–140.

151. Mason A.W. On subgroups of GL(n,A) which are generated by commutators. II. *J.reine angew.Math.*, 1981, Bd.**322**, S.118–135.

152. Mason A.W. A further note on subgroups of GL(n,A) which are generated by commutators. *Arch.Math.*, 1981, Bd.**37**, N.5, S.401–405.

153. Mason A.W. Nonstandard, normal subgroups and nonnormal, standard subgroups of the modular group. *Canad.Math.Bull.*, 1989, V.**32**, N.1, P.109–113.

154. Mason A.W., Pride S.J. Normal subgroups of prescribed order and zero le-

vel of the modular group and related groups. *J.London Math. Soc.*, 1990, V.**42**, P.465–474.

155. Mason A.W., Stothers W.W. On subgroups of GL(n,A) which are generated by commutators. *Invent.Math.*, 1974, V.**23**, P.327–346.

156. Matsumoto H. Subgroups of finite index in certain arithmetic groups. – *Proc.Symp.Pure Math.*, 1966, V.**9**, P.99–103.

157. Matsumoto H. Sur les sous-groupes arithmétiques des groupes semi-simples deployés. *Ann.Sci.Ecole Norm.Sup.*, $4^{\text{éme}}$ sér., 1969, T.**2**, P.1–62.

158. McCrimmon K. The Freudenthal-Springer-Tits construction of exceptional Jordan algebras. *Trans.Amer.Math.Soc.*, 1969, V.**139**, P.495–510.

159. McDonald B.R. Automorphisms of $GL_n(R)$. *Trans.Amer.Math.Soc.*, 1978, V.**246**, P.155–171.

160. McDonald B.R. *Geometric algebra over local rings*. Marcel Dekker: N.Y. 1976, 421P.

161. McDonald B.R. *Linear algebra over commutative rings*. Marcel Dekker: N.Y. 1984, 544P.

162. Meyberg K. Eine Theorie der Freudenthalschen Trippelsysteme. I,II. *Proc. Nederl.Akad.Wetensch.*, Ser.A, 1968, V.**30**, P.162–174; 175–190.

163. Milnor J. *Introduction to algebraic K-theory*. Princeton Univ. Press: Princeton, N.J., 1971, 184P.

164. Mizuno K. The conjugate classes of Chevalley groups of type E_6. *J.Fac. Sci.Univ.Tokyo*, 1977, V.**24**, N.3, P.525–563.

165. Mizuno K. The conjugate classes of unipotent elements of the Chevalley groups E_7 and E_8. *Tokyo J.Math.*, 1980, V.**3**, N.2, P.391–458.

166. Morita J. Coverings of generalized Chevalley groups associated with affine Lie algebras. *Tsukuba J.Math.*, 1982, V.**6**, N.1, P.1–8.

167. O'Meara O.T. *Lectures on linear groups*. Conf.Board of the Math.Sci. Reg. Conf.Ser.in Math., 1974, N.22, 87P.

168. O'Meara O.T. *Lectures on symplectic groups*. Amer.Math.Soc.Surveys, 1978, N.16, 122P.

169. Orlik P., Solomon R. Singularities II; automorphisms of forms. *Math. Ann.*, 1978, Bd.**231**, S.229–240.

170. Plotkin E.B. On the net subgroups of twisted Chevalley groups. *Latvian Math.J.*, 1984, V.**28**, P.179–193 (in Russian).

171. Plotkin E.B. *On the stabilization of K_1-functor for Chevalley groups.*

Manuscript deposited to VINITI 03.12.1984, N.7648-84, 34P. (in Russian).

172. Plotkin E.B. Chevalley-Matsumoto type decomposition for twisted Chevalley groups. In: *Topological spaces and their maps*. Riga, 1985, P.108-119 (in Russian).

173. Plotkin E.B. *Net subgroups of Chevalley groups and stability problems for K_1-functor*. Ph.D. Thesis. Leningrad State Univ., 1985, 118P. (in Russian).

174. Plotkin E.B. Surjective stability of K_1-functor for Chevalley groups of normal and twisted types. *Russian Math.Surveys*, 1989, V.**44**, N.2, P.239-240.

175. Plotkin E.B. Surjective stability of K_1-functor for some exceptional Chevalley groups. Submitted to *Zap.nauch.seminarov LOMI.*

176. Proctor R.A. Classical Bruhat orders and lexicographic shellability. *J.Algebra*, 1982, V.**77**, N.1, P.104-126.

177. Proctor R.A. Bruhat lattices, plane partition generating functions, and minuscule representations. *Europ.J.Combinatorics*, 1984, V.**5**, P.331-350.

178. Proctor R.A. A Dynkin diagram classification theorem arising from a combinatorial problem. *Adv.Math.*, 1986, V.**62**, N.2, P.103-117.

179. Quillen D. Projective modules over polynomial rings. *Invent.Math.*, 1976, V.**36**, N.1, P.167-173.

180. Ree R. On some simple groups defined by C.Chevalley *Trans.Amer. Math. Soc.*, 1957, V.**84**, N.2, P.392-400.

181. Ree R. A family of simple groups associated with the simple Lie algebras of type (F_4). *Amer.J.Math.*, 1961, V.**83**, P.401-420.

182. Ree R. A family of simple groups associated with the simple Lie algebras of type (G_2). *Amer.J.Math.*, 1961, V.**83**, P.432-462.

183. Ree R. Construction of certain semi-simple groups. *Canad.J.Math.*, 1964, V.**16**, P.490-508.

184. Ronan M.A., Smith S.D. Sheaves on buildings and modular representations of Chevalley groups. *J.Algebra*, 1985, V.**96**, N.2, P.319-346.

185. Satake I. *Classification theory of semi-simple algebraic groups.* Marcel Dekker: N.Y., 1971, 149P.

186. Schupp P.E. Embeddings into simple groups. *J.London Math.Soc.*, 1976, V.**13**, P.90-94.

187. Seitz G.M. The maximal subgroups of classical algebraic groups. *Mem.*

Amer. Math.Soc., 1987, V.**67**, N.365, 286P.

188. Seitz G.M. Maximal subgroups of exceptional exceptional groups. *Mem. Amer. Math.Soc.*, 1991, V.**90**, N.441, 197P.

189. Seligman G.B. On automorphisms of Lie algebras of classical type. III - *Trans.Amer.Math.Soc.*

190. Seligman G. *Modular Lie algebras*. Springer: Berlin et al., 1967, 165P.

191. Seshadri C.S. Geometry of G/P. I. Standard monomial theory for minuscule P - In: *C.P.Ramanujan: a tribute*. Tata Press: Bombay, 1978, P.207–239.

192. Shoji T. The conjugacy classes of Chevalley groups of type (F_4) over finite fields of characteristic $p \neq 2$. *J.Fac.Sci.Univ.Tokyo*, 1974, V.**21**,N.1, P.1–17.

193. Splitthoff S. Finite presentability of Steinberg groups and related Chevalley groups. *Contemp.Math.*, 1986, V.**55**, Part.II, P.635–687.

194. Springer T.A. The projective octave plane. I,II. *Proc.Nederl. Akad. Wetensch.*, ser.A, 1960, V.**63**, P.74–101.

195. Springer T.A. Characterization of a class of cubic forms. *Proc. Nederl. Akad.Wetensch.*, ser.A, 1962, V.**65**, P.259–265.

196. Springer T.A. On the geometric algebra of the octave plane. *Proc. Nederl. Akad.Wetensch.*, ser.A, 1962, V.**65**, P.451–468.

197. Springer T.A. *Jordan algebras and algebraic groups*. Springer: N.Y. Heidelberg, 1973, 169P.

198. Springer T.A. *Linear algebraic groups*. 2nd ed. Birkhäuser: Boston et al., 1981, 320P.

199. Springer T.A. Linear algebraic groups. *Fundam.trends in Math.*, 1989, V.**55**, P.5–136. (in Russian, English translation to appear in Springer).

200. Springer T.A., Veldkamp V.D. On Hjelmslev-Moufang planes. *Math.Z.*, 1968, Bd.**107**, S.249–263.

201. Stanley R.P. Weyl groups, the hard Lefschetz theorem, and the Sperner property. *SIAM J.Alg.Discr.Methods*, 1980, V.**1**, P.168–184.

202. Stein M.R. Relativizing functors on rings and algebraic K-theory. *J.Algebra*, 1971, V.**19**, N.1, P.140–152.

203. Stein M.R. Generators, relations and coverings of Chevalley groups over commutative rings. *Amer.J.Math.*, 1971, V.**93**, N.4, P.965–1004.

204. Stein M.R. Surjective stability in dimension 0 for K_2 and related functors - *Trans.Amer.Math.Soc.*, 1973, V.**178**, N.1, P.165–191.

205. Stein M.R. Matsumoto's solution of the congruence subgroup problem and stability theorems in algebraic K-theory. *Proc. 19th Meeting Algebra Section Math.Soc.Japan*, 1983, P.32–44.

206. Stein M.R. Stability theorems for K_1, K_2 and related functors modeled on Chevalley groupes. *Japan J.Math.*, 1978, V.**4**, N.1, P.77–108.

207. Steinberg R. Variations on a theme of Chevalley. *Pacific J.Math.*, 1959, V.**9**, N.5, P.875–891.

208. Steinberg R. Générateurs, relations et revêtements des groupes algébriques. *Colloque sur la théorie des groupes algébriques (Bruxelles, 1962)*. Paris: Gauthier-Villars, 1962, P.113–127.

209. Steinberg R. *Lectures on Chevalley groups*. Yale University, 1967.

210. Steinberg R. Some consequences of the elementary relations in SL_n. *Contemp.Math.*, 1985, V.**45**, P.335–350.

211. Stensholt E. An application of the Steinberg construction of twisted groups. *Pacific J.Math.*, 1974, V.**55**, P.595–618.

212. Stensholt E. Certain embeddings among finite groups of Lie type. *J.Algebra*. 1978. V.**53**, N.1. P.136–187.

213. Stepanov A.V. *Stability conditions in the theory of linear groups over rings*. Ph.D. Thesis. Leningrad State Univ., 1987, 112P.

214. Stepanov A.V. A ring of finite stable rank is not necessarily finite in the sense of Dedekind. *Soviet Math.Doklady*, 1988, V.**36**, N.2, P.301–304.

215. Strecker G. Unitäre Gruppen über beliebigen lokalen Ringen. *J.Algebra*, 1979, V.**57**, N.2, P.258–270.

216. Suslin A.A. Projective modules over polynomial rings are free. – *Soviet Math.Doklady*, 1976, V.**229**, N.5, P.1063–1066.

217. Suslin A.A. On a theorem of Cohn. *J.Sov.Math.*, 1981, V.**17**, N.2.

218. Suslin A.A. On the structure of the general linear group over polynomial rings. *Soviet Math.*Izv., 1977, V.**41**, N.2, P.503–516.

219. Suslin A.A. Algebraic K-theory. *J.Sov. Math.*, 1985, V.**28**, N.6, P.870–923.

220. Suslin A.A., Kopeiko V.I. Quadratic modules and orthogonal groups over polynomial rings. *J.Sov.Math.*, 1982, V.**20**, N.6, P.2665–2691.

221. Suslin A.A., Tulenbaev M.S. A theorem on stabilization for Milnor's K_2-functor. *J.Sov.Math.*, 1981, V.**17**, P.1804–1819.

222. Suzuki K. On normal subgroups of twisted Chevalley groups over local

rings. *Sci.Repts.Tokyo Kyoiku Daigaku*, 1977, V.**13**, P.237–249.

223. Suzuki K. On the automorphoisms of Chevalley groups over $\mathfrak{p}$-adic integer rings. *Kumamoto J.Sci. (Math.)*, 1984, V.**16**, N.1, P.39–47.

224. Swan R. Generators and relations for certain special linear groups. *Adv.Math.*, 1971, V.**6**, P.1–77.

225. Taddei G. Invariance du sous-groupe symplectique élémentaire dans le groupe symplectique sur un anneau. *C.R.Acad.Sci.Paris*, sér.I, 1982, T.**295**, N.2, P.47–50.

226. Taddei G. *Schémas de Chevalley-Demazure, fonctions représentatives et théorème de normalité*. Thèse, Univ.de Genève, 1985, 58P.

227. Taddei G. Normalité des groupes élémentaire dans les groupes de Chevalley sur un anneau. *Contemp.Math.*, 1986, V.**55**, Part II, P.693–710.

228. Tavgen' O.I. Bounded generation of Chevalley groups over rings of algebraic numbers. *Izv.Akad Nauk U.S.S.R.*, 1990, V.**54**, N.1, P.97–122. (in Russian, English translation to appear in *Soviet Math. Izv.*).

229. Testerman D.M. Irreducible subgroups of exceptional algebraic groups. – *Mem.Amer.Math.Soc.*, 1989, V.**75**, N.390, 190P.

230. Testerman D.M. A construction of certain maximal subgroups of the algebraic groups E_6 and F_4. *Comm.Algebra*, 1989, V.**17**, N.4, P.1003–1016.

231. Tits J. Le plan projective des octaves et les groupes de Lie exceptionnels. *Acad.Roy.Belg.Bull.Cl.Sci.*, 1953, V.**39**, P.309–329.

232. Tits J. Le plan projective des octaves et les groupes exceptionnels E_6 et E_7. *Acad.Roy.Belg.Bull.Cl.Sci.*, 1954, V.**40**, P.29–40.

233. Tits J. Sur les constantes de structure et le théorème d'existence des algèbres de Lie semi-simples. *Publ.Math.Inst.Hautes Et.Sci.*, 1966, N.**31**, P.21–58.

234. Tits J. Normalisateurs de tores. I. Groupes de Coxeter étendus. *J.Algebra*, 1966, V.**4**, N.1, P.96–116.

235. Tits J. Classification of algebraic semi-simple groups. *Proc.Symp. Pure Math.*, 1966, V.**9**, P.33–62.

236. Tits J. Buildings of spherical type and finite BN-pairs. *Lecture Notes Math.*, 1974, V.**288**, 299P.

237. Tits J. Systèmes générateurs de groupes de congruence. *C.R.Acad. Sci.Paris*, sér.A-B, 1976, V.**283**, N.9, P.693–695.

238. Tits J. A local approach to buildings. In: *The Geom.vein.* Springer: N.Y.

et al., 1981, P.519–547.

239. Tulenbaev M.S. The Schur multiplier of the group of elementary matrices of finite order. *J.Sov.Math.*, 1981, V.**17**, N.4.

240. Vaserstein L.N. K_1-theory and the congruence subgroup problem. *Math.Notes*, 1969, V.**5**, P.233–244.

241. Vaserstein L.N. On the stabilization of the general linear group over a ring. *Math. U.S.S.R. Sbornik*, 1969, V.**8**, P.383–400.

242. Vaserstein L.N. Stabilization of unitary and orthogonal groups over a ring. - *Math. U.S.S.R. Sbornik*, 1970, V.**10**, P.307–326.

243. Vaserstein L.N. The stable rank of rings and the dimension of topological spaces. *Funct.Anal.Applic.*, 1971, V.**5**, N.2, P.17–27.

244. Vaserstein L.N. On normal subgroups of GL_n over a ring. *Lecture Notes Math.*, 1981, V.**854**, P.456–465.

245. Vaserstein L.N. Classical groups over rings. *Canad.Math.Soc.Conf. Proc.*, 1984, V.**4**, P.131–140.

246. Vaserstein L.N. The subnormal structure of general linear groups. *Math. Proc.Cambridge Philos.Soc.*, 1986, V.**99**, P.425–431.

247. Vaserstein L. On normal subgroups of Chevalley groups over commutative rings. *Tôhoku Math.J.*, 1986, V.**38**, P.219–230.

248. Vaserstein L. Normal subgroups of orthogonal groups over commutative rings. - *Amer.J.Math.*, 1988, V.**110**, N.5, P.955–973.

249. Vaserstein L.N. Normal subgroups of symplectic groups over rings. *K-theory*, 1989, V.**2**, N.5, P.647–673.

250. Vaserstein L.N. The subnormal structure of general linear groups over rings. *Math.Proc.Cambridge Philos.Soc.*, 1990, V.**108**, N.2, P.219–229.

251. Vaserstein L.N., Suslin A.A. Serre's problem on projective modules over polynomial rings and algebraic K-theory. *Soviet Math.Izv.*, 1976, V.**40**, N.5, P.993–1054.

252. Vavilov N.A. Parabolic subgroups of Chevalley groups over a commutative ring. *J.Sov.Math.*, 1984, V.**26**, N.3, P.1848–1860.

253. Vavilov N.A. On solution of linear homogeneous systems of equations. In: *Algebra and discrete Math.*, Latvian Univ.Press, 1986, P.37–42 (in Russian).

254. Vavilov N.A. *Subgroups of split classical groups*. Dr.of Sci. thesis (Habilitazionschrift), Leningrad State Univ., 1987, 334P (in Russian).

255. Vavilov N.A. Weight elements of Chevalley groups. *Soviet Math.Doklady*, 1988, vol.37, N.1, P.92–95.

256. Vavilov N.A. Structure of split classical groups over commutative rings. *Soviet Math.Doklady*, 1988, V.**37**, P.550–553.

257. Vavilov N.A. Bruhat decomposition of the semisimple long root elements in Chevalley groups. In: *Rings and modules. Limit theorems of probability theory, II*, Leningrad Univ.Press, 1988, P.18–39. (in Russian, English transl. to appear in AMS transl.series).

258. Vavilov N.A. On the geometry of long root subgroups in Chevalley groups. *Vestnik Leningr.Univ., Math.*, 1988, N.1, P.8–11.

259. Vavilov N.A. On the problem of normality of the elementary subgroup in a Chevalley group. *Algebraic and discrete systems.* Ivanovo Univ., 1988, P.7–25. (in Russian).

260. Vavilov N.A. On the interrelation of a long and a short root subgroups in a Chevalley group. *Vestnik Leningr.Univ., Math.*, 1989, N.1, P.3–7.

261. Vavilov N.A. A note on the subnormal structure of general linear groups. – *Math.Proc.Cambridge Phil.Soc.*, 1990, V.**107**, N.2, P.193–196.

262. Vavilov N.A. On subgroups of the split classical groups. *Trudy Mat. Inst. Steklov*, 1990, V.**183**, P.29–41 (in Russian, English translation to appear in Proc.Math. Inst.Steklov).

263. Vavilov N.A. Subgroups of Chevalley groups containing a maximal torus. *Trudy Leningrad Math.Soc.*, 1990, V.**1**, P.64–109 (in Russian, English translation to appear in Proc.Leningrad Math. Soc.).

264. Vavilov N.A., Plotkin E.B. Net subgroups of Chevalley groups. I,II. *J.Sov. Math.*, 1982, V.**19**, N.1, P.1000–1006; 1984, V.**27**, N.4, P.2874–2885.

265. Vavilov N.A., Plotkin E.B., Stepanov A.V. Calculations in Chevalley groups over commutative rings. *Soviet Math.Doklady*, 1990, V.**40**, N.1, P.145–147.

266. Veldkamp V.D. Collineation groups in Hjelmslev-Moufang planes. *Math.Z.*, 1968, Bd.**108**, N.1, S.37–52.

267. Verma D.-N. Möbius inversion for the Bruhat ordering on a Weyl group. *Ann. Sci.Ecole Norm.Sup.*, 4éme sér., 1971, T.**4**, P.393–398.

268. Vorst T. The general linear group of polynomial rings over regular rings. *Comm.Algebra*, 1981, V.**9**, N.5, P.499–509.

269. Wagner A. On the classification of the classical groups. *Math.Z.*, 1967, Bd.**97**, N.1, S.66–76.

270. Waterhouse W.C. *Introduction to affine group schemes.* Springer: Berlin et al., 1979, 164P.

271. Wilson J.S. The normal and subnormal structure of general linear groups. *Proc.Cambridge Phil.Soc.*, 1972, V.**71**, P.163–177.

272. Zalesskii A.E. Semisimple root elements of algebraic groups. *Preprint Inst. Math.Bielorussian Acad.Sci.*, Minsk, 1980, N.13, 23P (in Russian).

273. Zalesskii A.E. Linear groups. *J.Sov.Math.*, 1985, V.**31**, N.3.

274. Zalesskii A.E. Linear groups. *Fundam.trends in Math.*, 1989, V.**50**, P.5–120. (in Russian, English translation to appear in Springer).

275. Zarhin Yu.G. Weights of semi-simple Lie algebras in the cohomology of algebraic varieties. *Izv.Akad.Nauk U.S.S.R.*, 1984, V.**48**, N.2, P.264–304.

Construction of Lie Algebras from Associative Triple Systems and Generalized J-ternary Algebras

Kiyosi YAMAGUTI

Department of Mathematics, Faculty of School Education
Hiroshima University, Shinonome 3-chome, Minami-ku
Hiroshima 734, Japan

Abstract

Let A, A' be the commutative associative triple systems. Let a quadruple $(J, U(\varepsilon), \sigma, f)$ be a generalized J-ternary algebra which is a generalization of J-ternary algebra due to B. N. Allison and W. Hein. It is shown that using tensor products, from A, A' and a generalized J-ternary algebra, a general Lie triple system T is constructed under certain binary product and ternary product, from which a Lie algebra is obtained as the standard imbedding of T.

1. Introduction

Among the constructions of Lie algebras from non-associative algebraic systems, it seems that there are two remarkable constructions, that is, the constructions via Lie triple systems and the constructions via general Lie triple systems. The constructions of Lie algebras by H. Freudenthal [2] and I. L. Kantor [4] and the others are the former case and the constructions by B. N. Allison and W. Hein and the others are the latter case.

B. N. Allison [1] and W. Hein [3] defined a J-ternary algebra and constructed a Lie algebra from it. Let a quadruple $(J, U(\varepsilon), \sigma, f)$ be a generalized J-ternary algebra, where J is a Jordan triple system, $U(\varepsilon), \varepsilon = 1$ or -1, is a Freudenthal-Kantor triple system, σ is a special representation of the Jordan triple system J into a vector space $U(\varepsilon)$, and f is an alternative bilinear mapping of $U(\varepsilon)$ into J, satisfying some conditions. Let A, A' be the commutative associative triple systems due to O. Loos [5]. Then, a general Lie triple system is constructed from A, A' and the generalized J-ternary algebra, which is an algebraic system with a binary product and a ternary product. This is a generalization of a result in [11]. It is assumed that any vector space considered here is a finite dimensional vector space over a field of characteristic different from 2.

2. A construction of general Lie triple systems

DEFINITION(O. Loos [5]). A triple system A with a trilinear product $\langle abc\rangle := l(a,b)c := m(a,c)b$ is called an associative triple system if

$$\langle ab\langle cde\rangle\rangle = \langle a\langle dcb\rangle e\rangle = \langle\langle abc\rangle de\rangle,$$

and an associative triple system is said to be commutative if

$$m(a,b) = m(b,a),$$

$a, b, c, d \in A$.

DEFINITION [15, 14]. An endomorphism m^* of a commutative associative triple system A is said to satisfy the condition (K) if

$$\begin{aligned} m^* l(a,b) = l(b,a)m^* &= m(m^*a, b),\\ m^*m(a,b) &= l(m^*a, b),\\ m(a,b)m^* &= l(a, m^*b). \end{aligned}$$

An endomorphism $m(a,b)$ satisfies the condition (K) and also we have the relations $m(a,b)m(c,d)m(e,f) = m(e,f)m(c,d)m(a,b)$, $m(m(a,b)c,d) = m(l(d,c)a, b) = m(a, l(d,c)b)$, $a, b, c, d, e, f \in A$.

Assume that the endomorphisms m^*, n^* of A satisfy the condition (K), then the following hold:

$$\begin{aligned} m^*n^*m(a,b) &= m(m^*n^*a, b) = m(a, m^*n^*b),\\ m(a,b)n^*m^* &= m(m^*n^*a, b) = m(m^*n^*b, a),\\ m^*m(a,b)n^* &= m(n^*a, m^*b) = m(n^*b, m^*a). \end{aligned}$$

DEFINITION. For $\varepsilon = 1$ or -1, a triple system $U(\varepsilon)$ with a trilinear product $\langle abc\rangle := L(a,b)c := M(a,c)b$ is called a Freudenthal-Kantor triple system if

$$\text{(U1)} \qquad [L(a,b), L(c,d)] = L(L(a,b)c,d) + \varepsilon L(c, L(b,a)d)$$

and

$$\text{(U2)} \qquad K(K(a,b)c,d) = L(d,c)K(a,b) - \varepsilon K(a,b)L(c,d),$$

$a,b,c,d \in U(\varepsilon)$, where $K(a,b)$ is a linear operator defined as $K(a,b) := M(a,b) - M(b,a)$.

If $\varepsilon = -1$, then $U(-1)$ is a generalized Jordan triple system of second order due to I. L. Kantor [4], and a Jordan triple system is a generalized Jordan triple system satisfying the condition $K(a,b) = 0$. For $\varepsilon = 1$ the algebraic system $U(1)$ is defined in [10, p.4].

The following is a generalization of J-ternary algebras due to B. N. Allison [1] and W. Hein [3].

DEFINITION. For $\varepsilon = 1$ or -1, a quadruple $(J, U(\varepsilon), \sigma, f)$ is called a generalized J-ternary algebra if

J is a Jordan triple system with a triple product $\langle xyz\rangle := L(x,y)z$,

$U(\varepsilon)$ is a Freudenthal-Kantor triple system with a product $\langle abc\rangle := L(a,b)c$,

σ is a special representation of the Jordan triple system J into the vector space $U(\varepsilon)$, that is, σ is a linear mapping of J into Hom$U(\varepsilon)$ satisfying

$$\sigma(\langle xyz\rangle) = \sigma(x)\sigma(y)\sigma(z) + \sigma(z)\sigma(y)\sigma(x),$$

f is an alternative bilinear mapping of $U(\varepsilon)$ into J, and the following conditions are satisfied:

$$(1) \qquad \varepsilon\langle x\ f(a,b)\ y\rangle = f(\sigma(x)a, \sigma(y)b) - f(\sigma(x)b, \sigma(y)a),$$

$$(2) \qquad \langle x\ y\ f(a,b)\rangle = f(\sigma(x)\sigma(y)a, b) + f(a, \sigma(x)\sigma(y)b),$$

$$(3) \qquad \sigma f(a,b) = K(a,b),$$

$$(4) \qquad K(\sigma(x)a, b) - L(b,a)\sigma(x) + \varepsilon\sigma(x)L(a,b) = 0,$$

$$(5) \qquad K(a,b)\sigma(x) + L(a, \sigma(x)b) - L(b, \sigma(x)a) = 0,$$

$$(6) \qquad \varepsilon\sigma(x)K(a,b) + L(\sigma(x)a, b) - L(\sigma(x)b, a) = 0,$$

$$(7) \qquad f(L(a,b)c,d) + f(c, L(a,b)d) = f(K(c,d)b, a),$$

$x, y \in J$, $a, b, c, d \in U(\varepsilon)$.

Let $U(\varepsilon)$ be a Freudenthal-Kantor triple system, then the linear span J of $K(a,b)$'s is a Jordan triple system with respect to a product $\langle K(a,b)\ K(c,d)\ K(e,f)\rangle := K(a,b)K(c,d)K(e,f)+K(e,f)K(c,d)K(a,b)$ and the quadruple $(J, U(\varepsilon), Id, K)$ is a generalized J-ternary algebra.

DEFINITION [7]. A general Lie triple system is an algebraic system T with a bilinear product $xy := L(x)y$ and a trilinear product $[xyz] := L(x,y)z$ satisfying

$$xx = 0, \tag{1}$$

$$[xxy] = 0, \tag{2}$$

$$[xyz] + [yzx] + [zxy] + (xy)z + (yz)x + (zx)y = 0, \tag{3}$$

$$L(xy, z) + L(yz, x) + L(zx, y) = 0, \tag{4}$$

$$[L(x,y), L(z)] = L(L(x,y), z), \tag{5}$$

$$[L(x,y), L(u,v)] = L(L(x,y)u, v) + L(u, L(x,y)v), \tag{6}$$

$x, y, z, u, v \in T$.

A linear mapping D of T is called a derivation of T if $[D, L(x)] = L(Dx)$ and $[D, L(x,y)] = L(Dx,y) + L(x,Dy)$. The conditions (5) and (6) show that any left multiplication $L(x,y)$ is a derivation of T which is said to be inner.

Let $\mathcal{D}$ and $\mathcal{D}(T)$ be the Lie algebras generated by derivations and inner derivations respectively, then the direct sums $\mathcal{D} \oplus T$ and $\mathcal{L} = \mathcal{D}(T) \oplus T$ as the vector space become a Lie algebra, where the multiplication $[\ ,\ ]$ for $\mathcal{L}$ is defined as $[L(x,y), L(u,v)]$ by (6), $[L(x,y), z] = -[z, L(x,y)] = L(x,y)z$, $[x,y] = L(x,y) + xy$ [6; 7]. The Lie algebra $\mathcal{L}$ is called a standard imbedding of general Lie triple system T.

In a general Lie triple system T, if $\mathcal{D}(T) = (0)$, then the axioms stated above become the axioms of Lie algebra and if a mapping $L(x)$ is trivial for all x, then the axioms (1) through (6) become the axioms of Lie triple system. If M is a Mal'cev algebra with product xy, then M becomes a general Lie triple system with respect to the product xy and a trilinear product $[xyz] := x(yz) - y(xz) + (xy)z$ [8, 9, also cf. 12, 13].

Let A, A' be the commutative associative triple systems. Let m_i^*, n_i^* and $m_i^{*\prime}, n_i^{*\prime}$ $(i = 1,2,3)$ be the linear mappings of A and A' respectively, satisfying the condition (K) and assume that $m_1^* m_2^* m_3^* = m_3^* m_2^* m_1^*$ and $m_1^{*\prime} m_2^{*\prime} m_3^{*\prime} = m_3^{*\prime} m_2^{*\prime} m_1^{*\prime}$. Denote by M^*, $M^{*\prime}$ the sets of these mappings.

Consider the vector space $T := M^* \otimes J \otimes M^{*\prime} \oplus \overline{M^* \otimes J \otimes M^{*\prime}} \oplus A \otimes U(\varepsilon) \otimes A' \oplus \overline{A \otimes U(\varepsilon) \otimes A'}$, where $\overline{X}$ means the copy of X. An element of T is denoted as

$$(m^* \otimes x \otimes m^{*\prime}, n^* \otimes y \otimes n^{*\prime}, p \otimes a \otimes p', q \otimes b \otimes q')$$

or

$$\begin{pmatrix} m^* \otimes x \otimes m^{*\prime} \\ n^* \otimes y \otimes n^{*\prime} \\ p \otimes a \otimes p' \\ q \otimes b \otimes q' \end{pmatrix}$$

in a column vector form.

Define a bilinear product in T as

$$\begin{pmatrix} m_1^* \otimes x_1 \otimes m_1^{*\prime} \\ n_1^* \otimes y_1 \otimes n_1^{*\prime} \\ p_1 \otimes a_1 \otimes p_1' \\ q_1 \otimes b_1 \otimes q_1' \end{pmatrix} \begin{pmatrix} m_2^* \otimes x_2 \otimes m_2^{*\prime} \\ n_2^* \otimes y_2 \otimes n_2^{*\prime} \\ p_2 \otimes a_2 \otimes p_2' \\ q_2 \otimes b_2 \otimes q_2' \end{pmatrix}$$
$$= \begin{pmatrix} -\varepsilon m(q_1, q_2) \otimes f(b_1, b_2) \otimes m(q_1', q_2') \\ m(p_1, p_2) \otimes f(a_1, a_2) \otimes m(p_1', p_2') \\ n_1^* q_2 \otimes \sigma(y_1) b_2 \otimes n_1^{*\prime} q_2' - n_2^* q_1 \otimes \sigma(y_2) b_1 \otimes n_2^{*\prime} q_1' \\ m_1^* p_2 \otimes \sigma(x_1) a_2 \otimes m_1^{*\prime} p_2' - m_2^* p_1 \otimes \sigma(x_2) a_1 \otimes m_2^{*\prime} p_1' \end{pmatrix}.$$

A trilinear product $[\ ,\ ,\ ]$ in T is defined as follows.

If $X = (m_1^* \otimes x_1 \otimes m_1^{*\prime}, 0, 0, 0)$, $Y = (0, n_2^* \otimes y_2 \otimes n_2^{*\prime}, 0, 0)$, then for $Z = (m^* \otimes x \otimes m^{*\prime}, n^* \otimes y \otimes n^{*\prime}, p \otimes a \otimes p', q \otimes b \otimes q')$

$$[X\ Y\ Z] = \begin{pmatrix} m_1^* n_2^* m^* \otimes L(x_1, y_2) x \otimes m_1^{*\prime} n_2^{*\prime} m^{*\prime} \\ -n_2^* m_1^* n^* \otimes L(y_2, x_1) y \otimes n_2^{*\prime} m_1^{*\prime} n^{*\prime} \\ -n_2^* m_1^* p \otimes \sigma(y_2) \sigma(x_1) a \otimes n_2^{*\prime} m_1^{*\prime} p' \\ m_1^* n_2^* q \otimes \sigma(x_1) \sigma(y_2) b \otimes m_1^{*\prime} n_2^{*\prime} q' \end{pmatrix}.$$

If $X = (0, 0, p_1 \otimes a_1 \otimes p_1', 0)$, $Y = (0, 0, 0, q_2 \otimes b_2 \otimes q_2')$, then for $Z = (m^* \otimes x \otimes m^{*\prime}, n^* \otimes y \otimes n^{*\prime}, p \otimes a \otimes p', q \otimes b \otimes q')$

$$[X\ Y\ Z] = \begin{pmatrix} \varepsilon m(m^* p_1, q_2) \otimes f(\sigma(x) a_1, b_2) \otimes m(m^{*\prime} p_1', q_2') \\ m(n^* q_2, p_1) \otimes f(\sigma(y) b_2, a_1) \otimes m(n^{*\prime} q_2', p_1') \\ l(p_1, q_2) p \otimes L(a_1, b_2) a \otimes l(p_1', q_2') p' \\ \varepsilon l(q_2, p_1) q \otimes L(b_2, a_1) b \otimes l(q_2', p_1') q' \end{pmatrix},$$

and $L(X,Y) = -L(Y,X)$ and the other triple products are trivial. A general left multiplication is, by definition, a linear extension of above multiplications.

THEOREM. Let A, A' be the commutative associative triple systems and a quadruple $(J, U(\varepsilon), \sigma, f)$ be a generalized J-ternary algebra. Then, the direct sum $M^* \otimes J \otimes M^{*\prime} \oplus \overline{M^* \otimes J \otimes M^{*\prime}} \oplus A \otimes U(\varepsilon) \otimes A' \oplus \overline{A \otimes U(\varepsilon) \otimes A'}$ becomes a general Lie triple system with respect to the bilinear product and trilinear product defined above.

ACKNOWLEDGEMENT. This paper is supported in part by the Grant in Aid for Fundamental Scientific Research of the Ministry of Education, Science and Culture (C) 62540050.

References

[1] B.N. Allison, A construction of Lie algebras from J-ternary algebras, Amer. J. Math. **98**, 285-294 (1976).

[2] H. Freudenthal, Beziehungen der E_7 und E_8 zur Oktavenebene, I, II, Nederl. Akad. Wetensch. Proc. Ser. A, **57**, 218-230; 363-368 (1954).

[3] W. Hein, A construction of Lie algebras by triple systems, Trans. Amer. Math. Soc. **205**, 79-95 (1975).

[4] I.L. Kantor, Models of exceptional Lie algebras, Soviet Math. Dokl. **14**, 254-258 (1973).

[5] O. Loos: Jordan pairs, Lecture Notes in Math. **460**, Springer-Verlag, 1975.

[6] K. Nomizu, Invariant affine connections on homogeneous spaces, Amer. J. Math. **76**, 33-65 (1954).

[7] K. Yamaguti, On the Lie triple system and its generalization, J. Sci. Hiroshima Univ. Ser. A **21**, 155-160 (1958).

[8] K. Yamaguti, Note on Malcev algebras, Kumamoto J. Sci. Ser. A **5**, 203-207 (1962).

[9] K. Yamaguti, On the theory of Malcev algebras, Kumamoto J. Sci. Ser. A **6**, 9-45 (1963).

[10] K. Yamaguti, Remarks on characterizations of the points and symplecta in the metasymplectic geometry, Mem. Fac. Gen. Ed. Kumamoto Univ. Ser. Nat. Sci. No. 11, 1-8 (1976).

[11] K. Yamaguti, An algebraic system arising from metasymplectic geometry. III, Bull. Fac. Sch. Ed., Hiroshima Univ. II, **9**, 65-73 (1986).

[12] K. Yamaguti, Constructions of Lie (super)algebras from Freudenthal-Kantor triple system $U(\varepsilon, \delta)$,. Proceedings of the 14th ICGTMP (Ed. by Y.M. Cho), 222-225, World Scientific Publ. Co., Singapore, 1986.

[13] K. Yamaguti, Constructions of Lie (super)algebras from triple systems, Lecture Notes in Physics **313** (Ed. by H.-D. Doebner, J.-D. Hennig, T.D. Palev), 190-197, Springer-Verlag, 1988.

[14] K. Yamaguti, A construction of Lie (super)algebras from associative triple systems and Freudenthal-Kantor (super)triple systems, Gauss Symposium held at Guarujá, SP, Brazil, July 24-27 (1989).

[15] K. Yamaguti and H. Tanabe, A construction of generalized J-(super)ternary pairs from generalized J-(super)ternary pairs, Bull. Fac. Sch. Ed. Hiroshima Univ. Part II, **10**, 43-61 (1987).

SYMPLECTIC TRIPLE SYSTEMS AND GRADED LIE ALGEBRAS

OSAMI YASUKURA

Department of Mathematics, Yokohama City University
Seto 22-2, Kanazawa-ku, Yokohama 236, Japan

ABSTRACT

Yamaguti-Asano's correspondence between complex simple symplectic triple systems and complex simple graded Lie algebras of contact type, is reviewed.

§1. Symplectic Triple Systems.

In [14], Professor K. Yamaguti and Professor H. Asano constructed a complex simple Lie algebra from a complex simple symplectic triple system, which is defined also in [14]. In [1] and [2], Professor H. Asano showed that all complex simple Lie algebras of rank non less than two are constructed in this manner. In fact, the correspondence is one to one up to isomorphisms.

1.1. DEFINITION. Let $\mathcal{T}$ be a vector space over complex numbers $\mathbf{C}$, with a trilinear product [xyz] $\in \mathcal{T}$ for $(x, y, z) \in \mathcal{T} \times \mathcal{T} \times \mathcal{T}$, and a skew-symmetric bilinear form $< x, y > \in \mathbf{C}$ for $(x, y) \in \mathcal{T} \times \mathcal{T}$. A triple system $\mathcal{T} = (\mathcal{T}, [\quad], <,>)$ is called a *symplectic triple system* if the following four conditions are satisfied:

$$\text{(ST.0)} \qquad < [uvx], y > + < x, [uvy] >= 0,$$
$$\text{(ST.1)} \qquad [xyz] - [yxz] = 0,$$
$$\text{(ST.2)} \qquad [xyz] - [xzy] =< x, z > y- < x, y > z + 2 < y, z > x,$$
$$\text{(ST.3)} \qquad [uv[xyz]] = [[uvx]yz] + [x[uvy]z] + [xy[uvz]].$$

The First condition (ST.0) is induced from (ST.2) and (ST.3) (cf. [14], [1], [2]). In general, $\mathcal{T}$ is called a *semi-symplectic triple system*, if the above three conditions (ST.0), (ST.1), and (ST.2) are satisfied.

1.2. PROPOSITION (YAMAGUTI-ASANO[14]). *A semi-symplectic triple system* $(\mathcal{T}, [\quad], <,>)$ *is simple, i.e. having non-trivial tri-ideal, if and only if* $<,>$ *is a symplectic form, i.e. being non-degenerate.*

PROOF: From the conditions (ST.0) and (ST.2), the result is obtained.

1.3. EXAMPLE. Let $C_{n+1} = \mathbf{C}^{2n}$ be the complex 2n-space. For $x = (x_1, ..., x_{2n})$ and $y = (y_1, ..., y_{2n}) \in \mathbf{C}^{2n}$, denote $< x, y >= \sum_{i=1}^{n}(x_i y_{i+n} - x_{i+n} y_i)$, being a symplectic form on $\mathcal{T}$. Then define a tri-linear product on C_{n+1} by the following formula:

$$\text{(CT.1)} \qquad [xyz] =< x, z > y+ < y, z > x,$$

which canonically induces the conditions (ST.0), (ST.1), (ST.2), (ST.3), and

$$\text{(CT.0)} \qquad < [xxx], x >= 0.$$

Then C_{n+1} is called the *simple symplectic triple system of type* C_{n+1}. If $(\mathcal{T}', [\ \ ]', <,>')$ is another simple symplectic triple system of $\dim_{\mathbf{C}} \mathcal{T}' = 2n$ satisfying the condition (CT.1), then there exists a linear isomorphism $f\colon C_{n+1} \longrightarrow \mathcal{T}'$ such that

$$\text{(MT.0)} \qquad < x, y >=< f(x), f(y) >',$$

$$\text{(MT.1)} \qquad f([xyz]) = [f(x)f(y)f(z)]',$$

since all symplectic forms are equivalent up to base-changes.

1.4. DEFINITION. Let $\mathcal{T}$ and $\mathcal{T}'$ be semi-symplectic triple systems. A linear mapping $f\colon \mathcal{T} \longrightarrow \mathcal{T}'$ is called a *homomorphism*, if f satisfies the condition (MT.0) and (MT.1). The condition (MT.1) implies (MT.0) by putting $z = x$ in (ST.2). A homomorphism $f\colon \mathcal{T} \longrightarrow \mathcal{T}'$ is called an *isomorphism* if f is bijective. If there exists an isomorphism $f\colon \mathcal{T} \longrightarrow \mathcal{T}'$, then $\mathcal{T}$ and $\mathcal{T}'$ are called *isomorphic.*

1.5. REMARK. The condition (ST.2) means that the commutator of $[xyz]$ with respect to y, z is same as that of the corresponding *canonical symplectic triple system* defined by (CT.1) on the same underlying vector space with the same skew-symmetric bilinear form.

1.6. THEOREM (ASANO[3]). *Let $(\mathcal{T}, [\ \ ], <,>)$ be a simple semi-symplectic triple system satisfying the condition (CT.0). Then $\mathcal{T}$ satisfies the condition (CT.1), so that $\mathcal{T}$ is isomorphic to the simple symplectic triple system of type C_{n+1} for $n = \dim_{\mathbf{C}} \mathcal{T}/2$.*

PROOF: Linearizing the condition (CT.0), one has that (1) $< y, [xxx] > + 2 < x, [yxx] > + < x, [xxy] > = 0$. Because of (ST.1) and (ST.2), one has that (2) $[yxx] = [xyx] = [xxy] - 3 < x, y > x$. So that (3) $< x, [yxx] > = < x, [xxy] > = < y, [xxx] >$, where the last equation follows from (ST.0). Back to (1) from (3), $< y, [xxx] > = 0$. Because of 1.2. Theorem, one has that (4) $[xxx] = 0$. Linearizing (4), (5) $2[yxx] + [xxy] = 0$. Back to (2) from (5), $3[xxy] - 6 < x, y > x = 0$, so that (6) $[xxy] = 2 < x, y > x$. Linearizing (6), one has the condition (CT.1).

§2. Graded Lie Algebras.

Let $\mathcal{G}$ be a complex simple Lie algebra with a Cartan subalgebra $\mathcal{H}$, the Killing form B, and the root system $\mathcal{R}$ in the dual space $\mathcal{H}^*$ of $\mathcal{H}$. Fix a lexicographic ordering $>$ in $\mathcal{H}^*$ and let $\mathcal{R}_+$ (or $\mathcal{R}_-$) be the set of positive (resp. negative) roots. Denote $\gamma \in \mathcal{R}_+$ be the highest root. Let $\alpha \in \mathcal{H}^*$. Denote the α -eigen space by $\mathcal{G}_\alpha = \{x \in \mathcal{G}; [H, x] = \alpha(H)x \text{ for all } H \in \mathcal{H}\}$. Define $H_\alpha \in \mathcal{H}$ by $B(H_\alpha, H) = \alpha(H)$ for all $H \in \mathcal{H}$. Denote the *Dynkin's form* by $D(x, y) = B(H_\gamma, H_\gamma) \cdot B(x, y)/2$ and $K_\alpha = 2H_\alpha / B(H_\alpha, H_\alpha)$. Let $\mathcal{G}'$ be another complex simple Lie algebra with the Dynkin's form D'. If $f : \mathcal{G} \longrightarrow \mathcal{G}'$ is a non zero homomorphism, then there is unique constant j_f such that $D'(f(x), f(y)) = j_f \cdot D(x, y)$ for all $x, y \in \mathcal{G}$. This j_f is an integer (cf. Dynkin[6], Atiyah-Hitchin-Singer[4], Yasukura[15]), so-called the *index* of f.

2.1. PROPOSITION (WOLF[12], TASAKI[11]). *For $\beta \in \mathcal{R}$, the following three conditions are all equivalent: (1) β is a long root.*
(2) There is a lexicographic ordering such that β is the highest root.
(3) The eigenvalue of $ad(H_\beta)$ is 0, $\pm B(H_\beta, H_\beta)/2$, or $\pm B(H_\beta, H_\beta)$ with multiplicity 1 for $\pm B(H_\beta, H_\beta)$.

PROOF: For $\alpha, \beta \in \mathcal{R}$, the possibilities of $B(H_\alpha, H_\alpha)/ B(H_\beta, H_\beta)$ are 0, 1, 2, 3, 1/2, 1/3. Then $\#\{B(H_\alpha, H_\alpha); \alpha \in \mathcal{R}\} \leq 2$, i.e. the length of roots are at most two kinds: *long* or *short*. Assume the condition (1). Then $B(H_\beta, H_\beta) \geq B(H_\alpha, H_\alpha)$ for all $\alpha \in \mathcal{R}$. So that $B(H_\alpha, H_\beta)^2 \leq B(H_\beta, H_\beta)^2$, where the equality follows when $H_\alpha = \pm H_\beta$, i.e. $\alpha = \pm\beta$. Let $H_\beta, H_2, ..., H_r$ be a basis of $\mathcal{H}$ and $>$ be the lexicographic ordering with respect to this ordering basis. Since $\beta(H_\beta) \geq \alpha(H_\beta)$ where the equality follows when $\beta = \alpha$, then β is the highest root with respect to this ordering. Assume the condition (2). Then $\alpha + \beta$, $\alpha - 2\beta \in \mathcal{H}^* \backslash \mathcal{R}$ for all $\alpha \in \mathcal{R}_+ \backslash \{\beta\}$. So the eigenvalue of $ad(2H_\beta / B(H_\beta, H_\beta))$ on $\mathcal{G}_\alpha$ is $2B(H_\beta, H_\alpha)/B(H_\beta, H_\beta) = 0$, or ± 1 (for $\alpha \in \mathcal{R} \backslash \{\pm\beta\}$), ± 2 (for $\alpha = \pm\beta$). Assume the condition (3). If there exists $\alpha \in \mathcal{R}$ such that $B(H_\alpha, H_\alpha) > B(H_\beta, H_\beta)$, then $\alpha \neq \beta$ and $| 2B(H_\beta, H_\alpha)/B(H_\alpha, H_\alpha) | < 1$, so $B(H_\beta, H_\alpha) = 0$. But there is a root λ in the Weyl chamber containing β such that $B(H_\alpha, H_\alpha) = B(H_\lambda, H_\lambda)$. Then $0 = B(H_\beta, H_\lambda) > 0$ (cf. Goto-Grosshans [7; p.349, (7.5.3)]). It is a contradiction. So that $B(H_\beta, H_\beta) \geq B(H_\alpha, H_\alpha)$ for all $\alpha \in \mathcal{R}$.

2.2. DEFINITION. Let $\mathcal{G}_i$ be the $i-$eigen space of $ad(K_\gamma)$. Then one has that

$$\text{(CLA.0)} \qquad \mathcal{G} = \sum_{i \in \mathbf{Z}} \mathcal{G}_i \ , [\mathcal{G}_i, \mathcal{G}_j] \subset \mathcal{G}_{i+j} \ ,$$

$$\text{(CLA.1)} \qquad \mathcal{G}_i = \{0\} \text{ for } | i | \geq 3 \ , \dim_{\mathbf{C}} \mathcal{G}_{\pm 2} = 1 \ , \mathcal{G}_{\pm 1} \neq \{0\}.$$

In general, a *contact Lie algebra* is by definition a complex simple Lie algebra $\mathcal{G}$ with a **Z**-graded structure of contact type: (CLA.0) and (CLA.1). For two contact Lie algebras $\mathcal{G}$ and $\mathcal{G}'$, a homomorphism $f : \mathcal{G} \longrightarrow \mathcal{G}'$ is called *graded*, if $f(\mathcal{G}_i) \subset \mathcal{G}'_i$ for all $i \in \{\pm 2, \pm 1, 0\}$.

2.3. PROPOSITION. *Let $\mathcal{G}$ be a contact Lie algebra. Then one has the following eight results:*
(1) There is the unique element $K \in \mathcal{G}_0$ such that $[K, x] = ix$ for all $x \in \mathcal{G}_i$ and $i \in \{0, \pm 1, \pm 2\}$.
(2) Let $\mathcal{H}$ be a maximal abelian subalgebra of $\mathcal{G}_0$ containing K, then $\mathcal{H}$ is a Cartan subalgebra of $\mathcal{G}$ and $\mathcal{G}_0$.
(3) Let $H_1, ..., H_r$ be a basis of $\mathcal{H}$ such that $H_1 = K$. Consider the lexicographic ordering on $\mathcal{H}^*$ with respect to this ordered basis. Let γ be the highest root, and $\mathcal{R}_+$ (resp. $\mathcal{R}_-$) be the set of positive (resp. negative) roots. Put $\mathcal{R}_i = \{\alpha \in \mathcal{R}; \mathcal{G}_\alpha \subset \mathcal{G}_i\}$. Then $\mathcal{R}_1 \subset \mathcal{R}_+ \backslash \{\gamma\}, \mathcal{R}_{-1} \subset \mathcal{R}_- \backslash \{-\gamma\}, \mathcal{R}_{\pm 2} = \{\pm\gamma\}, \mathcal{G}_{\pm 1} = \sum_{\alpha \in \mathcal{R}_{\pm 1}} \mathcal{G}_\alpha$, and $\mathcal{G}_{\pm 2} = \mathcal{G}_{\pm\gamma}$.
(4) If $i, j \in \{\pm 1, \pm 2\}$, then $[\mathcal{G}_i, \mathcal{G}_j] = \mathcal{G}_{i+j}$.
(5) For $\alpha \in \mathcal{R}_1$ (or $\mathcal{R}_{-1}$) and $x \in \mathcal{G}_\alpha$, if $[x, \mathcal{G}_1] = \{0\}$ (resp. $[x, \mathcal{G}_{-1}] = \{0\}$), then $x = 0$.
(6) $K = K_\gamma = 2H_\gamma / B(H_\gamma, H_\gamma)$.
(7) Let $\mathcal{G} = \sum_i \mathcal{G}_i$ and $\mathcal{G}' = \sum_i \mathcal{G}'_i$ be contact Lie algebras. A graded homomorphism f from $\mathcal{G}$ into $\mathcal{G}'$ is of index 1. Conversely, let f be a homomorphism of index 1, then there is an inner automorphism g on $\mathcal{G}'$ such that gf from $\mathcal{G}$ into $\mathcal{G}'$ is graded.
(8) Two contact Lie algebras are graded isomorphic when they are isomorphic.

PROOF: (cf. Kobayashi-Nagano[8], Tanaka[10], Cheng[5]) (1) Let f be a linear transformation on $\mathcal{G}$ such that $f(x) = ix$ if $x \in \mathcal{G}_i$ for $i = 0, \pm 1, \pm 2$. Then f is a Lie algebraic automorphism of $\mathcal{G}$. Since $\mathcal{G}$ is simple, there is unique $K \in \mathcal{G}$ such that $f = ad(K)$. Represent $K = \sum_{i=-2}^{2} y_i$; $y_i \in \mathcal{G}_i$. If $x \in \mathcal{G}_j$, then $jx = [K, x] = \sum [y_i, x] \in \mathcal{G}_j$, so that $[y_0, x] = [K, x]$. Hence, $K = y_0 \in \mathcal{G}_0$. (2) Since $\mathcal{G}$ is simple, a maximal abelian subalgebra is a Cartan subalgebra. (3) For $\alpha, \beta \in \mathcal{H}^*, \alpha > \beta$ if and only if there is $i \in \{1,...,r\}$ such that $\alpha(H_1) = \beta(H_1), ..., \alpha(H_{i-1}) = \beta(H_{i-1}), \alpha(H_i) > \beta(H_i)$. (4) If $i', j' \in \{\pm 1, \pm 2\}$, and $i' + j' = i + j$, then $\{i, j\} = \{i', j'\}$. So that $\sum_{k \neq i+j} \mathcal{G}_k + [\mathcal{G}_i, \mathcal{G}_j]$ is an ideal of $\mathcal{G}$. (5) Let $\alpha \in \mathcal{R}_1$ and $0 \neq x \in \mathcal{G}_\alpha$. If $[x, \mathcal{G}_1] = 0$, then $B(x, \mathcal{G}_{-1}) = B(x, [\mathcal{G}_1, \mathcal{G}_{-2}]) = B([x, \mathcal{G}_1], \mathcal{G}_{-2}) = 0$. Since $B(\mathcal{G}_\alpha, \mathcal{G}_{-\alpha}) \neq 0$, it is a contradiction. So that $[x, \mathcal{G}_1] \neq 0$. Similar for $\mathcal{R}_{-1}$. (6) Because of (3), $ad(K) = ad(K_\gamma)$ on $\mathcal{G}_{\pm 2}$. If $\alpha \in \mathcal{R}_+ \backslash \{\gamma\}$, then $\alpha(K_\gamma) \in \{0,1\}$. For each $\alpha \in \mathcal{R}_1$, put $0 \neq x_\alpha \in \mathcal{G}_\alpha$ and $y_\alpha = \sum_{\beta \neq \alpha} y_\beta \in \mathcal{G}_1$ such that $0 \neq [x_\alpha, y_\alpha] \in \mathcal{G}_2$. If $\alpha(K_\gamma) = 0$, then $[x_\alpha, [K_\gamma, y_\alpha]] = [K_\gamma, [x_\alpha, y_\alpha]] = 2[x_\alpha, y_\alpha]$. Since non-zero $[x_\alpha, y_\beta]'s$ for $\beta's$ are linearly independent, and since $[K_\gamma, y_\beta] = 0$ or y_β, it is a contradiction. So that $\alpha(K_\gamma) \neq 0$. Hence $\alpha(K_\gamma) = 1$, and $ad(K_\gamma) = ad(K)$ on $\mathcal{G}_1$. Similarly on $\mathcal{G}_{-1}$. So does on $\mathcal{G}_0 = [\mathcal{G}_1, \mathcal{G}_{-1}]$. Hence $K_\gamma = K$. (7) Let f be a graded non-zero homomorphism. Let $E_\pm \in \mathcal{G}_{\pm\gamma}$ be $[E_+, E_-] = K_\gamma$. Since $f(E_\pm) \in \mathcal{G}'_{\pm\gamma'}$, there is a constant c such that $f(K_\gamma) = cK'_{\gamma'}$. Since $ad(f(K_\gamma)) = ad(K'_{\gamma'})$ on $\mathcal{G}'_{\pm 2}$, one has that $f(K_\gamma) = K'_{\gamma'}$. Since $D(K_\gamma, K_\gamma) = D'(K'_{\gamma'}, K'_{\gamma'}) = 2$, the index of f is 1. Conversely, let f be of index 1. Then, there is a long root $\beta' \in \mathcal{R}'$ such that $f(K_\gamma) = K'_{\beta'}$ by 2.1.Prop. and Dynkin [6; Thm. 2.4.], which is the highest root with respect to a suitable ordering. Let g be an inner automorphism of $\mathcal{G}'$ such that $g(K'_{\beta'})$ and $K'_{\gamma'}$ are in the same Weyl chamber. Then $D'(g(K'_{\beta'}), K'_{\gamma'}) = 1$ or 2 by 2.1.Prop.(3). If $D'(g(K'_{\beta'}), K'_{\gamma'}) = 1$, then the cosine of the angle between β' and γ' is 1/4, which is impossible, so $D'(g(K'_{\beta'}), K'_{\gamma'}) = 2$, and $g(K'_{\beta'}) = K'_{\gamma'}$ by 2.1.Prop.(3). Then gf is a graded homomorphism. (8) follows from (7), since each isomorphism is of index 1.

2.4. DEFINITION. Let $\mathcal{C}$ be the category of all contact Lie algebras of rank ≥ 2 up to isomorphisms as objects, with graded homomorphisms up to equivalences by isomorphisms as morphisms. Let $\mathcal{B}$ be the category of all simple symplectic triple systems up to isomorphisms as objects, with homomorphisms up to equivalences by isomorphisms as morphisms.

2.5. DEFINITION (YAMAGUTI-ASANO[14]). For a symplectic triple system $\mathcal{T}$, define $\mathfrak{a}(\mathcal{T}) = \sum_{i=-2}^{2} \mathcal{G}_i; \mathcal{G}_{-1} = \{0\} \times \mathcal{T}, \mathcal{G}_1 = \mathcal{T} \times \{0\}, \mathcal{G}_{-2} = \{cW; c \in \mathbf{C}\}, \mathcal{G}_2 = \{cV; c \in \mathbf{C}\}, \mathcal{G}_0 = \{cK; c \in \mathbf{C}\} \oplus \{L(x,y); x, y \in \mathcal{T}\}$, where $L(x,y), K, V$, and W are linear transformations on $\mathcal{T}^2 = \mathcal{G}_1 \oplus \mathcal{G}_{-1}$ such that $L(x,y)(v,w) = ([xyv], [xyw])$, $K(v,w) = (v, -w)$, $V(v,w) = (w, 0)$, $W(v,w) = (0, v)$ for $(v,w) \in \mathcal{T}^2$, and a bilinear product $[\ ,\ ]$ on $\mathfrak{a}(\mathcal{T})$ is given as follows:

$$\begin{aligned}
&[K,V] = 2V, [K,W] = -2W, [V,W] = K,\\
&[L(x,y), K] = [L(x,y), V] = [L(x,y), W] = 0,\\
&[L(x,y), L(v,w)] = L([xyv], w) + L(v, [xyw]),\\
&[K,(v,w)] = (v,-w), [V,(v,w)] = (w,0), [W,(v,w)] = (0,v),\\
&[L(x,y),(v,w)] = ([xyv],[xyw]),\\
&[(x,y),(v,w)] = L(x,w) - L(v,y) + (< v,y > - < x,w >)K\\
&\qquad\qquad + 2 < x,v > V - 2 < y,w > W.
\end{aligned}$$

2.6. THEOREM (YAMAGUTI-ASANO[14]). *The algebra $(\mathfrak{a}(\mathcal{T}), [\ ,\ ])$ is a well-defined complex Lie algebra, which is simple if and only if $\mathcal{T}$ is simple.*

2.7. REMARK. Since $\dim_{\mathbf{C}} \mathfrak{a}(\mathcal{T}) > 3$, rank $\mathfrak{a}(\mathcal{T}) \geq 2$. If $f\colon \mathcal{T} \longrightarrow \mathcal{T}'$ is a non-zero homomorphism, then $\mathfrak{a}(\mathrm{f})\colon \mathfrak{a}(\mathcal{T}) \longrightarrow \mathfrak{a}(\mathcal{T}')$; $\mathfrak{a}(f)(cW + dV + eK + (x,y) + L(v,w)) = cW' + dV' + eK' + (f(x), f(y)) + L'(f(v), f(w))$, is a well-defined homomorphism, because $\mathfrak{a}(f(\mathcal{T}))$ is simple (cf. Satake [9; p.39, Prop. 9.1.]). In fact, if $\sum L(v_i, w_i) = 0$, then $\sum L(f(v_i), f(w_i)) = 0$ in $\mathfrak{a}(f(\mathcal{T}))$. Hence, it defines a *functor* $\mathfrak{a}\colon \mathcal{B} \longrightarrow \mathcal{C}$.

2.8. THEOREM (ASANO[1], [2]). *There is a functor $\mathfrak{s}\colon \mathcal{C} \longrightarrow \mathcal{B}$ such that the composition $\mathfrak{a} \circ \mathfrak{s}$ equals to the identity functor on the category $\mathcal{C}$.*

PROOF: Let $\mathcal{G} = \sum \mathcal{G}_i$ be a contact Lie algebra. Let $E_\pm \in \mathcal{G}_{\pm 2} = \mathcal{G}_{\pm\gamma}$ be such that $B(E_+, E_-) = 2/B(H_\gamma, H_\gamma)$. Then $K_\gamma = B(E_+, E_-)H_\gamma = [E_+, E_-]$ and $[K_\gamma, E_\pm] = \pm 2E_\pm$. On $\mathcal{G}_1$, define a tri-linear product $[\quad]$ and a skew-symmetric bilinear form $<,>$ as

$$\begin{aligned}
&[xyz] = [[x,[E_-,y]] + [y,[E_-,x]], z]/2,\\
&2 < x,y > E_+ = [x,y].
\end{aligned}$$

Then, $\mathfrak{s}(\mathcal{G}) := (\mathcal{G}_1, [\ \ ], <,>)$ is a simple symplectic triple system (cf. [1], [2]). If one choices another $E'_\pm$, then there is a constant $k \in \mathbf{C}$ such that $E'_+ = k^2 E_+$ and $E'_- = E_-/(k^2)$, and that the resulting symplectic triple system $\mathfrak{s}'(\mathcal{G}) = (\mathcal{G}_{-1}, [\ \ ]', <,>')$ is isomorphic to $\mathfrak{s}(\mathcal{G})$ by $F(x) = kx \in \mathfrak{s}'(\mathcal{G}) = \mathcal{G}_{-1}$ for $x \in \mathfrak{s}(\mathcal{G}) = \mathcal{G}_{-1}$. If f: $\mathcal{G} \longrightarrow \mathcal{G}'$ is a graded homomorphism, then there is a constant $k \in \mathbf{C}$ such that $f(E_+) = E'_+/(k^2)$. Since the index of f is 1 (cf. 2.3.Prop.(7)), and $D(E_+, E_-) = D'(E'_+, E'_-) = 1$, one has that $f(E_-) = k^2 E'_-$. Then $\mathfrak{s}(f) = k(f \mid \mathcal{G}_1)$: $\mathcal{G}_1 \longrightarrow \mathcal{G}'_1$ is a homomorphism. Put κ: $\mathfrak{a}(\mathfrak{s}(\mathcal{G})) = \mathfrak{a}(\mathcal{G}_1) \longrightarrow \mathcal{G}$; $\kappa(cV + dW + eK + (x,y) + L(v,w)) = cE_+ + dE_- + eK_\gamma + x + [E_-, y] + ([[v, E_-], w] + [[w, E_-], v])/2$, which is a well-defined isomorphism (cf. [1], [2], [9]). Similarly put an isomorphism κ':$\mathfrak{a}(\mathfrak{s}(\mathcal{G}')) = \mathfrak{a}(\mathcal{G}'_1) \longrightarrow \mathcal{G}'$. If $f : \mathcal{G} \longrightarrow \mathcal{G}'$ is a graded homomorphism with $f(E_+) = E'/(k^2_+)$, then $f \circ \kappa = \mu \circ \kappa' \circ \mathfrak{a}(\mathfrak{s}(f))$, where μ is an automorphism of $\mathcal{G}'$ such that $\mu(x) = k^{-i}x$ for $x \in \mathcal{G}'_i (i = 0, \pm 1, \pm 2)$. So $\mathfrak{a}(\mathfrak{s}(f))$: $\mathfrak{a}(\mathcal{G}_1) \longrightarrow \mathfrak{a}(\mathcal{G}'_1)$ is equivalent to f.

2.9. THEOREM. *The composition $\mathfrak{s} \circ \mathfrak{a}$: $\mathcal{B} \longrightarrow \mathcal{B}$, is the identity functor on the category $\mathcal{B}$.*

PROOF: Let $\mathcal{T}$ be a simple symplectic triple system. Then $\mathfrak{a}(\mathcal{T}) = \sum_{i=-2}^{2} \mathcal{G}_i$ is a contact Lie algebra, and $K = K_\gamma$ by 2.3.Proposition and 2.5.Definition. Putting $E_+ = V$ and $E_- = W$ in 2.6.Theorem, the symplectic triple system $\mathfrak{s}(\mathfrak{a}(\mathcal{T})) = \mathcal{T} \times \{0\}$ has the product $[(x,0)(y,0)(z,0)] = [[(x,0), [W,(y,0)]] + [(y,0), [W, (x,0)]], (z,0)]/2 = ([xyz], 0)$ and the alternating form $< (x,0), (y,0) > = < x, y >$ by 2.5.Definition. So that $\mathfrak{s}(\mathfrak{a}(\mathcal{T}))$ is isomorphic to $\mathcal{T}$. If f from $\mathcal{T}$ into $\mathcal{T}'$ is a homomorphism, then $\mathfrak{a}(f)(V) = V'$ and $\mathfrak{s}(\mathfrak{a}(f)) = \mathfrak{a}(f) \mid \mathcal{T} \times \{0\} = f \times id$, which is equivalent to f.

REFERENCES

[1] H. Asano, On triple systems (in Japanese), *Yokohama City Univ. Ronso* **Ser. Nat. Sci. 27** (1975), 7-31.

[2] H. Asano, Symplectic triple systems and simple Lie algebras (in Japanese), *Kyoto Univ. RIMS Kokyuroku* **308** (1977), 41-54.

[3] H. Asano, Private communication at Yokohama City University, May 8, 1989.

[4] M.F. Atiyah, N. Hitchin and I. Singer, Self-duality in four-dimensional Riemannian geometry, *Proc. R. Soc. Lond.* **A. 362** (1978), 425-461.

[5] J.- H. Cheng, *Graded Lie algebras of the second kind*, Ph.D. Thesis, University of Notre Dame, 1983.

[6] E.B. Dynkin, Semisimple subalgebras of semisimple Lie algebras, *American Math. Soc. Translations* **6** (1957), 111-244.

[7] M. Goto and F.D. Grosshans, *Semisimple Lie algebras*, Dekker, 1978.

[8] S. Kobayashi and T. Nagano, On filtered Lie algebras and geometric structures I, *J. Math. Mech.* **13** (1964), 875-908.

[9] I. Satake, *Algebraic structures of symmetric domains*, Iwanami Shoten and Princeton Univ. Press, 1980.

[10] N. Tanaka, On non-degenerate real hypersurfaces, graded Lie algebras and Cartan connections, *Japan J. Math.* **2** (1976), 131-190.

[11] H. Tasaki, Quaternionic submanifolds in quaternionic symmetric spaces, *Tôhoku Math.J.* **38** (1986), 513-538.

[12] J.A. Wolf, Complex homogeneous contact manifolds and quaternionic symmetric spaces, *J. Math. Mech.* **14** (1965), 1033-1047.

[13] K. Yamaguti, Metasymplectic geometry and triple systems (in Japanese), *Kyoto Univ. RIMS Kokyuroku* **308** (1977), 55-92.

[14] K. Yamaguti and H. Asano, On the Freudenthal's construction of exceptional Lie algebras, *Proc. Japan Academy* **51** (1975), 253-258.

[15] O. Yasukura, On subalgebras of type A_1 in simple Lie algebras, *Algebras, Groups and Geometries* **5** (1988), 359-368.

Orbit types of the compact Lie group E_6 in the complex exceptional Jordan algebra $\mathfrak{J}^C$

Ichiro YOKOTA

Department of Mathematics, Shinshu University
1-1, Asahi, Matsumoto, Nagano, Japan

Abstract

Let $\mathfrak{J}$ be the exceptional Jordan algebra, then the simply connected compact exceptional Lie group F_4 acts on $\mathfrak{J}$ and F_4 has three orbit types which are one point, $F_4/Spin(9)$ and $F_4/Spin(8)$ ([2], p.313). In this paper we determine the orbit types of the simply connected compact Lie group E_6 in the complex exceptional Jordan algebra $\mathfrak{J}^C$. As result, E_6 has five orbit types which are one point, E_6/F_4, $E_6/Spin(10)$, $E_6/Spin(9)$ and $E_6/Spin(8)$.

1. Orbits of the group F_4 in the algebra $\mathfrak{J}$

Let $\mathfrak{C}$ be the division Cayley algebra and $\mathfrak{J} = \{X \in M(3, \mathfrak{C}) \mid X^* = X\}$ be the exceptional Jordan algebra with the multiplication $X \circ Y = \frac{1}{2}(XY + YX)$, the inner product $(X, Y) = \mathrm{tr}(X \circ Y)$ and the Freudenthal multiplication $X \times Y = \frac{1}{2}(2X \circ Y - \mathrm{tr}(X)Y - \mathrm{tr}(Y)X + (\mathrm{tr}(X)\mathrm{tr}(Y) - (X, Y))E)$ (where E is the 3×3 unit matrix). Now

$$\begin{aligned} F_4 &= \{\alpha \in \mathrm{Iso}_{\boldsymbol{R}}(\mathfrak{J}) \mid \alpha(X \circ Y) = \alpha X \circ \alpha Y\} \\ &= \{\alpha \in \mathrm{Iso}_{\boldsymbol{R}}(\mathfrak{J}) \mid \alpha(X \times Y) = \alpha X \times \alpha Y\} \end{aligned}$$

(where $\boldsymbol{R}$ is the field of real numbers) is the simply connected compact Lie group of type F_4 ([1], [5]).

LEMMA 1.1 ([1]). *Any element $X \in \mathfrak{J}$ can be transformed to a diagonal form by a certain $\alpha \in F_4$:*

$$\alpha X = \begin{pmatrix} \xi_1 & 0 & 0 \\ 0 & \xi_2 & 0 \\ 0 & 0 & \xi_3 \end{pmatrix} \quad \text{(which is briefly written by } (\xi_1, \xi_2, \xi_3)).$$

Where $\xi_1, \xi_2, \xi_3 \in \boldsymbol{R}$ *are uniquely determined by* X *up to permutations (independent of the choice of* $\alpha \in F_4$*).*

As a direct result of Lemma 1.1, we have

THEOREM 1.2. *There exists a one-to-one correspondence between the set of all orbits of the group* F_4 *in* $\mathfrak{J}$ *and the set of all triples* $(\xi_1, \xi_2, \xi_3) \in \boldsymbol{R}^3$ *(up to permutations).*

To prove the following Theorem 1.4, we use

LEMMA 1.3. (1)([5]) *For* $\alpha \in F_4$, *we have* $\alpha E = E$ *and* $\mathrm{tr}(\alpha X) = \mathrm{tr}(X)$ *for* $X \in \mathfrak{J}$.

(2)([3]) *The group* F_4 *has subgroups* $Spin(9)$, $Spin(8)$ *as*

$$\begin{aligned} Spin(9) &= \{\alpha \in F_4 \mid \alpha E_1 = E_1\}, \\ Spin(8) &= \{\alpha \in F_4 \mid \alpha E_i = E_i, \ i = 1,2,3\}, \end{aligned}$$

where $E_1 = (1,0,0)$, $E_2 = (0,1,0)$, $E_3 = (0,0,1)$.

THEOREM 1.4. *The orbit types of the group* F_4 *in* $\mathfrak{J}$ *are as follows.*

(1) *The orbit through* (ξ_1, ξ_2, ξ_3) *(where* ξ_1, ξ_2, ξ_3 *are distinct) is diffeomorphic to* $F_4/Spin(8) \simeq \{X \in \mathfrak{J} \mid 2X \times (X \times X) = X, \ \mathrm{tr}(X) = 0, \ X \neq 0\}$ $(=_{put} \mathfrak{C}Q)$.

(2) *The orbit through* (ξ_1, ξ, ξ) *(where* $\xi_1 \neq \xi$ *) is diffeomorphic to* $F_4/Spin(9) \simeq \{X \in \mathfrak{J} \mid X^2 = X, \ \mathrm{tr}(X) = 1\}$ $(=_{put} \mathfrak{C}P_2)$.

(3) *The orbit through* (ξ, ξ, ξ) *is one point.*

PROOF. (1) Obviously the group F_4 acts transitively on the orbit $F_4(\xi_1, \xi_2, \xi_3)$. We shall determine the isotropy subgroup of F_4 at (ξ_1, ξ_2, ξ_3). If $\alpha \in F_4$ satisfies $\alpha(\xi_1, \xi_2, \xi_3) = (\xi_1, \xi_2, \xi_3)$, then $\alpha(\eta_1, \eta_2, 0) = (\eta_1, \eta_2, 0)$ (where $\eta_1 = \xi_1 - \xi_3 \neq 0, \eta_2 = \xi_2 - \xi_3 \neq 0$) because $\alpha E = E$ (Lemma 1.3). Since $\alpha((\eta_1 E_1 + \eta_2 E_2)^{\times 2}) = (\eta_1 E_1 + \eta_2 E_2)^{\times 2}$, that is, $\alpha(\eta_1 \eta_2 E_3) = \eta_1 \eta_2 E_3$, we have $\alpha E_3 = E_3$. Similarly $\alpha E_1 = E_1$, $\alpha E_2 = E_2$. Hence $\alpha \in Spin(8)$ (Lemma 1.3). Thus we have $F_4/Spin(8) \simeq F_4(\xi_1, \xi_2, \xi_3)$. Next we shall show $F_4/Spin(8) \simeq \mathfrak{C}Q$. The group F_4 acts on $\mathfrak{C}Q$ (Lemma 1.3). This action is transitive. In fact, for a given $X \in \mathfrak{C}Q$, choose $\alpha \in F_4$ such that $\alpha X = (\lambda_1, \lambda_2, \lambda_3)$ (Lemma 1.1). From the condition $2\alpha X \times (\alpha X \times \alpha X) = \alpha(2X \times (X \times X)) = \alpha X$, $\mathrm{tr}(\alpha X) = \mathrm{tr}(X)$ (Lemma 1.3) $= 0$, we have

$\lambda_i({\lambda_{i+1}}^2 + {\lambda_{i+2}}^2) = \lambda_i$, $i = 1,2,3$, and $\lambda_1 + \lambda_2 + \lambda_3 = 0$, hence ${\lambda_i}^2 = {\lambda_{i+1}}^2 = 1$, $\lambda_{i+2} = 0$, that is, $\alpha X = \pm(E_i - E_{i+1})$, $i = 1,2,3$, and these are transformed to $E_1 - E_2 \in \mathfrak{C}Q$ by some $\beta \in F_4$, respectively. Thus the action is transitive. The isotropy subgroup of F_4 at $E_1 - E_2$ is $Spin(8)$ as is shown in the above. Thus we have $F_4/Spin(8) \simeq \mathfrak{C}Q$.

(2) Obviously the group F_4 acts transitively on the orbit $F_4(\xi_1, \xi, \xi)$. We shall determine the isotropy subgroup of F_4 at (ξ_1, ξ, ξ). If $\alpha \in F_4$ satisfies $\alpha(\xi_1, \xi, \xi) = (\xi_1, \xi, \xi)$, then $\alpha(\xi_1 - \xi, 0, 0) = (\xi_1 - \xi, 0, 0)$ because $\alpha E = E$ (Lemma 1.3). Therefore $\alpha E_1 = E_1$, that is, $\alpha \in Spin(9)$. (Lemma 1.3). Hence we have $F_4/Spin(9) \simeq F_4(\xi_1, \xi, \xi)$. Next we shall show $F_4/Spin(9) \simeq \mathfrak{C}P_2$. The group F_4 acts on $\mathfrak{C}P_2$ (Lemma 1.3). This action is transitive. In fact, for a given $X \in \mathfrak{C}P_2$, choose $\alpha \in F_4$ such that $\alpha X = (\lambda_1, \lambda_2, \lambda_3)$ (Lemma 1.1). From the condition $(\alpha X)^2 = \alpha(X^2) = \alpha X$, $\operatorname{tr}(\alpha X) = \operatorname{tr}(X)$ (Lemma 1.3) $= 1$, we have ${\lambda_i}^2 = \lambda_i, i = 1,2,3$ and $\lambda_1 + \lambda_2 + \lambda_3 = 1$, hence $\lambda_i = 1$, $\lambda_{i+1} = \lambda_{i+2} = 0$, that is, $\alpha X = E_i, i = 1,2,3$, and these are transformed to $E_1 \in \mathfrak{C}P_2$ by some $\beta \in F_4$, respectively. Thus the action is transitive. The isotropy subgroup of F_4 at E_1 is $Spin(9)$ (Lemma 1.3). Thus we have $F_4/Spin(9) \simeq \mathfrak{C}P_2$ (which is the Cayley projective plane ([1])). (In (2), (3), the indexes are considered as mod 3).

(3) It is trivial because $\alpha E = E$ for $\alpha \in F_4$ (Lemma 1.3).

2. Orbits of the group E_6 in the algebra $\mathfrak{J}^C$

Let $C = \boldsymbol{R}^C$ and $\mathfrak{J}^C$ be the complexifications of $\boldsymbol{R}$ and $\mathfrak{J}$, respectively. The complex conjugations in C and $\mathfrak{J}^C$ are denoted by τ. (For $\xi \in C$, we denote $\sqrt{(\tau\xi)\xi}$ by $|\xi|$). In $\mathfrak{J}^C$, the determinant $\det X = \frac{1}{3}(X, X \times X)$ and the Hermitian inner product $\langle X, Y\rangle = (\tau X, Y)$ are defined. Now

$$\begin{aligned} E_6 &= \{\alpha \in \mathrm{Iso}_C(\mathfrak{J}^C) \mid \det \alpha X = \det X,\ \langle \alpha X, \alpha Y\rangle = \langle X, Y\rangle\} \\ &= \{\alpha \in \mathrm{Iso}_C(\mathfrak{J}^C) \mid \tau\alpha\tau(X \times Y) = \alpha X \times \alpha Y,\ \langle \alpha X, \alpha Y\rangle = \langle X, Y\rangle\} \end{aligned}$$

is the simply connected compact Lie group of type E_6 ([4]).

LEMMA 2.1. *Any element $X \in \mathfrak{J}^C$ can be transformed to a diagonal form by* *a certain* $\alpha \in E_6$:

$$\alpha X = \begin{pmatrix} \xi_1 & 0 & 0 \\ 0 & \xi_2 & 0 \\ 0 & 0 & \xi_3 \end{pmatrix}, \quad \xi_1, \xi_2, \xi_3 \in C.$$

Moreover we can choose $\alpha \in E_6$ *so that two of* ξ_1, ξ_2, ξ_3 *are non-negative real numbers.*

Proof is in [4]. The last statement of Lemma 2.1 follows from

LEMMA 2.2 ([4]). *For* $\lambda \in C$, $|\lambda| = 1$, *the mapping* $\alpha_{12}(\lambda) : \mathfrak{J}^C \to \mathfrak{J}^C$ *defined by*

$$\alpha_{12}(\lambda) \begin{pmatrix} \xi_1 & x_3 & \overline{x}_2 \\ \overline{x}_3 & \xi_2 & x_1 \\ x_2 & \overline{x}_1 & \xi_3 \end{pmatrix} = \begin{pmatrix} \lambda^2\xi_1 & x_3 & \lambda\overline{x}_2 \\ \overline{x}_3 & \lambda^{-2}\xi_2 & \lambda^{-1}x_1 \\ \lambda x_2 & \lambda^{-1}\overline{x}_1 & \xi_3 \end{pmatrix}$$

belongs to E_6. *Similarly we can define elements* $\alpha_{23}(\lambda), \alpha_{31}(\lambda)$ *of* E_6.

To prove the following Theorem 2.4, we use

LEMMA 2.3 ([4], [5]). *The group* E_6 *has subgroups* F_4, $Spin(10)$ *as*

$$F_4 = \{\alpha \in E_6 \mid \alpha E = E\},$$
$$Spin(10) = \{\alpha \in E_6 \mid \alpha E_1 = E_1\}.$$

THEOREM 2.4. *The orbit types of the group* E_6 *in* $\mathfrak{J}^C$ *are as follows.*

(1) *The orbit through* (ξ_1, ξ_2, ξ_3) (*where* $|\xi_1|, |\xi_2|, |\xi_3|$ *are distinct*) *is diffeomorphic to* $E_6/Spin(8)$.

(2) *The orbit through* (ξ_1, ξ_2, ξ_3) (*where* $|\xi_1| \neq |\xi_2| = |\xi_3|$, $\xi_3 \neq 0$) *is diffeomorphic to* $E_6/Spin(9) \simeq \{X \in \mathfrak{J}^C \mid 2\tau X \times (X \times X) = X, \ \langle X, X\rangle = 2\}$ $(=_{put} \mathfrak{W}_9)$.

(3) *The orbit through* $(\xi_1, 0, 0)$ (*where* $\xi_1 \neq 0$) *is diffeomorphic to* $E_6/Spin(10) \simeq \{X \in \mathfrak{J}^C \mid X \times X = 0, \ \langle X, X\rangle = 1\}$ $(=_{put} \mathfrak{W}_{10})$.

(4) *The orbit through* (ξ_1, ξ_2, ξ_3) (*where* $|\xi_1| = |\xi_2| = |\xi_3| \neq 0$) *is diffeomorphic to* $E_6/F_4 \simeq \{X \in \mathfrak{J}^C \mid \det X = 1, \ \langle X, X\rangle = 3\}$ $(=_{put} EIV)$.

(5) *The orbit through* $(0, 0, 0)$ *is one point.*

PROOF. Obviously the group E_6 acts transitively on the orbit $E_6(\xi_1, \xi_2, \xi_3)$. We shall determine the isotropy subgroup $(E_6)_{(\xi_1, \xi_2, \xi_3)}$ of E_6 at (ξ_1, ξ_2, ξ_3). We may determine the isotropy subgroup $(E_6)_{\beta(\xi\xi_1,\xi\xi_2,\xi\xi_3)}$ (for some $\beta \in E_6$, $\xi \in C$, $\xi \neq 0$) instead of $(E_6)_{(\xi_1,\xi_2,\xi_3)}$.

(1)(a) Case $|\xi_1|, |\xi_2|, |\xi_3|$ are distinct and $\xi_1\xi_2\xi_3 \neq 0$. Let $\alpha \in (E_6)_{(\xi_1,\xi_2,\xi_3)}$, that is,

$$\alpha(\xi_1 E_1 + \xi_2 E_2 + \xi_3 E_3) = \xi_1 E_1 + \xi_2 E_2 + \xi_3 E_3. \tag{i}$$

From $\tau\alpha\tau((\xi_1 E_1 + \xi_2 E_2 + \xi_3 E_3)^{\times 2}) = (\alpha(\xi_1 E_1 + \xi_2 E_2 + \xi_3 E_3))^{\times 2} = (\xi_1 E_1 + \xi_2 E_2 + \xi_3 E_3)^{\times 2}$, we have

$$\alpha(\tau(\xi_2\xi_3)E_1 + \tau(\xi_3\xi_1)E_2 + \tau(\xi_1\xi_2)E_3) = \tau(\xi_2\xi_3)E_1 + \tau(\xi_3\xi_1)E_2 + \tau(\xi_1\xi_2)E_3. \tag{ii}$$

Take $\xi_1^{-1}\times$(i)$-\tau(\xi_2\xi_3)\times$(ii), then $\alpha(\eta_2 E_2 + \eta_3 E_3) = \eta_2 E_2 + \eta_3 E_3$, where $\eta_2 = \xi_1^{-1}\xi_2 - (\tau\xi_2)^{-1}(\tau\xi_1) \neq 0$, $\eta_3 = \xi_1^{-1}\xi_3 - (\tau\xi_3)^{-1}(\tau\xi_1) \neq 0$. From $\tau\alpha\tau((\eta_2 E_2 + \eta_3 E_3)^{\times 2}) = (\alpha(\eta_2 E_2 + \eta_3 E_3))^{\times 2} = (\eta_2 E_2 + \eta_3 E_3)^{\times 2}$, we have $\alpha(\tau(\eta_2\eta_3)E_1) = \tau(\eta_2\eta_3)E_1$, hence $\alpha E_1 = E_1$. Similarly $\alpha E_2 = E_2$, $\alpha E_3 = E_3$. Therefore $\alpha \in Spin(8)$ (Lemma 1.3, Lemma 2.3). Conversely $Spin(8) \subset (E_6)_{(\xi_1,\xi_2,\xi_3)}$. Thus we have $E_6/Spin(8) \simeq E_6(\xi_1, \xi_2, \xi_3)$.

(1)(b) Case $|\xi_2| \neq |\xi_3|$, $\xi_1 = 0$, and $\xi_2\xi_3 \neq 0$. Let $\alpha \in (E_6)_{(0,\xi_2,\xi_3)}$, that is,

$$\alpha(\xi_2 E_2 + \xi_3 E_3) = \xi_2 E_2 + \xi_3 E_3. \tag{iii}$$

As in (1)(a), we have $\alpha E_1 = E_1$. From $\tau\alpha\tau(E_1 \times (\xi_2 E_2 + \xi_3 E_3)) = \alpha E_1 \times \alpha(\xi_2 E_2 + \xi_3 E_3) = E_1 \times (\xi_2 E_2 + \xi_3 E_3)$, we have

$$\alpha((\tau\xi_3)E_2 + (\tau\xi_2)E_3) = (\tau\xi_3)E_2 + (\tau\xi_2)E_3. \tag{iv}$$

Take $\xi_3^{-1}\times$(iii)$-(\tau\xi_2)^{-1}\times$(iv), then $\alpha(\eta E_2) = \eta E_2$, where $\eta = \xi_3^{-1}\xi_2 - (\tau\xi_2)^{-1}(\tau\xi_3) \neq 0$, hence $\alpha E_2 = E_2$. Similarly $\alpha E_3 = E_3$. Hence $\alpha \in Spin(8)$. Thus we have $E_6/Spin(8) \simeq E_6(0, \xi_2, \xi_3)$.

(2) Case $|\xi_1| \neq |\xi_2| = |\xi_3|$, $\xi_3 \neq 0$. We may assume that $\xi_2 = \xi_3 = 1$ (Lemma 2.2). Let $\alpha \in (E_6)_{(\xi_1,1,1)}$, that is,

$$\alpha(\xi_1 E_1 + E_2 + E_3) = \xi_1 E_1 + E_2 + E_3.$$

As in (1), we have $\alpha E_1 = E_1$ and hence $\alpha(E_2 + E_3) = E_2 + E_3$. Therefore $\alpha \in Spin(9)$ (Lemma 1.3, Lemma 2.3). Conversely $Spin(9) \subset (E_6)_{(\xi_1,1,1)}$. Thus we have $E_6/Spin(9) \simeq E_6(\xi_1, 1, 1)$. Next we shall show $E_6/Spin(9) \simeq \mathfrak{W}_9$. The group E_6 acts on $\mathfrak{W}_9$. This action is transitive. In fact, for a given $X \in \mathfrak{W}_9$, choose $\alpha \in E_6$ such that $\alpha X = (\lambda_1, \lambda_2, \lambda_3)$ (Lemma 2.1). From the condition $(\lambda_1, \lambda_2, \lambda_3) \in$

$\mathfrak{W}_9$, we have $\lambda_i(|\lambda_{i+1}|^2 + |\lambda_{i+2}|^2) = \lambda_i$, $i = 1, 2, 3$ and $|\lambda_1|^2 + |\lambda_2|^2 + |\lambda_3|^2 = 2$. Hence $\lambda_i = 0$ and $|\lambda_{i+1}| = |\lambda_{i+2}| = 1$, and this element αX is transformed to $E_2 + E_3 \in \mathfrak{W}_9$ by some $\beta \in E_6$ (Lemma 2.2). Thus the action is transitive. The isotropy subgroup of E_6 at $E_2 + E_3$ is $Spin(9)$ as is shown in the above. Thus we have $E_6/Spin(9) \simeq \mathfrak{W}_9$.

(3) Case $\xi_1 \neq 0$ and $\xi_2 = \xi_3 = 0$. Let $\alpha \in (E_6)_{(\xi_1,0,0)}$, then $\alpha E_1 = E_1$, hence $\alpha \in Spin(10)$ (Lemma 2.3). Thus we have $E_6/Spin(10) \simeq E_6(\xi_1, 0, 0)$. Next we shall show $E_6/Spin(10) \simeq \mathfrak{W}_{10}$. The group E_6 acts on $\mathfrak{W}_{10}$. This action is transitive. In fact, for a given $X \in \mathfrak{W}_{10}$, choose $\alpha \in E_6$ such that $\alpha X = (\lambda_1, \lambda_2, \lambda_3)$ (Lemma 2.1). From the condition $(\lambda_1, \lambda_2, \lambda_3) \in \mathfrak{W}_{10}$, we have $\lambda_1\lambda_2 = \lambda_2\lambda_3 = \lambda_3\lambda_1 = 0$ and $|\lambda_1|^2 + |\lambda_2|^2 + |\lambda_3|^3 = 1$. Hence $\lambda_i = \lambda_{i+1} = 0$, $|\lambda_{i+2}| = 1$, and this element αX is transformed to $E_1 \in \mathfrak{W}_{10}$ by some $\beta \in E_6$. Thus the action is transitive. The isotropy subgroup of E_6 at E_1 is $Spin(10)$ (Lemma 2.3). Thus we have $E_6/Spin(10) \simeq \mathfrak{W}_{10}$.

(4) Case $|\xi_1| = |\xi_2| = |\xi_3| \neq 0$. We may assume that $\xi_1 = \xi_2 = \xi_3 (= \xi)$ (Lemma 2.2). Now let $\alpha \in (E_6)_{(\xi,\xi,\xi)}$, then $\alpha E = E$, hence $\alpha \in F_4$ (Lemma 2.3). Thus we have $E_6/F_4 \simeq E_6(\xi, \xi, \xi)$. Next we shall show $E_6/F_4 \simeq EIV$. The group E_6 acts on EIV. This action is transitive. In fact, for a given $X \in EIV$, choose $\alpha \in E_6$ such that $\alpha X = (\lambda_1, \lambda_2, \lambda_3)$ (Lemma 2.1). From the condition $(\lambda_1, \lambda_2, \lambda_3) \in EIV$, we have $\lambda_1\lambda_2\lambda_3 = 1$ and $|\lambda_1|^2 + |\lambda_2|^2 + |\lambda_3|^2 = 1$. Hence $|\lambda_1| = |\lambda_2| = |\lambda_3| = 1$ and $\lambda_1\lambda_2\lambda_3 = 1$, and this element αX is transformed to $E \in EIV$ by some $\beta \in E_6$ (Lemma 2.3). Thus the action is transitive. The isotropy subgroup of E_6 at E is F_4 (Lemma 2.3). Thus we have $E_6/F_4 \simeq EIV$ (which is the compact symmetric space of type EIV ([4])).

REMARK. The author does not know how to realize $E_6/Spin(8)$ in $\mathfrak{J}^C$ as $E_6/Spin(9)$, $E_6/Spin(10)$, E_6/F_4 of Theorem 2.4.

References

[1] H. Freudenthal, *Oktaven, Ausnahmegruppen und Oktavengeometrie*, Math. Inst. Rijksuniv. te Utrecht, 1951.

[2] F. R. Harvey, *Spinors and Calibrations*, Academic Press, 1990.

[3] Y. Matsushima, *Some remarks on the exceptional simple Lie group F_4*, Nagoya Math. J. **4** (1954), 83-88.

[4] I. Yokota, *Simply connected compact simple Lie group $E_{6(-78)}$ of type E_6 and its involutive automorphisms*, J. Math. Kyoto Univ., **20** (1980), 448-473.

[5] I. Yokota, *Realizations of involutive automorphisms σ and G^σ of exceptional linear Lie groups G, part I, $G = G_2$, F_4, and E_6*, Tsukuba J. Math., **14** (1990), 185-223.

List of Talks

H. Asano	Generalized Jordan triple systems
K. Atsuyama	Simple Lie algebras and projective spaces in a wider sense
R. Farnsteiner	Cohomology of Frobenius extensions
T. Ikeda	Derivations and central extensions of a generalized Witt algebra
N. Kamiya	On (ε, δ)-Freudenthal-Kantor triple systems
M. Kikkawa	Projectivity of homogeneous left loops
A. Koulibaly	Sur la cohomologie des alèbres de Malcev
F. Kubo	An identity on symmetric algebras of Lie algebras
S. Maeda	Real hypersurfaces of a complex projective space
A. Mitsukawa	On Lie algebras in which the Frattini subalgebra is equal to the derived ideal
J. Morita	Schur multipliers of Kac-Moody groups and associated K_2
M. Nishikawa	Exponential images in real linear groups
K. Nôno	Linearizations of a differential operator of second order and its function theory

R.H. OEHMKE	Commutative algebras and affine connections
E. PLOTKIN	Stability for K_1-functor for Chevalley groups
N. VAVILOV	Structure of Chevalley groups over commutative rings
K. YAMAGUTI	Construction of Lie algebras from associative triple systems and generalized J-ternary algebras
E.I. ZEL'MANOV	Non-associative superalgebras

List of Participants

Hiroshi ASANO	Department of Mathematics, Yokohama City University, Kanazawa-ku, Yokohama 236, Japan
Kenji ATSUYAMA	Department of Mathematics, Kumamoto Institute of Technology, Ikeda, Kumamoto 860, Japan
Rolf FARNSTEINER	Department of Mathematics, University of Wisconsin, Milwaukee, WI 53201, USA
Toshiharu IKEDA	Department of Mathematics, Faculty of Science, Hiroshima University, Naka-ku, Hiroshima 730, Japan
Akiyoshi KAI	Department of Mathematics, Faculty of Science, Hiroshima University, Naka-ku, Hiroshima 730, Japan
Yoshiaki KAKIICHI	Department of Mathematics, Faculty of Engineering, Toyo University, Kawagoe City, Saitama 350, Japan
Noriaki KAMIYA	Department of Mathematics, Faculty of Science, Shimane University, Matsue 690, Japan
Naoki KAWAMOTO	Department of Mathematics, Maritime Safety Academy, Wakaba-cho, Kure 737, Japan
Michihiko KIKKAWA	Department of Mathematics, Faculty of Science, Shimane University, Matsue 690, Japan
Akry KOULIBALY	Institut de Mathématiques et de Physique, Université de Ouagadougou, B.P. 7021, Ouagadougou, Burkina Faso

Fujio KUBO	Department of Mathematics, Kyushu Institute of Technology, Tobata-ku, Kitakyushu 804, Japan
Sadahiro MAEDA	Department of Mathematics, Nagoya Institute of Technology, Shôwa-ku, Nagoya 466, Japan
Osamu MARUO	Department of Mathematics, Faculty of Education, Hiroshima University, Higashi-Hiroshima 724, Japan
Atsushi MITSUKAWA	Department of Mathematics, Faculty of Science, Hiroshima University, Naka-ku, Hiroshima 730, Japan
Jun MORITA	Institute of Mathematics, University of Tsukuba, Tsukuba, Ibaraki 305, Japan
Mitsuru NISHIKAWA	Department of Mathematics, Fukuoka University of Education, Munakata, Fukuoka 811-41, Japan
Kiyoharu NÔNO	Department of Mathematics, Fukuoka University of Education, Munakata, Fukuoka 811-41, Japan
Robert H. OEHMKE	Department of Mathematics, The University of Iowa, Iowa City, IA 52242-1466, USA
Eugene PLOTKIN	Department of Mathematics, Riga Technical University, Riga, Latvia, USSR
Takanori SAKAMOTO	Department of Mathematics, Fukuoka University of Education, Munakata, Fukuoka 811-41, Japan
Manabu SANAMI	Department of Mathematics, Toba National College of Maritime Technology, Ikegami-cho, Toba, Mie-ken 517, Japan
Nikolai VAVILOV	Department of Mathematics and Mechanics, Leningrad State University, 198904, Leningrad, USSR

Kiyosi YAMAGUTI — Department of Mathematics, Faculty of School Education, Hiroshima University, Minami-ku, Hiroshima, 734, Japan

Osami YASUKURA — Department of Mathematics, Yokohama City University, Kanazawa-ku, Yokohama 236, Japan

Efim I. ZEL'MANOV — Institute of Mathematics, Siberian Branch of the Academy of Sciences of the USSR, 630090, Novosibirsk, USSR